Sustainability Sciences in Asia and Africa

Sustainable Agriculture and Food Security

Series Editor
Rajeev K. Varshney,
WA State Agricultural Biotechnology Centre,
Centre for Crop and Food Innovation,
Food Futures Institute Murdoch University,
Murdoch, WA, Australia

Sustainability as a scientific endeavor requires broad understanding of the interconnections in our global environment. It is therefore positioned at the crossroads of research in natural sciences, technology and human sciences. This book series supports the global efforts by providing timely coverage of the progress, opportunities, and challenges towards achieving sustainable development goals (SDGs) in Asia and Africa. It fosters the research in transdisciplinary academic fields spanning across agriculture and global food security, health and diseases, access to clean water and affordable energy, climate resilience and effects of global climate change on biodiversity, water and land. This series is planned to narrate the stories of scientific development and progression towards SDGs from the regions of Asia and Africa. The overall aim of the series is to provide a comprehensive publication platform and act as a knowledge engine in the growth of sustainability sciences with a special focus on developing nations. The series constitute of four sub-series each focusing on a specific theme closely linked to a set of SDGs. Each sub-series is leaders and experts in agricultural sciences, environmental sciences, biomedicine and medical sciences respectively. Books in this series are useful for environmentalists, biotechnologists, plant scientists, medical practitioners, conservation biologists, policy makers, NGOs and scientific bodies working in the areas of sustainability sciences in the developing regions.

This book series support the global efforts towards sustainability by providing timely coverage of the progress, opportunities, and challenges of sustainable food production and consumption in Asia and Africa. The series narrates the success stories and research endeavors from the regions of Africa and Asia on issues relating to SDG 2: Zero hunger. It fosters the research in transdisciplinary academic fields spanning across sustainable agriculture systems and practices, post- harvest and food supply chains. It will also focus on breeding programs for resilient crops, efficiency in crop cycle, various factors of food security, as well as improving nutrition and curbing hunger and malnutrition. The focus of the series is to provide a comprehensive publication platform and act as a knowledge engine in the growth of sustainability sciences with a special focus on developing nations. The series will publish mainly edited volumes but some authored volumes. These volumes will have chapters from eminent personalities in their area of research from different parts of the world.

Bhagirath Singh Chauhan • Khawar Jabran
Editors

Biology and Management of Weeds in Asia and Africa

Springer

Editors
Bhagirath Singh Chauhan
The Centre for Crop Science
Queensland Alliance for Agriculture
and Food Innovation (QAAFI)
The University of Queensland
Gatton, QLD, Australia

Khawar Jabran
Department of Plant Production
and Technologies
Nigde Omer Halisdemir University
Nigde, Türkiye

ISSN 2730-6771 ISSN 2730-678X (electronic)
Sustainability Sciences in Asia and Africa
ISSN 2730-6798 2730-6801 (electronic)
Sustainable Agriculture and Food Security
ISBN 978-981-95-7313-4 ISBN 978-981-95-7314-1 (eBook)
https://doi.org/10.1007/978-981-95-7314-1

This Springer imprint is published by the registered company Springer Nature Singapore Pte Ltd.
The registered company address is: 152 Beach Road, #21-01/04 Gateway East, Singapore 189721, Singapore

Series Preface

The UN's second Sustainable Development Goal (SDG) aims to end hunger, achieve food security and improved nutrition, and promote sustainable agriculture by 2030. Yet over 700 million people still suffer from undernourishment, meaning that global food production systems will need to expand their capacities and produce almost twice their existing levels to feed the world's ever-growing population.

This poses a significant challenge, and while farmers, researchers, and policymakers the world over have been tireless in their efforts to address the global food security gap, we're a long way off from where we need to be to meet SDG 2: Zero Hunger. In recent years, the adoption of new research approaches has led to improvements in food access, farmer autonomy, and the stabilization of agriculture's global carbon footprint. While the advent of these approaches, such as genomics-assisted breeding, has dramatically improved crop breeding programs the world over, the challenge to reduce hunger remains pronounced in the Global South, particularly Asia and Africa, which contains the largest proportion of people suffering from malnutrition and other nutrition-related issues. Furthermore, the agri-food systems in Asia and Africa are severely constrained by the subsistence nature of farming, barriers to accessing new research tools and technologies, and the increasing impacts of environmental pollution, soil and biodiversity degradation.

This book series, Sustainable Agriculture and Food Security, aims to support global efforts toward sustainability by providing timely coverage of the progress, opportunities, and challenges of sustainable food production and consumption in Asia and Africa. The series narrates the success stories and research endeavors from the regions of Africa and Asia on issues relating to SDG 2: Zero Hunger. It fosters research in transdisciplinary academic fields spanning sustainable agriculture systems and practices, pre-seed and post-harvest innovations, and food supply chains. The focus of the series is to provide a comprehensive publication platform and to act as a repository of knowledge in sustainability sciences with a particular focus on developing nations.

In addition to growing challenges facing agriculture, this series looks to address persistent and pervasive threats to global food security. One such threat is the prevalence of weeds, which continues to be an ever-present and costly issue facing farmers in both the developed and developing world. Weeds compete with crops for water, sunlight, nutrients, and space, and even moderate weed pressure can drastically reduce yields. Moreover, some weed species are prolific seed producers, with seeds that can persist in the soil seedbank for decades, and, concerningly, the number of herbicide-resistant weeds has exploded in the past 25 years.

This book, *Biology and Management of Weeds in Asia and Africa*, edited by Dr. Bhagirath Chauhan, explores weed prevalence and mitigation strategies in 10 different countries in Africa and Asia. It presents 10 chapters written by experts and practitioners from across the globe, each containing up-to-date and important insights into this important topic. I commend the Editors for including multidisciplinary and global perspectives and would like to congratulate all the authors for their valuable contributions.

I wish to extend my sincere thanks and gratitude to the Springer staff, particularly Sanchi Bhimrajka, Editor (Books, Medicine and Life Sciences, Asia), and Nick Melchior, Editorial Director (Books, Medicine and Life Sciences, Asia), for their constant support in the accomplishment of this compendium. The cooperation received from my supervisors and research leaders, such as Prof. Andrew Deeks (Vice Chancellor and President), Prof. Peter Eastwood (Deputy Vice Chancellor- Research and Innovation), Prof. Peter Davies (Pro Vice Chancellor, Food Futures Institute), and my laboratory colleagues—Prof. Reyazul Rouf Mir, Anu Chitikineni from Murdoch University (Australia), are also gratefully acknowledged. I would like to thank my family members—Monika Varshney, Prakhar Varshney, and Preksha Varshney, for their love and support in discharging my duties as Series Editor.

Perth, WA, Australia Rajeev K. Varshney

Preface

Weeds are among the three most important pests of field crops, along with disease pathogens and insect pests. The potential global crop losses due to weeds are approximately 37% in soybeans, 23% in wheat, 30% in potatoes, 40% in maize, 36% in cotton, and 37% in rice. Understanding the country-wise situation of weed infestations is crucial for developing effective weed management strategies. This book aims to present a comprehensive overview of weed biology and ecology, weed infestations, their impacts on crops, and management practices across different countries. It compiles information on the extent of weed problems, economic losses caused by weeds, and the major challenges influencing weed control. Each of the ten countries in Asia and Africa has been covered in a separate chapter, detailing weed issues in cropped areas, yield losses, and management strategies. Herbicide-resistant weeds and their management represent a key focus, with chapters discussing the prevalence of resistance and the solutions adopted in different countries. The book also highlights non-chemical weed control methods currently practiced and the state-of-the-art chemical control approaches used for major crops. Integrated weed management (IWM) practices implemented in various countries are discussed in detail, along with the influence of climate change on weeds and weed management. Overall, the book provides an in-depth understanding of weed-related challenges and management strategies at the global and regional levels.

Gatton, QLD, Australia — Bhagirath Singh Chauhan
Nigde, Turkey — Khawar Jabran
18 October 2025

Contents

Series Editor and Editorial Board Members

About the Series Editor

Rajeev Kumar Varshney, FRS, is a renowned agricultural scientist with over two decades of experience in genomics and molecular breeding. Currently, he holds multiple prestigious positions at Murdoch University in Australia. He has made a significant impact on global food security by creating genomic resources for major "orphan" tropical crops, using these resources to identify genetic loci and candidate genes for drought and pest tolerance in staple crops in sub-Saharan Africa and India, and spearheading international programs that deliver superior crop varieties to some of the world's most impoverished farmers. He is an accomplished author, with more than 600 publications to his name, and has been recognized as a Clarivate Highly Cited Researcher for ten consecutive years (2014–2023).

Food Futures Institute, Centre for Crop and Food Innovation, State Agricultural Biotechnology Centre, Agriculture and Food Security, Murdoch University, Perth, WA, Australia.

Editorial Board Members

About the Editors

Bhagirath Singh Chauhan leads the cropping weeds team at the University of Queensland (UQ), based at the university's Gatton campus. He works closely with growers in the northern grain region of Australia and leads national research projects focused on weed ecology, biology, and management. Before joining the UQ, he worked at the International Rice Research Institute (IRRI), where he conducted research across several major rice-growing countries, including Bangladesh, India, the Philippines, Sri Lanka, and Vietnam. Professor Chauhan has published peer-reviewed journal articles, book chapters, and review papers that have collectively advanced the global understanding of weed ecology and management.

Khawar Jabran is an associate professor at the Department of Plant Production and Technologies at Niğde Ömer Halisdemir University, Turkey, and is actively involved in research, teaching, and scientific publication. The main areas of his research work include weed science, crop management, and ecological weed control. Alongside carrying several research projects, he has contributed significantly to scientific writing through his roles as an author, editor, and a reviewer. In addition to several scientific article, he has also authored and edited several high-quality books such as *Potato Production*, *Non-Chemical Weed Control* and *Crop Protection under Climate Change*.

Contributors

Gabriel O. Adesina Department of Crop and Environmental Protection, Ladoke Akintola University of Technology, Ogbomoso, Nigeria

Taseer Ahmad Department of Plant Production and Technologies, Nigde Omer Halisdemir University, Nigde, Turkey

Muhammad Saiful Ahmad-Hamdani Department of Crop Science, Faculty of Agriculture, Universiti Putra Malaysia, Serdang, Selangor, Malaysia

Sharif Ahmed International Rice Research Institute, Bangladesh Office, Dhaka, Bangladesh

Oyebanji O. Alagbo Department of Crop Production and Protection, Faculty of Agriculture, Obafemi Awolowo University, Ile-Ife, Nigeria

S. Anusha ICAR-Sugarcane Breeding Institute, Veerakeralam, Coimbatore, Tamil Nadu, India

Farhena Aslam Department of Agronomy, University of Agriculture, Faisalabad, Pakistan

Chapter 1
Biology and Management of Weeds in Bangladesh

Sharif Ahmed and Bhagirath Singh Chauhan

Abstract Weeds pose a significant threat to agricultural productivity. Bangladesh is a tropical country where weeds severely affect different crops, leading to substantial economic losses for farmers. Understanding the biology of a particular crop's most prevalent weed species is crucial to developing targeted management strategies. This chapter explains the biology and management of the most common weed species in five important crops: rice, corn, jute, sugarcane, and onion, considering their economic value and weed management significance in Bangladesh. Rice is a staple crop, covering over 75% of cultivable land, making it one of the most important crops in Bangladesh. The yield losses from weed infestation in rice range from 15% to 100%, depending on the season, establishment methods, types of weed infestation, land position, and farmers' weed management practices. Chemical weed control with herbicides is becoming popular in Bangladesh for its cost-effectiveness and labor efficiency. However, relying solely on chemicals is not a sustainable long-term strategy. Farmers' understanding of weed biology can enhance integrated management approaches for better sustainability. Corn is currently the most promising crop in Bangladesh, but weed infestation can lead to yield losses of 30%–50%, posing a threat to crop productivity. The yield losses in jute, sugarcane, and onion due to weeds have been 90%, 20%–70%, and 40%–80%, respectively. This chapter outlines the biology of major weeds and various management practices, including integrated approaches for specific crops. It serves as a valuable resource for farmers, researchers, and policymakers aiming to understand and manage weeds in key Bangladeshi crops.

S. Ahmed (✉)
International Rice Research Institute, Bangladesh Office, Dhaka, Bangladesh
e-mail: sharif.ahmed@cgiar.org

B. S. Chauhan
The Centre for Crop Science, Queensland Alliance for Agriculture and Food Innovation (QAAFI), The University of Queensland, Gatton, QLD, Australia

B. S. Chauhan, K. Jabran (eds.), *Biology and Management of Weeds in Asia and Africa*, Sustainability Sciences in Asia and Africa,
https://doi.org/10.1007/978-981-95-7314-1_1

Keywords Weed infestation · Weed biology · Weed management · Integrated weed management · Sustainable agriculture

1.1 Introduction

Agriculture is the largest employment sector in Bangladesh and serves as the backbone of the country's economic system, employing 40% of the total labor force and contributing 12% to the gross domestic product (Anonymous 2023). The agriculture sector has an overwhelming impact on employment generation, poverty alleviation, human resources development, and food security. Agro-ecological conditions are very complex in most parts of Bangladesh and are based on the physiography, soils, land levels in relation to flooding, and agro-climatology. Bangladesh has been divided into 30 agro-ecological zones (AEZ). Due to diverse agro-ecological conditions in Bangladesh, diverse crops are grown here, but crop intensity and diversity greatly vary across the country (Nasim et al. 2017). The population of Bangladesh is increasing consistently; however, agricultural land is decreasing. Therefore, to meet the food demands of the increasing population, agricultural productivity and production need to be increased vertically. This includes the development of high-yielding and problem-oriented crop varieties, excellence in agronomy, and reduction of yield loss due to pests (Nasim et al. 2021). Among the pests that are responsible for yield loss in agriculture, weeds are the number one, and global yield losses due to weed infestation are around 34% (Oerke 2006).

In Bangladesh, weeds are one of the major biological constraints that limit crop productivity (Ahmed et al. 2021). The prevailing climate and edaphic conditions of Bangladesh are highly favorable for luxuriant weed growth (Ahmed and Chauhan 2014). Based on the crop type, growing seasons, topography, and management practices, yield losses due to weeds greatly vary in the country (Ahmed et al. 2016). Besides causing yield losses in crops, weeds also impair the quality of produce and pose health and environmental hazards (Anwar et al. 2021). The total economic losses will be much higher if the indirect effects of weeds on health, losses of biodiversity, nutrient depletion, and grain quality are taken into consideration.

In most crops, weeds must be managed to maintain the optimum yield and quality of the products because weeds compete with crops for essential resources such as sunlight, water, and nutrients. Effective weed management strategies vary by region and depend on several factors. Weed management strategies generally include preventive, cultural, manual, mechanical, biological, and chemical methods. Manual weeding is a traditional and effective weed management strategy. However, it is labor-intensive, and due to labor shortages in agriculture, as well as the drudgery nature of this work, manual weed management is becoming less sustainable.

Chemical weed control using herbicides is becoming popular day by day due to its cost-effectiveness, ease of use, and ability to control a wide range of weed species (Ahmed et al. 2021; Gianessi 2013). It often leads to quicker results compared to manual or mechanical methods. Additionally, it is very effective for no-till and reduced-till systems (Kumar et al. 2013). Herbicides are a blessing for our agriculture, which has made a revolution in farming, but we should not forget that this is a two-edged sword (Kudsk and Streibig 2003). Some herbicides can have adverse effects on non-target plants, wildlife, and water quality if not used properly, potentially causing ecological damage (Norsworthy et al. 2012). Certain herbicides can also pose health risks to humans and animals if they are improperly applied or if there is drift (Ghazi et al. 2023). Relying too much on herbicides can lead to the emergence of herbicide-resistant weed populations, making weed management in the future more challenging and expensive (Ahmed and Chauhan 2015). Therefore, weed management needs integrated approaches considering weed biology and ecology with all possible options (Chauhan et al. 2015).

Weed biology is a comprehensive scientific field in biological science that covers weeds' classification, life cycles, establishment, growth, and reproductive behaviors (Bhowmik 1997). The biology of weeds is essential to know how weeds adapt and thrive in various environments (Ngo et al. 2017). Understanding the biological response of weeds to environmental factors such as temperature, soil depth, light, salinity, and residue management is fundamental for effective weed management in different crops (Anwar et al. 2021).

Insights into seed germination patterns help in anticipating weed pressure, which helps in decision weed management (Travlos et al. 2020). Temperature influences germination rates and growth patterns, allowing farmers to time their control efforts more effectively; for instance, certain weeds thrive in warmer conditions, prompting targeted interventions during peak growth periods (Ahmed et al. 2015).

Light availability plays a critical role because many weeds require specific light conditions to germinate or grow. Therefore, adjusting plant density and using cover crops can suppress these light-loving species (Sias et al. 2021). Salinity levels in the soil can also influence weed species composition, as some weeds are more saline-tolerant than crops (Cirillo et al. 2018). Seed burial depth affects root development, which can determine a weed's competitive ability; managing soil depth through various cultivation practices can enhance crop resilience (Chamara et al. 2018). Proper residue management can inhibit weed seed germination and reduce their establishment by creating a physical barrier or altering micro-environmental conditions (Nichols et al. 2015). By considering these aspects of weed biology in management practices, farmers can effectively reduce weed populations. Effective and economic weed management strategies are essential for fostering agricultural sustainability. This book chapter provides a broad overview of the biology and management of weeds of some major crops, such as rice (*Oryza sativa* L.), corn (*Zea mays* L.), jute (*Corchorus capsularis* L.), sugarcane (*Saccharum officinarum* L.), and onion (*Allium cepa* L.) in Bangladesh, which contribute to the country's economy, and weed management has great economic importance for these crops.

1.2 Weed Biology

Weed biology is the study of the natural life processes, growth patterns, and reproduction of weed plants. It explores how weed species originate, develop, reproduce, disperse, and survive in diverse environmental conditions. Weed biology is the discipline that combines elements of plant physiology, ecology, and agronomy to better understand the behavior and characteristics of weeds. Knowledge of weed biology is essential for developing effective weed management strategies, which helps farmers and land managers to predict weed emergence, prevent their spread, and optimize control measures.

1.2.1 The Major Focus of Weed Biology

Weed biology encompasses several key aspects, including germination, growth habits, reproduction, seed dispersal, and dormancy.

1.2.1.1 Germination and Emergence

This refers to the processes by which weed seeds sprout and develop into seedlings. Factors influencing germination include light, temperature, moisture, and seed dormancy mechanisms. For example, some weed seeds require exposure to light to germinate, while others need specific temperature ranges (Fig. 1.1).

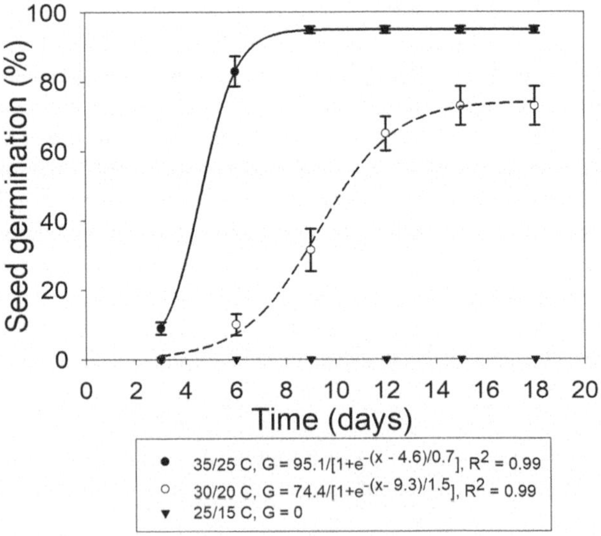

Fig. 1.1 Seed germination of Doveweed [*Murdannia nudiflora* (L.) Brenan] as affected by alternating day/night temperatures (25/15, 30/20, and 35/25 °C) and light (light/dark and dark). (Adopted from Ahmed et al. (2015))

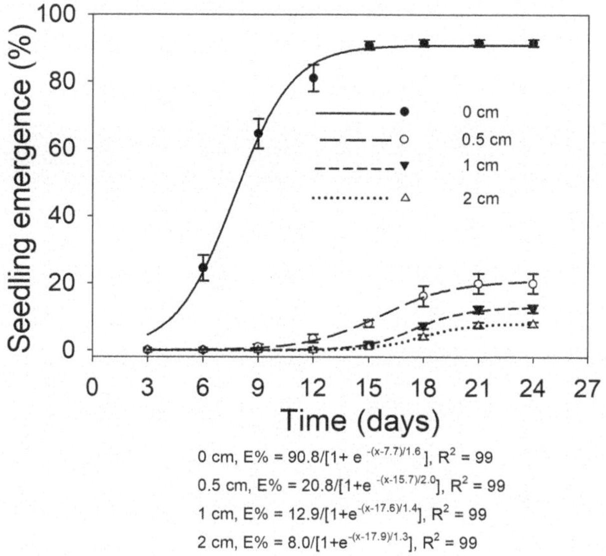

Fig. 1.2 Seedling emergence of *Murdannia nudiflora* as affected by different burial depths (cm). (Adopted from Ahmed et al. (2015))

Seed burial depth is a crucial factor that affects weed seed germination, making it essential for effective weed management. Seeds buried too deeply often struggle to use their energy reserves to reach the soil surface, leading to lower germination rates. Conversely, seeds placed too close to the surface may face unfavorable conditions, such as excessive exposure to the elements, drying out, or being eaten by wildlife. The optimal burial depth varies among different weed species, with factors like seed size and morphological characteristics playing important roles. For instance, larger seeds generally can germinate from greater depths, while smaller seeds tend to require a shallower position. By understanding these dynamics, farmers and land managers can adopt strategies such as careful tillage and the use of cover crops, which help suppress unwanted vegetation and promote healthier crops. Figure 1.2 shows the seedling emergence of *Murdannia nudiflora* as influenced by burial depth.

1.2.1.2 Growth Habit

Weeds exhibit various growth forms—annuals, biennials, and perennials (Table 1.1). Annuals complete their life cycle within 1 year, biennials take 2 years, and perennials persist for multiple years. Understanding these growth forms assists in timing control practices effectively.

Table 1.1 Biological characteristics of major weeds in rice, corn, jute, sugarcane, and onion crops

Weed species		Life cycle and type	Method(s) of reproduction	Flowering time	Dormancy	Seed production/plant
Common name	Scientific name					
Grass weed						
Bermuda grass	*Cynodon dactylon* (L.) Pers.	Perennial-C_4	Rhizomes, stolons, seeds	60–80 days	No dormancy, seeds survive 50 days of submergence	1000–5000 seeds
Crowfoot grass	*Dactyloctenium aegyptium* (L.) Willd.	Annual-C_4	Seeds	30–40 days	Low temperature (5 °C) prolonged the dormancy	Up to 66,000 seeds
Southern crabgrass	*Digitaria ciliaris* (Retz.) Koeler	Annual-C_4	Seeds	60–80 days	Variable, up to 7 months	Up to 150,000 seeds
Hairy crabgrass	*Digitaria sanguinalis* (L.) Scop.	Annual-C_4	Seeds	40–60 days	Low dormancy levels under optimal conditions	Approximately 10,000 seeds per plant
Jungle rice	Echinochloa colona (L.) link.	Annual-C_4	Seeds, stolons	30–45 days	Low or none; light requirement for germination	3500–111,000 seeds
Barnyard grass	*Echinochloa crus-galli* (L.) P. Beauv.	Annual-C_4	Seeds	Within 60 days	Variable, up to 4 months	Up to 40,000 seeds
Indian goosegrass	*Eleusine indica* (L.) Gaertn.	Annual-C_4	Seeds	30 days	Some, but usually short	Up to 120 thousand seeds
Southern cutgrass	*Leersia hexandra* SW.	Perennial-C_3	Rhizomes, seeds	70–100 days	Dormancy period of 6 months	10,000–100,000 seeds
Red sprangletop	*Leptochloa chinensis* (L.) Nees	Perennial-C_4	Seeds, plant fragments	60–90 days	Low or none	10,000–30,000 seeds
Fall panicgrass	*Panicum dichotomiflorum* Michx.	Perennial-C_4	Rhizomes, seeds	60–90 days	Several months to a couple of years	10,000 to over 100,000 seeds
Torpedo grass	*Panicum repens* L.	Perennial-C_4	Rhizomes, seeds	50–60 days	Several months to a couple of years	Up to 50,000 seeds
Knotgrass	*Paspallum distichum* L.	Perennial-C_4	Stolons, seeds, and rhizomes	50–80 days	Short-term dormancy	Around 10,000–30,000 seeds
Wild foxtail millet	*Setaria viridis* (L.) P.Beauv.	Annual-C_4	Seeds and rhizomes	40–60 days	Prolonged seed dormancy	Up to 6000 seeds/ panicle

Broadleaf						
Indian jointvetch	*Aeschynomene indica* L.	Annual or perennial-C_3	Seeds, Stam, and nodes	45–60 days	Short	1000–10,000 seeds
Sessile joyweed	*Alternanthera sessilis* (L) R.Br. ex DC.	Perennial-C_3	Seeds, stolons, stem fragment	30–60 days	Low seed dormancy	Over 20,000 seeds
Goatweed	*Ageratum conyzoides* L.	Annual-C_3	Seeds	Within 60 days	50% of seeds can germinate immediately, light required for germination	Up to 40,000 seeds per plant
Spiny amaranth	*Amaranthus spinosus* L.	Annual-C_4	Seeds	50–70 days	Variable, none to 4 months; long viability; no light	Up to 235,000 seeds
Sheperd's clock	*Anagalis arvensis* L.	Annual-C_3	Seeds	20–40 days	40–60 days	1000–20,000 seeds
Plumed cockscomb	Celosia argentea L.	Annual-C_4	Seeds	60–90 days	Do not exhibit significant dormancy	Up to 100,000 seeds
Lamb's quarters	*Chenopodium album* L.	Annual-C_3	Seeds	40–70 days	Short dormancy	10,000–100,000 seeds
Fringed spider flower	*Cleome rutidosperma* DC.	Annual-C_3	Seeds	45–60 days	Shot dormancy	1000–2500 seeds
Tropical spiderwort	*Commelina benghalensis* L.	Perennial-C_3	Seeds, stolons	30–40 days	Minimum dormancy period of 6 months	1500–20,000 seeds
Bindweed	*Convolvulus arvensis* L.	Perennial-C_4	Seeds, roots, and stem	60–90 days	Long dormancy	500–1000 seeds
Creeping cradle plant	*Cyanotis axillaries* (L.) D.Don ex Sweet	Perennial-C_3	Seeds and stem	60–120 days	Do not have significant dormancy	1000–10,000 seeds
Digera	*Digera arvensis* (L.) Mart.	Annual-C_4	Seeds	30–60 days	Exhibit a degree of dormancy	1000–10,000 seeds
False daisy	*Eclipta prostrata* L.	Annual-C_3	Seeds	20–40 days	None; light required for germination	5000–25,000 seeds

(continued)

Table 1.1 (continued)

Weed species		Life cycle and type	Method(s) of reproduction	Flowering time	Dormancy	Seed production/plant
Common name	Scientific name					
Asthma-plant	*Euphorbia hirta* L.	Annual-C_4	Seeds	30–60 days	Some degree of dormancy	20,000–100,000 seeds
Wild poinsettia	*Euphorbia geniculate* L.	Annual-C_4	Seeds	30–90 days	Short dormancy	No information
Shaggy soldier	*Galinsoga ciliate* (Rafin.) S.F.Blake	Annual-C_3	Seeds	20–40 days	Remain viable in the soil until conditions are favorable	1000–10,000 seeds
Morning glory	Ipomoea spp. L.	Perennial-C_3	Seeds	60–90 days	Do not exhibit significant dormancy	100–1000 seeds
Linear leaf water primrose	*Ludwigia hissopifolia* (G.Don) Excell	Annual-C_3	Seeds and vegetative parts	40–50 days	1–2 years	Up to 250,000 seeds per plant
Sweet clover	*Meliotus indica* (L.) all	Biennial-C_3	Seeds	60–90 days	May exhibit some dormancy	1000–10,000 seeds
Heartshape false pickerelweed	*Monochoria vaginalis* (Burm.f) C.Presl	Perennial-C_3	Seeds, rhizomes	Within 60 days	Need long anaerobic period to germinate	500–10,000 seeds
Doveweed	*Murdannia nudiflora* (L.) Brenan.	Perennial-C_3	Seeds, rhizomes	50–70 days	Several months to years	1000–5000 seeds
Dwarf waterclover	(Marsilea minuta L.)	Perennial-C_3	Seeds, rhizomes	45–70 days	Long-dormancy	100–10,000 seeds
Tall wood-sorrel	*Oxalis europea* (f. villicaulis) Wiegand	Perennial-C_3	Seeds, rhizomes, bulbs	40–60 days	Short	100–10,000 seeds
Parthenium weed	*Parthenium hysterophorus L.*	Annual-C_3	Seeds	30–45 days	Do not exhibit significant dormancy	10,000–25,000 seeds
Stone-breaker	*Phyllanthus niruri* L.	Annual-C_3	Seeds	40–70 days	Several months to years	10,000–20,000 seeds
Clammy groundcherry	*Physalis heterophylla* Nees	Annual-C_3	Seeds	35–50 days	Several months	100–1000 seeds
Common purslane	*Portulaca oleracea* L.	Annual-C_4	Seeds, stem	30–60 days	Can remain viable for a period and do not exhibit significant dormancy	10,000–100,000 seeds

Lesser arrow-head	*Sagittaria guyanensis* Kunth	Perennial-C_3	Seeds, tubers	40–60 days	Long dormancy	1000–10,000 seeds
Black nightshade	*Solanum nigrum* L.	Annual-C_3	Seeds	60–90 days	May exhibit some degree of dormancy	1000–10,000 seeds
Gooseweed	*Sphenoclea zeylanicam* Gaertn.	Annual-C_3	Seeds	40–60 days	Several months or even over a year	1000–15,000 seeds
Chickweed	*Stellaria media* (L.) Vill.	Annual-C_3	Seeds	30–60 days	Do not exhibit significant dormancy	1000–30,000 seeds
Desert horsepurslane	*Trianthema portulacastrum* L.	Perennial-C_4	Seeds, stem	60–90 days	May not exhibit significant dormancy	1600–10,000 seeds per plant
Sedge weed						
Smallflower umbrella-sedge	*Cyperus difformis* L.	Annual-C_3	Seeds	20–30 days	Yes	3000–25,000 seeds
Rice flatsedge	*Cyperus iria* L.	Annual-C_4	Seeds	20–30 days	Can germinate about 75 days after shedding	3000–5000 seeds per plant
Purple nutsedge	*Cyperus rotundus* L.	Perennial-C_4	Tubers, rhizomes	About 60–90	Yes, apical dominance in tubers	No seed
Lesser fimbristylis	*Fimbristylis miliacea* (L.) Vahl	Perennial-C_4	Seeds	30–40 days	None; light requirement for germination	5000–15,000 seeds per plant
Bulrush	*Scirpus juncoides* Roxb.	Annual-C_3	Seeds	60–90 days	2–3 months	1500 seeds per plant
Saltmarsh bulrush	*Scirpus mucronatus* L.	Perennial-C_3	Seed, rhizomes, and stolons	70–100 days	Does not have a prolonged seed dormancy	Several thousand to over 50,000 seeds

All information in this table was collected from different books, published articles, and websites

1.2.1.3 Reproduction

Weeds mainly reproduce through seeds, though many can also propagate vegetatively via rhizomes, stolons, or tubers. The reproductive capacity of a weed influences its ability to spread and persist. Most of the weeds produce a large number of seeds (Table 1.1).

1.2.1.4 Seed Dispersal

Weeds disperse their seeds through various methods, including wind, water, animals, and human activities, which impact the spread of weed populations across fields and landscapes.

1.2.1.5 Dormancy

Some weed seeds can remain dormant for varying periods, sometimes for years, before germinating. Dormancy mechanisms ensure survival under unfavorable conditions, complicating management efforts.

1.2.2 Relationship of Weed Biology with Weed Ecology

Weed biology is closely linked to weed ecology, which studies how weeds interact with their environment and other organisms. While weed ecology explores broader interactions within ecosystems, weed biology provides a detailed understanding of weed life cycles and physiological traits that influence these ecological dynamics. For instance, understanding seed dormancy and germination (biological aspects) helps explain how weeds establish in specific environments (ecological aspects). Weed ecology involves several fundamental aspects, including weed characteristics and classification. Key characteristics that contribute to a plant's weediness include rapid seed germination, fast growth, the ability to absorb and utilize large amounts of nutrients, prolific seed production, seed traits that facilitate dispersal, mechanisms for seed dormancy, continuous germination cycles, adaptability to various environmental conditions, and high tolerance to stress. Weeds can be classified based on factors related to population dynamics, such as habitat, growth form, life cycle, reproduction, and seed type. Understanding the basic ecology of a weed is essential for implementing effective management strategies that can prevent the plant from producing seeds. Additionally, weed ecology encompasses aspects like seed production—its quantity and form—as well as seed dispersal through both abiotic and biotic factors. The ecology of weeds is closely related to the composition of plant communities, the evolution of weeds (including factors such as herbicide resistance), allelopathy, and competition. Both fields inform management practices by revealing factors influencing weed success and persistence, such as soil conditions, climate, competition with crops, and responses to control measures.

1.2.3 Weed Biology of Major Weeds in Rice, Corn, Jute, Sugarcane, and Onion

The detailed information in Table 1.1 characterizes the biology of major weeds impacting rice, corn, jute, sugarcane, and onion, including the life cycle, which categorizes weeds as either annual or perennial, and their metabolic types, classified as either C3 or C4 plants. Furthermore, it explores the reproduction methods used by these weeds, the flowering times necessary for producing viable seeds, and the dormancy characteristics of their seeds or propagation materials. It also emphasizes the remarkable seed production capacity of each plant. A thorough understanding of these biological traits not only enhances effective weed management practices but also aids in anticipating the potential effects of weeds on crop yield and agricultural sustainability.

1.3 Economic Significance of Rice, Corn, Jute, Sugarcane, and Onion in Bangladesh

Rice (*Oryza sativa* L.) holds paramount economic importance in Bangladesh, being the staple food and a vital component of its agricultural landscape. Rice is cultivated here on 11.7 million hectares of land, with an annual production of around 39 million tons (BBS 2023). The rice sector ensures food security for a large population and plays a vital role in the rural economy, where about 40% of the labor force is engaged in agriculture (Anonymous 2023). The cultivation and processing of rice significantly contribute to the country's gross domestic product (GDP), accounting for approximately 70% of the agricultural GDP (Saha et al. 2021). Economically, rice serves as a primary source of income for millions of farmers, directly connecting their livelihoods to this essential crop. Additionally, the rice industry sustains various supporting sectors such as milling, transportation, generating numerous employment opportunities, and fostering rural development.

Rice is grown in Bangladesh almost year-round and there are three rice growing seasons including *aus* (pre-monsoon rice), *aman* (monsoon rice), and *boro* (dry season irrigated rice). In the rainy season, from July to October, over 90% of the crop fields are covered by wet-season rice. In the dry season, *boro* rice is the major crop, although several non-rice crops such as potato, corn, onion, mustard, lentil, and different vegetables are also grown at this time. In the pre-monsoon season, either *aus* rice or jute is the major crop. In recent years, Bangladesh has already earned self-sufficiency in rice production, but still, Bangladesh needs to import Basmati rice. On the other hand, the export potential of aromatic and premium rice varieties positions Bangladesh favorably in international markets. The demand for zinc- and iron-enriched biofortified rice and low-glycemic index rice is increasing day by day. Additionally, rice scientists in Bangladesh are also focusing their research on climate-smart rice production (Ahmed et al. 2025).

Corn (*Zea mays* L.) the second most important crop in Bangladesh, holds significant economic importance in the country, emerging as one of its key agricultural commodities over the past few decades. Corn has gained prominence due to its high yield potential, flexible growing period where it can be grown throughout the year because of its C_4 characteristics, low water requirement, climate resilience, use in poultry and livestock industries, and versatile use. The corn sector makes a significant contribution to the national economy, constituting approximately 3%–4% of the total agricultural GDP (BBS 2023). Currently, corn is produced in 1.3 million ha of land with an annual production of 5.4 million metric tons (BBS 2023). Advances in agricultural technology and the availability of high-yielding and hybrid corn varieties have led to increased productivity, making corn an attractive option for farmers seeking high returns on investment. With the increasing demand for meat and dairy products, driven by population growth and evolving dietary preferences, the importance of corn has intensified. Additionally, the crop's adaptability to various climatic conditions makes it a resilient option for farmers in the face of climate change. The annual demand for corn in Bangladesh is approximately 7.0 million metric tons, and 25% of this demand has to be fulfilled through imports (BBS 2023). Farmers grow mostly winter corn, purchasing hybrid seeds from the market, but summer corn is also gaining popularity.

Jute (*Corchorus* spp.) is a cash crop of immense economic significance in Bangladesh. It is often called the "Golden Fiber" due to its golden-brown color and significant economic impact. As one of the largest global producers of jute, Bangladesh plays a crucial role in the jute industry, supporting the livelihoods of millions of farmers and laborers (Akter et al. 2020; Moniruzzaman et al. 2024). Jute cultivation occupies a significant portion of agricultural land, primarily in the south-western and north-western regions of Bangladesh, making it vital for the economy and the sustainable development of rural communities. The jute sector is crucial for the country's agricultural GDP and significantly contributes to generating foreign exchange through exports. Bangladesh exports a variety of jute products, including raw jute, jute sacks, and eco-friendly jute textiles to international markets, thereby enhancing its economic standing (Moniruzzaman et al. 2024). As global demand for sustainable and biodegradable materials has increased, jute has received renewed attention as a viable alternative to synthetic fibers, resulting in heightened investments in jute processing and value-added products. Additionally, jute cultivation is essential for environmental conservation as it enhances soil health and decreases carbon emissions, thus aiding the country's efforts to mitigate climate change. From 2010 to 2024, the average raw jute production ranged from 7 to 9 million bales (Moniruzzaman et al. 2024). Jute is mostly grown in the country from March to June.

Sugarcane (*Saccharum officinarum* L.) is a crucial economic crop in Bangladesh, serving as the second most important cash crop after jute and supporting many farmers' livelihoods. Historically, sugarcane has been cultivated for producing various traditional sweeteners like *gur* (jaggery) and sugar. Sugarcane is a valuable source of food (sucrose and syrups), fiber (cellulose), fodder (green tops, bagasse, and molasses), fuel, and chemicals (molasses, and alcohol). It is also considered one of the best converters of solar energy into biomass and sugar. The sugarcane industry plays a significant role in supporting food security and promoting economic growth due to the increasing domestic demand for sugar and sugar-based products. Furthermore, the exploration of alternative uses for sugarcane, particularly its potential in the renewable energy sector for biofuel production, highlights its importance in the context of sustainable development. Despite its importance, the sugarcane industry in Bangladesh faces challenges, including fluctuating market prices, inadequate infrastructure, and competition from imported sugar. Bangladesh has a yearly production of 5–6 million metric tons of sugarcane that comes from the cultivation of 0.7–1.2 million acres of land (BBS 2023). The sugar sector plays a critical role in generating employment opportunities across the value chain, from cultivation and harvesting to processing and marketing, thus bolstering rural economies. As an annual crop, sugarcane occupies the land year-round, preventing crop intensification and diversification.

Onion (*Allium cepa* L.) is a major spice in Bangladesh and holds significant importance as it is a fundamental ingredient in most meals, contributing flavor to the local cuisine. It is cooked in many ways, for example, by preparing curries and other delicious foods. As a widely consumed vegetable, onions are essential for flavoring a variety of traditional dishes, making their demand consistently high among households, restaurants, and food service establishments. There are various types of onions, including red, yellow, white, and green, each with its unique flavor. Onion farming provides a viable income source for rural communities due to its low input costs and short cultivation cycle. However, challenges such as price volatility, pest infestations, diseases, and inadequate storage facilities can hinder the growth of the onion sector. Addressing these challenges with effective agricultural policies, infrastructure investment, and research can enhance the economic significance of onion production. Approximately 0.5 million acres of land in the country are under onion cultivation, resulting in a production of about 2.5 million tons (BBS 2023). Bangladesh needs to import a huge amount of onion every year. Onion is mostly grown in Bangladesh from mid-October to mid-April.

1.4 Importance of Weed Management in Different Crops

1.4.1 Rice

The yield losses due to weeds in various rice cultivation systems in Bangladesh range from 15% to 40% for transplanted rice to up to 100% for directly seeded rice (Islam et al. 2021; Ahmed and Chauhan 2014; Ahmed et al. 2021). Weed infestation during the *aus* (pre-monsoon season) and *aman* (monsoon/wet season) is significantly higher than in the *boro* (dry season). This is mainly due to consistently high temperatures throughout the growing season, which enhance weed seed germination and growth. During the initial growth stage of *boro* rice, temperatures remain low due to which weeds cannot germinate and grow until the temperature rises in early to mid-February (Ahmed and Chauhan 2014). In addition, farmers grow their *boro* with more intensive care than *aus* and *aman* due to its higher yield potential and lower risk. Weed species found in rice fields depend on various factors such as season, site, cropping systems, and weed management practices. The common weed species found in rice fields of Bangladesh are presented in Table 1.2. The extent of yield loss in crops depends on their management practices.

The traditional methods of weed control practices in rice fields in Bangladesh are manual hand weeding in the wet field and weed pulling using a *hoe* in the dry field (Ahmed et al. 2021). Usually, two or three hand weedings are done for growing a rice crop, depending upon the nature of the weeds, their intensity of infestation and the nature of the crop grown. Chemical weed control using herbicides is gaining popularity in rice cultivation due to its rapid effects and lower costs compared to traditional methods. During the last decade, herbicide use has tremendously increased in Bangladesh, and now, the common weed management practices in transplanted rice are pre-emergence herbicide application followed by (fb) one or two cycles of hand weeding (Table 1.3). Farmers mostly apply pre-emergence herbicides 3–7 days after transplanting (DAT) with a mix of urea fertilizer in standing water conditions. Post-emergence herbicides for transplanted rice are not very common in Bangladesh because of the unavailability of post-emergence herbicides for rice in the market. The best integrated weed management strategies for rice from previous studies were herbicide-based IWM options with mefenacet + bensulfuron-methyl as pre-emergence, followed by either bispyribac-sodium or penoxsulam as post-emergence, followed by one hand-weeding (Ahmed et al. 2021).

1.4.2 Corn

Weeds are one of the most important factors that limit corn yield in Bangladesh. Corn is more sensitive to weed competition during its early growth period because of slow initial growth, wide space among crop rows, and heavily applied fertilizer. Maximum weed competition in corn occurs during the period of 2–6 weeks after

Table 1.2 Major weed species in rice, corn, jute, sugarcane, and onion in Bangladesh and similar environments (Ahmed and Chauhan 2014; Ahmed et al. 2021; Nasir et al. 2024; Parvez et al. 2013; Hossain et al. 2019; Islam 2014; Ghorai et al. 2013; Rathika et al. 2023; Khan et al. 2021; Chanda et al. 2021)

Crop name	Associated weeds
Rice	*Echinochloa colona, Cyperus rotundus, Panicum repens, Leersia hexandra, Cyperus difformis, Ludwigia hissopifolia, Leptochloa chinensis, Echinochloa crus-galli, Cynodon dactylon, Alternanthera sessilis, Fimbristylis miliacea, Digitaria ciliaris, Scirpus juncoides, Cyperus iria, Eclipta prostrata, Dactyloctenium aegyptium, Eleusine indica, Anagalis arvensis, Commelina benghalensis, Monochoria vaginalis, Sphenoclea zeylanica, Cleome rutidosperma, Galinsoga ciliate, Phyllanthus niruri, Celosia argentea, Murdannia nudiflora, Marsilea minuta, Oxalis europea, Aeschynomene indica, Ageratum conyzoides, Sagittaria guyanensis, Physalis heterophylla*
Corn	*Echinochloa colona, Celosia argentea, Cynotis axillaris,Euphorbia hirta, Chenopodium album, Cyperus rotundus, Euporbia geniculate, Portulaca oleracea, Digitaria* spp., *Ageratum* spp., *Euphorbia prostrata, Solanum nigrum, Parthenium hysterophorous*
Jute	*Cyperus rotundus, Cynodon dactylon, Echinochloa colona, Digitaria* spp., *Physalis heterophylla, Paspallum distichum, Eleusine indica, Euphorbia hirta, Phyllanthus niruri, Scirpus mucronatus, Setaria viridis, Alternanthera sessile, Portulaca oleracea, Trianthema* spp., *Commelina benghalensis, Leptochloa chinensis, Cyanotis axillaries, Amaranthus spinosus*
Sugarcane	*Digitaria* spp., *Panicum repens, Panicum dichotomiflorum, Cynodon dactylon, Eleusine indica, Amaranthus* spp., *Alternanthera* spp., *Chenopodium album, Portulaca oleracea, Parthenium hysterophorus, Cyperus rotundus, Chenapodium album, Convolvulus arvensis, Commelina bengalensis, Trianthema portulacastrum, Dactylocternium aegyptium, Ipomoea* spp., *Digera arvensis, Euphorbia* spp., *Eclipta prostrata, Phyllanthus niruri*
Onion	*Cyperus rotundus, Cynodon dactylon, Parthenium hysterophorus, Digitaria sanguinalis, Convolvulus arvensis, Anagallis arvensis, Chenopodium album, Meliotus indica, Ageratum conyzoides, Stellaria media, Alternathera sessilis, Amaranthus viridis*

sowing, which suggests the importance of maintaining the fields weed-free during this critical period of weed-crop competition (Keller et al. 2014). The competition between weeds and corn at critical growth stages could reduce both the quality and quantity of corn and cause yield losses by up to 70% (Sharma et al. 2000; Malviya et al. 2012; Hossain et al. 2019). Mostly upland annual and perennial weeds are found in corn fields (Table 1.2). Weed control in corn in Bangladesh primarily relies on mechanical methods, such as hand spading and hand hoeing. While these techniques are effective for weed management, they are not cost-effective and have several disadvantages (Hossain et al. 2019). Heavy rains during the summer season create a serious issue as the wet soil makes weeding impossible by conventional methods. Chemical weed control using herbicides is available now for corn, and they are very cost-effective. The best economic weed control options for corn are presented in Table 1.3.

Table 1.3 Most economical weed management methods for managing weeds in rice, corn, jute, sugarcane, and onion

Crop	Weed management	Reference
Rice (transplanted)	Pre emergence mefenacet + bensulfuron-methyl at 1–3 days after transplanting (DAT) followed by (fb) either bispyribac-sodium or penoxsulam as postemergence at 15–25 DAT fb one hand-weeding at 45–50 DAT	Ahmed et al. (2021)
	Pre-emergence herbicide oxadiargyl 190 ml ha^{-1} fb one hand weeding at 50 days after transplanting (DAT)	Hasanuzzaman et al. (2008)
	Pre-emergence herbicide pretilachlor fb one hand weeding at 21 DAT	Parvez et al. (2013)
Rice (direct seeded)	Pre-emergence herbicide pendimethalin 1000 g ai ha^{-1} at 2 days after sowing (DAS) fb post-emergence ethoxysulfuron 20 g ai ha^{-1} at 20 DAS fb one HW at 40–45 DAS or pre-emergence herbicide pendimethalin 1000 g ai ha^{-1} at 1–3 DAS fb post-emergence bispyribac-sodium or penoxsulam at 15–25 DAS fb one HW at 40–45 DAS	Ahmed et al. (2015, 2025)
Corn	Pre-emergence herbicide pendimethalin fb mixture of post-emergence atrazine@0.625 kg ha^{-1} + 2,4-D@0.5 kg ha^{-1} fb need-based one HW	Hossain et al. (2019)
Jute	Fenoxaprop-P-ethyl 15 DAS fb 1 HW at 35–40 DAS or quizalofop-ethyl@1.5 to 2.0 ml/1 at 20–25 DAS fb one hand weeding at 35–40 DAS	Islam (2014), Ghorai et al. (2013)
Sugarcane	Pre-emergence herbicide pendimethalin 1000 g ai ha^{-1} or atrazine 2 kg or Oxyfluorfen 750 ml/ha fb 2,4-D@1.0–2.0 kg ha^{-1} at 35–40 DAS	Rathika et al. (2023)
Onion	Pendimethalin or oxyfluorfen as a pre-emergence fb quizalofop ethyl at 40 g ha^{-1} as post-emergence + one HW 40–45 DAS	Khan et al. (2021)

1.4.3 *Jute*

The humid weather conditions, coupled with intermittent rainfall during the jute-growing season in Bangladesh, encourage weed growth, leading to heavy infestations in jute fields (Islam 2014). The magnitude of yield loss due to weeds in jute is up to 75% under uncontrolled or poor weed management (Akram et al. 2021; Hossain et al. 2023). In jute, the maximum weed infestation in Bangladesh was found in the third–ninth weeks after sowing (Hossain et al. 2023). If weeds are not managed properly during this critical weed control period, jute yield performance is severely affected. In Bangladesh, weeds in jute are generally managed by raking and *niri* (hand weeding), and weeding with thinning operations involves about 50% or more of the labor cost (Islam 2014). Nowadays, the use of herbicides is increasing, which is economically viable (Table 1.3). At the initial stage, jute plants grow slowly; therefore, control of weeds in the early stage of the crop is most important. The major weed species found in the jute fields in Bangladesh and similar weather conditions are presented in Table 1.2.

1.4.4 *Sugarcane*

Being a long-duration crop with wide row spacing (60–90 cm), sugarcane creates ample opportunities for severe weed infestations in vacant spaces during the crop growing season. The initial growth of sugarcane is slow, requiring about 30–45 days for germination and an additional 60–80 days to develop full canopy cover. Due to its long growing duration, cultivation under abundant nutrient supply, and frequent ratooning with minimal preparatory tillage, weed problems are usually more severe in sugarcane. Weeds have been estimated to cause up to a 70% reduction in cane yield depending upon the severity of infestation and other associated factors (Islam et al. 2016). The first 90–120 days of crop growth are critical for weed competition in sugarcane; therefore, weed management practices should maintain a weed-free field during this period. All types of grasses, broadleaf weeds, and sedges—both annual and perennial—infest sugarcane fields. The major weed species found in the sugarcane field in Bangladesh and similar weather conditions are presented in Table 1.2. The common weed management practices in sugarcane in Bangladesh are manual and mechanical control of weeds practiced 2–3 times (i.e., before the cane sprouts and at tillering phases of the crop). The best economic weed control options for sugarcane were pre-emergence herbicide pendimethalin or oxyfluorfen fb 2,4-D at 35–40 days after sowing (Table 1.3).

1.4.5 *Onions*

Onions are more susceptible to weed competition than other crops due to several inherent characteristics such as slow growth, small stature, and narrow upright leaves, which do not effectively shade out competing weeds. Additionally, onions have shallow root systems and lack dense foliage, making them more vulnerable to weed invasion. So, early-season weed control is essential for successful crop production. Unmanaged weed growth reduces the onion bulb yield by up to 80% depending upon the weed infestation level and other factors (James and Harlen 2010; Dhananivetha et al. 2015). The major weed species found in the onion field in Bangladesh and similar weather conditions are presented in Table 1.2.

Hand weeding in onions is a common practice in Bangladesh, but it is a tedious, expensive, and time-consuming task due to the close spacing of the plants and their shallow root systems. Spraying pre-emergence herbicides helps maintain weed-free conditions during the early growth stages of the crop. However, the second flush of weeds at a later stage can negatively impact bulb formation. Hand weeding is effective in keeping the weed population below the economic threshold level throughout the growth period of the crop. Combining pre-emergence herbicides with hand weeding can be costly, particularly due to limited labor availability and high labor costs. After bulb formation, using manual or mechanical methods of weed control can damage the bulbs. Therefore, applying early post-emergence herbicides may be beneficial in reducing damage to the bulbs, minimizing weed competition, and lowering overall weed control costs.

1.5 Weed Biology for Integrated Weed Management

A thorough understanding of the biology of weeds is crucial for developing effective integrated weed management (IWM) strategies for crops. For example, seed ecology (germination and emergence) is affected by several environmental factors, such as temperature, light, seed burial depth, water stress, flooding depth, crop residue retention, etc. Understanding the germination response of a weed species to these factors can help develop effective weed management programs by either suppressing germination or encouraging it at times when control is easier (Chauhan and Johnson 2010). Crop residue is frequently utilized as a mulching material in the crop field, and its retention is a crucial aspect of conservation agriculture practices (Hossain et al. 2021b). The presence of residue significantly aids in suppressing weeds. For instance, Ahmed et al. (2015) reported that applying 4 tons/hectare or more of rice residue effectively reduces the emergence of *M. nudiflora* (Fig. 1.3).

Weed emergence is a crucial process that determines the number of weed plants and the timing of their appearance. Understanding weed emergence patterns can justify or refute the application of herbicides in crop management. In rice-based cropping systems, where rice is often rotated with non-rice crops, the cropping sequence follows different tillage practices. These differences in tillage operations and crop residue management significantly affect the conditions that influence weed emergence. For instance, Hossain et al. (2021b) reported substantial variations in weed emergence across different tillage systems and residue retention levels (Table 1.4), underscoring the importance of weed biology studies.

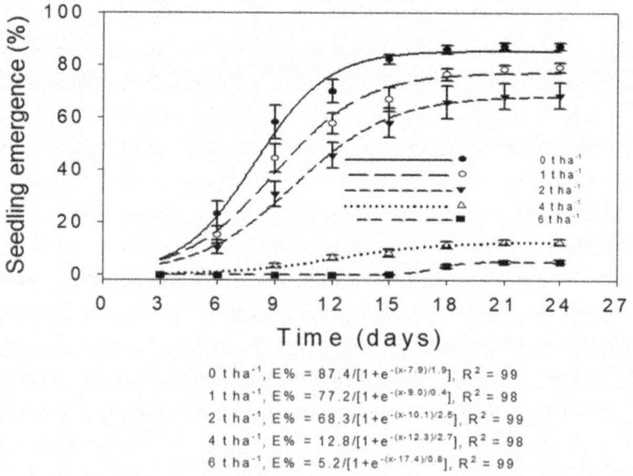

Fig. 1.3 Seedling emergence response of *Murdannia nudiflora*, as influenced by rice residue (t ha^{-1}). (Adopted from Ahmed et al. (2015))

Table 1.4 Composition of different weed species of the initial and final soil seed bank at 0–5 cm depth under different tillage systems and residue retention in Bangladesh

Weed species	Initial seed bank (seeds m^{-2})	Final seed bank (plants m^{-2})			
		CT R_0	CT R_{50}	ST R_0	ST R_{50}
Ageratum conyzoides	170	194	169	239	Absent
Alternanthera sessilis	173	203	184	181	168
Amaranthus viridis	118	157	136	Absent	Absent
Cyanotis axillaris	288	94	167	186	Absent
Cynodon dactylon	129	Absent	201	234	315
Cyperus difformis	348	381	337	148	137
Cyperus iria	209	123	Absent	Absent	110
Cyperus rotundus	161	259	112	303	381
Digitaria sanguinalis	137	154	134	Absent	231
Echinochloa cruss-galli	351	512	445	169	157
Echinochloa colona	Absent	131	114	Absent	184
Eclipta alba	98	166	145	Absent	Absent
Eichhornia crassipes	102	188	163	135	146
Eleusine indica	165	208	179	Absent	127
Euphorbia parviflora	104	150	129	Absent	Absent
Fimbristylis miliacea	275	30	258	253	214
Hedyotis corymbosa	144	202	175	Absent	Absent
Leersia hexandra	196	236	209	271	252
Lindernia hyssopifolia	103	113	Absent	Absent	209
Marsilea quadrifolia	122	152	Absent	110	102
Monochoria hastate	115	163	187	157	146
Nicotina plumbaginifolia	174	217	198	141	Absent
Panicum distichum	73	Absent	Absent	158	133
Physalis heterophylla	Absent	Absent	167	Absent	Absent
Pistia stratiotes	163	132	114	123	Absent
Polygonum coccineum	Absent	173	150	Absent	Absent
Solanum juncoides	Absent	99	86	137	Absent
Solanum nigrum	Absent	Absent	Absent	171	Absent

Adopted from Hossain et al. (2021a)
CT conventional tillage, *ST* strip tillage, R_0 no-residue, R_{50} 50% residue

1.6 Conclusion

This chapter highlights the biological traits and weed management of key weed species in five major crops in Bangladesh. Understanding the biology of major weeds that affect these crops—including their germination requirements, growth patterns, and responses to environmental stresses—is essential for developing targeted and sustainable weed control strategies. However, in Bangladesh, the weed biology studies are lacking, which presents a significant opportunity to enhance understanding of the major weed species. Policy interventions that support research on weed biology and develop the capacity of farmers and extension workers in timely and effective weed control practices are essential for the long-term success of sustainable weed management.

Acknowledgments The authors acknowledge the use of ChatGPT for language improvement.

References

Ahmed S, Chauhan BS (2014) Performance of different herbicides in dry-seeded rice in Bangladesh. Sci World J 2014:729418. https://doi.org/10.1155/2014/729418

Ahmed S, Chauhan BS (2015) Efficacy and phytotoxicity of different rates of oxadiargyl and pendimethalin in dry-seeded rice (*Oryza sativa* L.) in Bangladesh. Crop Prot 72:169–174

Ahmed S, Opeña JL, Chauhan BS (2015) Seed germination ecology of Doveweed (*Murdannia nudiflora*) and its implication for management in dry-seeded rice. Weed Sci 63(2):491–501. https://doi.org/10.1614/ws-d-14-00115.1

Ahmed S, Awan TH, Salim M, Chauhan BS (2016) Economics of nitrogen and integrated weed management in dry-seeded rice. J Anim Plant Sci 25:1675–1684

Ahmed S, Kumar V, Alam M, Dewan MR, Bhuiyan KA, Miajy AA, Saha A, Singh S, Timsina J, Krupnik TJ (2021) Integrated weed management in transplanted rice: options for addressing labor constraints and improving farmers' income in Bangladesh. Weed Technol 35(5):697–709

Ahmed S, Kumar V, Zaman AU, Dewan MR, Khatun A, Hossain K, Singh S, Timsina J, Krupnik TJ (2025) Dry direct-seeded and broadcast rice: a profitable and climate-smart alternative to puddled transplanted *aus* rice in Bangladesh. Field Crop Res 322:109739. https://doi.org/10.1016/j.fcr.2025.109739

Akram M, Ali MH, Islam MS, Masum SM, Malek M, Hasanuzzaman M (2021) Jute growth and yield influenced by integrated weed management. Bangladesh J Weed Sci 7(1&2):1–14

Akter S, Sadekin MN, Islam N (2020) Jute and jute products of Bangladesh: contributions and challenges. Asian Bus Rev 10(3):143. https://doi.org/10.18034/abr.v10i3.480

Anonymous (2023) Bangladesh's agriculture, natural resources, and rural development sector assessment and strategy. https://doi.org/10.22617/tcs230050. Accessed on 28 July 2024

Anwar MP, Islam AKMM, Yeasmin S, Rashid MH, Juraimi AS, Ahmed S, Shrestha A (2021) Weeds and their responses to management efforts in a changing climate. Agronomy 11(10):1921. https://doi.org/10.3390/agronomy11101921

BBS (2023) Bangladesh Bureau of Statistics. http://nsds.bbs.gov.bd/en/posts/111/Statistical%20Yearbook%202023. Accessed on 12 Sept 2024

Bhowmik PC (1997) Weed biology: importance to weed management. Weed Sci 45(3):349–356. https://doi.org/10.1017/s0043174500092973

Chamara BS, Marambe B, Kumar V, Ismail AM, Septiningsih EM, Chauhan BS (2018) Optimizing sowing and flooding depth for anaerobic germination-tolerant genotypes to enhance crop establishment, early growth, and weed management in dry-seeded rice (*Oryza sativa* L.). Front Plant Sci 9:1654. https://doi.org/10.3389/fpls.2018.01654

Chanda SC, Khan MJ, Sarkar SC, Sarwar AKMG, Paul SK (2021) Crop-weed association in different field crops at Sirajganj district in Bangladesh. J Bangladesh Agric Univ 19(2):215–222. https://doi.org/10.5455/JBAU.76123

Chauhan BS, Johnson DE (2010) The role of seed ecology in improving weed management strategies in the tropics. Adv Agron 105:221–262

Chauhan BS, Ahmed A, Awan TH, Jabran K, Manalil S (2015) Integrated weed management approach to improve weed control efficiencies for sustainable rice production in dry-seeded systems. Crop Prot 71:19–24

Cirillo V, Masin R, Maggio A, Zanin G (2018) Crop-weed interactions in saline environments. Euro J Agron 99:51–61. https://doi.org/10.1016/j.eja.2018.06.009

Dhananivetha M, Amanullah MM, Arthanari PM (2015) Effect of weed management methods on weed control and productivity of onion. Ann Soil Plant Res 7:301–303

Ghazi RM, Yusoff NRN, Halim NSA, Wahab IRA, Latif NA, Hasmoni SH, Zaini MAA, Zakaria ZA (2023) Health effects of herbicides and its current removal strategies. Bioengineered 14(1):2259526. https://doi.org/10.1080/21655979.2023.2259526

Ghorai A, De R, Chowdhury H, Majumdar B, Chakraborty A, Kumar M (2013) Integrated management of weeds in raw jute. Ind J Weed Sci 45(1):47–50

Gianessi LP (2013) The increasing importance of herbicides in worldwide crop production. Pest Manag Sci 69:1099–1105

Hasanuzzaman M, Islam MO, Bapari MS (2008) Efficacy of different herbicides over manual weeding in controlling weeds in transplanted rice. Aust J Crop Sci 2:18–24

Hossain A, Islam MT, Islam MS, Nurislam N, Ahmed S, Sarker KK, Gathala MK (2019) Chemical weed management in corn (*Zea mays* L.) under conservation agricultural systems: an outlook of the Eastern Gangetic Plains in South-Asia. In: IntechOpen eBooks. https://doi.org/10.5772/intechopen.89030

Hossain MM, Begum M, Hashem A, Rahman MM, Ahmed S, Hassan MM, Javed T, Shabbir R, Hadifa A, Sabagh AE, Bell RW (2021a) Strip tillage and crop residue retention decrease the size but increase the diversity of the weed seed bank under intensive rice-based crop rotations in Bangladesh. Agronomy 11(6):1164. https://doi.org/10.3390/agronomy11061164

Hossain A et al (2021b) Conservation agriculture: next-generation, climate resilient crop management practices for food security and environmental health. In: Jayaraman S, Dalal RC, Patra AK, Chaudhari SK (eds) Conservation agriculture: a sustainable approach for soil health and food security. Springer, Singapore. https://doi.org/10.1007/978-981-16-0827-8_28

Hossain MS, Sarker UK, Monira S, Ali MI, Hasan AK, Kaysar MS, Anwar MP, Begum M, Rashid MH, Rashid MH et al (2023) Determination of critical period for sustainable weed management and yield of jute (*Corchorus olitorius* L.) under sub-tropical conditions. Sustainability 15:9282. https://doi.org/10.3390/su15129282

Islam MM (2014) Research advances of jute field weeds in Bangladesh: a review. ARPN J Sci Technol 4:254–268

Islam N, Sohel MAT, Alam MN, Islam AKMR, Akhter S (2016) Performance of herbicides for weed control in sugarcane field. Bangladesh J Sugarcane 37:51–56

Islam AKMM, Nasir M, Mou MA, Yeasmin S, Islam MS, Ahmed S, Anwar MP, Hadifa A, Baazeem A, Iqbal MA et al (2021) Preliminary reports on comparative weed competitiveness of Bangladeshi monsoon and winter rice varieties under puddled transplanted conditions. Sustainability 13:5091. https://doi.org/10.3390/su13095091

James RL, Harlen MHV (2010) Multiplication of reduced rate herbicides for weed control in onion. Weed Tech 24:153–159

Keller M, Gantoli G, Möhring J, Gutjahr C, Gerhards R, Rueda-Ayala V (2014) Integrating economics in the critical period for weed control concept in corn. Weed Sci 62(4):608–618. https://doi.org/10.1614/ws-d-13-00184.1

Khan MA, Rahman MM, Mou SS (2021) Effect of integrated weed management practices on the growth, yield, quality and economic of onion (*Allium cepa* L.). Arch Agric Environ Sci 6(3):277–289. https://doi.org/10.26832/24566632.2021.060303

Kudsk P, Streibig JC (2003) Herbicides—a two-edged sword. Weed Res 43(2):90–102. https://doi.org/10.1046/j.1365-3180.2003.00328.x

Kumar V, Singh S, Chhokar RS, Malik RK, Brainard DC, Ladha JK (2013) Weed management strategies to reduce herbicide use in zero-till rice-wheat cropping systems of the Indo-Gangetic Plains. Weed Technol 27:241–254

Malviya A, Malviya N, Singh AK (2012) Integrated weed management in corn (*Zea mays* L.) under rainfed conditions. Ind J Dryland Agric Res Dev 27(1):70–73

Moniruzzaman S, Alam MJ, Sharma D, Begum IA, Patino MTO, McKenzie AM (2024) Value chain analysis of jute fiber in Bangladesh. Soc Sci Humanit Open 10:101088. https://doi.org/10.1016/j.ssaho.2024.101088

Nasim M, Shahidullah SA, Saha A, Muttaleb MA, Aditya TL, Ali MA, Kabir MS (2017) Distribution of crops and cropping patterns in Bangladesh. Bangladesh Rice J 21(2):1–55. http://www.brri.gov.bd/site/page/55d231c8-8ba2-46d9-93bf-a63d5aa129a4

Nasim M, Khatun A, Kabir MJ, Mostafizur ABM, Mamun MAA, Sarkar MAR, Salam MU, Kabir MS (2021) Intensification of cropping through utilization of fallow period and unutilized land resources in Bangladesh. Bangladesh Rice J 25(1):89–100. https://doi.org/10.3329/brj.v25i1.55181

Nasir M, Islam A, Islam M, Ahmed S, Salam M, Anwar M (2024) Weed competitiveness of popular corn varieties of Bangladesh. Fundam Appl Agric 9(4):294. https://doi.org/10.5455/faa.243245

Ngo TD, Boutsalis P, Preston C, Gill G (2017) Growth, development, and seed biology of feather fingergrass (*Chloris virgata*) in southern Australia. Weed Sci 65:413–425

Nichols V, Verhulst N, Cox R, Govaerts B (2015) Weed dynamics and conservation agriculture principles: a review. Field Crop Res 183:56–68. https://doi.org/10.1016/j.fcr.2015.07.012

Norsworthy J, Ward S, Shaw D, Llewellyn R, Nichols R, Webster T et al (2012) Reducing the risks of herbicide resistance: best management practices and recommendations. Weed Sci 60:31–62

Oerke EC (2006) Crop losses to pests. J Agric Sci 144(1):31–43. https://doi.org/10.1017/S0021859605005708

Parvez MS, Salam MA, Kato-Noguchi H, Mahfuza B (2013) Effect of cultivar and weeding regime on the performance of transplant aman rice. Int J Agric Crop Sci 6(11):654–666

Rathika S, Ramesh T, Jagadeesan R (2023) Weed management in sugarcane: a review. Pharma Innov J 12(6):3883–3887

Saha I, Durand-Morat A, Nalley LL, Alam MJ, Nayga R (2021) Rice quality and its impacts on food security and sustainability in Bangladesh. PLoS One 16(12):e0261118. https://doi.org/10.1371/journal.pone.0261118. PMID: 34972112; PMCID: PMC8719737

Sharma AR, Toor AS, Sur HS (2000) Effect of interculture operations and scheduling of atrazine application on weed control and productivity of rainfed corn (*Zea mays*) in Shiwalik foothills of Punjab. Ind J Agric Sci 70:757–761

Sias C, Wolters BR, Reiter MS, Flessner ML (2021) Cover crops as a weed seed bank management tool: a soil down review. Ital J Agron 16(4):1852. https://doi.org/10.4081/ija.2021.1852

Travlos I, Gazoulis I, Kanatas P, Tsekoura A, Zannopoulos S, Papastylianou P (2020) Key factors affecting weed seeds' germination, weed emergence, and their possible role for the efficacy of false seedbed technique as weed management practice. Front Agron 2:1. https://doi.org/10.3389/fagro.2020.00001

Chapter 2
Biology and Management of Weeds in India

Simerjeet Kaur, S. Anusha, Arjun Tayade, Murali Arthanari Palanisamy, and Bhagirath Singh Chauhan

Abstract Agriculture serves as the backbone of India's economy, playing a pivotal role in ensuring food security, providing employment, and contributing to overall economic development. The largest Indian population is engaged in agricultural activities growing a wide range of cereals, pulses, oilseeds, fruits, vegetables, and non-food crops. In this chapter, six crops, namely rice (*Oryza sativa* L.), wheat (*Triticum aestivum* L.), corn (*Zea mays* L.), pulses, cotton (*Gossypium* spp.) and sugarcane (*Saccharum officinarum* L.), are included for their weed management aspect. Weed infestation poses a significant threat to crop production, leading to substantial yield losses and economic challenges for farmers. By delving into weed biology and management techniques, farmers may be empowered with the knowledge and tools necessary to mitigate weed-related risks, optimize crop productivity, and promote sustainable agricultural practices. Traditionally, manual weeding and intercultural operations have been used to manage weeds, but changing rainfall pat-

S. Kaur (✉)
Department of Agronomy, Punjab Agricultural University, Ludhiana, Punjab, India
e-mail: simer@pau.edu

S. Anusha
ICAR-Sugarcane Breeding Institute, Coimbatore, Tamil Nadu, India
e-mail: anusha.s@icar.org.in

A. Tayade
Division of Crop Production, ICAR-Central Institute for Cotton Research, Nagpur, Maharashtra, India

M. A. Palanisamy
Department of Agronomy, Tamil Nadu Agricultural University, Coimbatore, Tamil Nadu, India
e-mail: agronmurali@tnau.ac.in

B. S. Chauhan
The Centre for Crop Science, Queensland Alliance for Agriculture and Food Innovation (QAAFI), The University of Queensland, Gatton, QLD, Australia
e-mail: b.chauhan@uq.edu.au

B. S. Chauhan, K. Jabran (eds.), *Biology and Management of Weeds in Asia and Africa*, Sustainability Sciences in Asia and Africa,
https://doi.org/10.1007/978-981-95-7314-1_2

terns, rising labor costs, and increasing agricultural expenses have made weed management more difficult. Though herbicides are the easiest, cheapest, and timely weed management approach, it is helpful and effective only for a short period. Majority of the research in India is based on herbicides, and very limited studies on weed biology and IWM are carried out on problematic weeds. The shift toward integrated weed management (IWM) highlights the growing need for more sustainable and effective weed control methods in crop production. Therefore, IWM research must focus on weed biology and ecology, combination of mechanical, cultural, and chemical methods, and potential benefits of weeds. New cutting-edge futuristic technologies such as machine learning, drones, and robotics may aid in the development of innovative and environmentally friendly weed management approaches.

Keywords Cotton · Corn · Pulses · Rice · Sugarcane · Integrated weed management · Wheat

2.1 Introduction

Agriculture, with its allied sectors, is unquestionably the largest livelihood provider in India, and remains vital to India's socioeconomic fabric, more so in the vast rural areas. India has diverse agroclimatic zones ranging from temperate to tropical and humid to arid regions (Anonymous 2025a). It is the seventh largest country by area and has only 2.4% of global land. Further, it is the most populous country in the world with over 1.42 billion population that is 17.76% (one sixth) of total world population. About 18% of Indian farmers own 1–2 ha of land, while over 67% own less than 1 ha. With an average land holding of less than 1.1 ha, small and marginal farmers made up approximately 86% of the population (Kareemulla et al. 2021). Moreover, there are different soil types, climatic conditions, and traditional cultivation methods in the nation's agricultural landscape, which produces crops that support livelihood of millions of people.

Major crops grown in India are rice (*Oryza sativa* L.), wheat (*Triticum aestivum* L. emend. Fiori and Paol.), corn (*Zea mays* L.), cotton (*Gossypium* spp.), sugarcane (*Saccharum officinarum* L.), tea (*Camellia sinensis* (L.) Kuntze), coffee (*Coffea arabica* L.), jute (*Corchorus capsularis* L.), pulses, oilseeds, and millets. Among cereal crops, rice, wheat, and corn are the first three important food crops (Anonymous 2025a). Pulses and corn seeds are very important dietary sources for humans and animals in the world. Particularly, pulses are consumed as the world's second-most food, which is an important source of proteins, whereas, corn is mostly consumed as an animal dietary system and is an important source of carbohydrates and proteins. The rice-wheat cropping system is the most dominant and profitable system in the Indo-Gangetic Plain region of north-west India. Corn-wheat is the third most important cropping system in India after rice-wheat and rice-rice that contributes about 3% in the national food basket.

In 2030, the population is expected to increase to 1.5 billion people, and around 345 million tons of food grains will be required to feed (NITI Aayog 2018). It would be an arduous task to produce food in the current agricultural scenario of shrinking agricultural land due to alternate land use, increasing threats of climate change, water scarcity, rapid urbanization, etc. to feed the burgeoning population (Kaur and Chauhan 2023). In India, per capita agricultural land is shrinking due to continued fragmentation (population pressure and inheritance laws), urbanization, and alternate land use. Targeting the yield gaps caused by weed competition is one approach to achieve this target.

Globally, in crop production, weeds cause great yield losses and are considered a major biotic stress in crop production. In economic terms, approximately 1800 weed species not only caused a 31.5% reduction in crop production but also amounted to $USD 32 billion per year in economic losses (Kubiak et al. 2022). Weeds compete with crops for light, water, and nutrients and can drastically reduce crop yields and quality. In a recent study, a total economic loss of about USD 11 billion was estimated due to weeds in 10 major field crops from 18 states in India (Gharde et al. 2018). Estimates of actual economic losses were the highest in the case of rice (worth $USD 4420 million), followed by wheat (worth $USD 3376 million) and soybean (*Glycine max* (L.) Merr.) (worth $USD 1559 million), respectively. Yaduraju (2012) also estimated the economic losses as approximately USD 13 billion when losses due to weeds were taken as 10%, which would amount to a loss of about 25 mt of total food grains in India. The density and duration of weed species are critical factors that define the intensity of crop-weed competition and the resultant losses in crop yield.

Weed flora in cropped fields is highly diverse, encompassing grasses, sedges, and broadleaved weeds. Weed species and its floristic composition present in any agricultural field reflect the history of crop production and/or protection activities in the field (Hofmeijer et al. 2021). Weed floristic composition in a particular crop is a function of agronomic practices such as cropping rotation, tillage systems, residue management, fertilization, crop density, geometry, water management, and herbicide use history, along with environmental and edaphic conditions. For example, rice was not the major crop of northwestern India before the green revolution, and perennial weeds like *Cynodon dactylon* (L.) Pers., *Sorghum halepense* (L.) Pers., *Cyperus rotundus* L., *Convolvulus arvensis* L. and *Cirsium arvense* (L.) Scop. were predominant during the summer season under such situations (Kaur et al. 2022). Change in tillage practices (puddling) for rice cultivation has resulted in a major shift in weed flora from aerobic summer-annual weeds to water-loving, semi-aquatic, and aquatic weeds viz.; *Echinochloa crus-galli* (L.) P. Beauv., *Paspalum distichum* L., *Cyperus iria* L., *Ischaemum rugosum* Salisb., *Cyperus difformis* L., *Fimbristylis miliacea* (L.) Vahl, *Leptochloa chinensis* (L.) Nees, *Ammannia baccifera* L., *Caesulia axillaris* Roxb., *Sphenoclea zeylanica* Gaertn., *Sagittaria guayanensis* Kunth., *Scirpus juncoides* Roxb., *Eleocharis acicularis* (L.) Roem & Schult., *Najas arguta* Kunth., *Hydrilla verticillata* (L. f.) Royle, and *Ipomoea aquatica* Forssk. Further due to canal irrigation system, typical weeds of hilly areas like

Ageratum conyzoides L. invaded the rice fields, watercourses, and bunds of the fields in the state.

2.2 Biology of Problematic Weeds

Phalaris minor Retz. is a self-pollinated, winter annual grass and is a satellite weed of wheat crops. It reproduces mainly through seeds, and at maturity, seed dispersal takes place after caryopses shatter from their glumes (Matus-Cádiz and Hucl 2002). During winter months, its emergence starts in November, and multiple cohorts emerge with subsequent irrigation given throughout the cropping season depending upon the available space (Singh 2007). During the initial 60 days, its dry biomass accumulates at slower speed, and it is faster between 60 and 90 days after sowing (DAS) (Malik and Singh 1995). Plant invests 27% of aboveground biomass in seed production, irrespective of plant size (Franke et al. 2007). The aboveground biomass failed to influence the thousand seed weight, and even small plants also produced seeds of equal individual seed weights; but, the numbers of seeds were reduced. The thousand seed weight of *P. minor* varies from 1.5 to 2.1 g (Bhan and Choudary 1976). Singh et al. (2025) observed that seed production per panicle was 309.1 ± 7.5, and almost 74% of seeds were retained on the panicle at the harvest, thus indicating the potential for use of harvest weed seed control tactics.

Avena spp. is one of the troublesome winter annual grass weeds of temperate agricultural landscapes and is a naturalized species in the Indian subcontinent (Holm et al. 1979). It is an annual grass with fibrous roots that reaches up to 150 cm high with profuse tillering (Tidemann et al. 2021). Like *P. minor*, the initial morphology similarities of this weed to wheat plant make it difficult to control with mechanical methods. Further, its seeds exhibit multiple dormancy mechanisms in the soil after shedding, which help in the enriched weed seed bank for several years (Holm et al. 2000). Singh et al. (2025) observed that the total number of seeds per panicle ranged from 52 to 61, and most of seeds were shed (only 10%–11% retention) by the time of wheat harvest.

Echinochloa colona (L.) Link, a C_4 plant native to tropical and subtropical Asia, now spreads across warm regions of Asia, Africa, and Australia, growing as a prostrate to erect, shallow-rooted annual or occasionally perennial plant up to 100 cm tall (Rao 2021). However, *E. crus-galli* is robust, tufted annual grass with an erect to decumbent 150 cm tall stem that is often branched at the base. *E. colona* can germinate throughout the growing season in several upland summer crops, but it is a less vigorous competitor than *E. crus-galli* (Zhang and Watson 1997). Many of the growth characteristics of *E. crus-galli* are like those of rice, making it a strong competitor to rice in fields across the world (Rao et al. 2007).

Digitaria sanguinalis (L.) Scop., features sparse tufts and weak, spreading stems, reaching up to 100 cm in length, with soft, flat leaf blades and spikelets occurring in pairs. *Digitaria* spp. are summer annuals; however, they can begin to exhibit growth of a perennial species due to extended durations of favorable growing conditions

(Jones et al. 2021). This is a C_4 plant, and vegetative growth is invariably dictated by light and temperature (Shetty et al. 1982). *Digitaria* spp. are determinate plants, monoecious, and mainly self-pollinated and can produce up to 150,000 seeds per plant.

Commelina benghalensis L., a perennial, succulent, and sprawling herb found in tropical Asia and Africa, produces aerial and underground flowers (Walker and Evenson 1985). It can form new roots when leaf nodes encounter soil. Moreover, rhizomes and underground flowers start developing 6 weeks after emergence, before the development of aerial flowers. This weed represents floral dimorphism and seed polymorphism (Maheshwari and Singh 1934).

Trianthema portulacastrum L. is an annual broadleaf weed belonging to the Aizoaceae family and is a widely distributed weed in summer and rainy season fields and horticultural crops in India. It is a strong competitor, reducing significant losses in corn and soybean yield (Kaur and Aggarwal 2017). It has high acclimatization capacity and produced more growth under elevated temperatures up to 4 °C (Mandal et al. 2017). The invasion hotspots for *T. portulacastrum* will be shifted towards the south and east of the country, including a few western parts like Gujarat, in future climatic scenarios (Singh et al. 2024a).

Parthenium hysterophorus L., considered one of the most aggressive invasive and problematic weeds, germinates in spring, produces flowers and seeds throughout its life cycle, and dies in late autumn, with optimal seed germination observed at 21/16 °C (day/night). The ability of *P. hysterophorus* to perpetuate in diverse climatic regimes, soil types, and disturbed environments enhances its potential to occupy a wide array of habitat ranges (Annapurna and Singh 2003). The phytotoxic ability of the weed has been proved against many agricultural crops. It exhibits diverse characteristics and adaptations, posing significant challenges to agriculture and biodiversity conservation efforts worldwide (Kaur et al. 2021). Ahmad et al. (2019) mapped current and potential distribution of *P. hysterophorus* in India and observed a significant shift in niche in response to climate changes.

Cyperus rotundus, native to India, is widely naturalized across Africa, America, South Asia, and southern/central Europe, significantly impacting subtropical and tropical agroecosystems in both hemispheres. This erect, glabrous, grass-like perennial sedge can reach heights of 7–40 cm, with prolific production of rhizomes and tubers. It germinates in warm seasons, has prolific rhizome and tuber production, and flowering is favored by favorable conditions. *C. rotundus*, a perennial weed, has been stated to be 60%–80% of total weed flora in sugarcane, causing more than 40% yield reduction (Chand et al. 2014; Giraldeli et al. 2021). It is a rapidly spreading perennial C_4 sedge reproducing extensively through underground tubers and rhizomes and rarely by seeds. More than 80% of tubers of *C. rotundus* occur in the upper 15 cm of soil. Rhizome penetration of *C. rotundus* will be more in light textured soils than heavy-textured soils. Under ideal soil conditions, a single tuber can produce more than 100 tubers within 90 days. This sedge can propagate best under high soil moisture, high fertility, and high temperature, making it a serious weed of sugarcane (Stoller and Sweet 1987).

Cynodon dactylon is a warm-season perennial known for its rapid spread via rhizomes and stolons, forming dense turf resilient to foot traffic and environmental challenges. Thriving in diverse soils, it favors heavy clay and alkaline conditions but tolerates varying pH levels. Its growth peaks in spring and summer, persisting year-round due to its ability to regenerate vegetatively. Its robust growth and adaptability allow it to outcompete crops for nutrients, water, and light, severely affecting agricultural productivity. While primarily spreading vegetatively, it also reproduces through seeds, germinating in warm, moist conditions. This weed thrives in fertile, moisture-retentive deep black cotton soils (vertisols, found in central India), where it establishes extensive underground rhizomes and stolons, making mechanical or manual control methods largely ineffective. It is a persistent and aggressive perennial grass weed, and a comprehensive approach involving integrated weed management strategies is required to achieve effective long-term control (Satao et al. 1995). Its robust growth and adaptability allow it to outcompete crops for nutrients, water, and light, severely affecting agricultural productivity. The infestation of *C. dactylon* in sugarcane reduces the cane yield as well as the number of subsequent ratoons (Dalley et al. 2013). *Cynodon dactylon* is a troublesome perennial stoloniferous and rhizomatous creeping C_4 grass with long runners. It is a serious weed of sugarcane throughout Asia due to its wide ecological tolerance and effective vegetative propagation through rhizomes and stolon fragments in addition to seed production (Ramakrishnan and Kumar 1971). It responds positively to high light intensities, and it is adversely affected by shading. Hence, competition of *C. dactylon* is more severe in the initial growing stages of sugarcane than in later stages with dense canopy (Horowitz 1972).

Rottboellia cochinchinensis (Lour.) Clayton is a noxious grass weed infesting sugarcane both in tropics and subtropics causing yield losses of 50%–70% in sugarcane (Millholon 1992). This weed is also a major problem in corn crop fields. It is a fast-growing annual or perennial C_4 grass reaching up to 4 m in height, and a single plant can produce 100 tillers with over 15,000 seeds (Correia 2016). These seeds germinate throughout the growing season, making it difficult to control. These seeds can remain dormant in the soil for up to 4 years. Manual removal of *R. cochinchinensis* from the infested field is very difficult due to the presence of rigid trichomes in leaf sheaths (Chauhan 2013).

Ipomoea spp. are the most bothersome broadleaved weeds in sugarcane cultivation in many countries, including India. Sugarcane crops can be badly damaged due to twining of this weed around sugarcane clumps, which can cause yield loss up to 25% (Singh et al. 2018). These weeds also interfere with harvesting operations. With the extension of sugarcane cultivation to rice fields, twining weeds, or creeper weeds, or binding weeds, have become a major problem. Management of creeper weeds poses serious problems, as they germinate after the earthing up, and it is difficult to use mechanical methods. Seeds of *Ipomoea* spp. buried under the soil during sugarcane planting exhibit very poor germination. These seeds will be scarified and brought to soil surface during earthing-up operations (Bhullar et al. 2012). Binding weeds like *Ipomoea* spp. germinate in sugarcane fields 120 days after cane planting and occupy the top plane of the cane, compete for sunlight, and also hinder cane harvesting (Kathiresan et al. 2004).

2.3 Weed Infestation in Major Crops

Rice For rice production, India ranks second in the world after China. India is the largest exporter of rice globally. Rice is the staple food crop in India and is grown over the largest area. It requires warm temperatures (above 25 °C) with high humidity and annual rainfall of above 100 cm. In rice, production agronomy affected the distribution of weed species comprising a population. Changes in tillage and crop establishment methods significantly affected the weed flora infestation level and its composition. The crop-weed competition is influenced by i. rice production systems (transplanting versus direct seeding); ii. relative heights of rice and weed; iii. Cropping density; iv. rice cultivar; and v. soil moisture and nutrient availability (Rao et al. 2007). Rice is mainly grown with the puddling and transplanting technique. In transplanted production systems, 25–30-day-old rice seedlings are transplanted (manually or mechanically) into puddled soil, and that plant height difference gives rice a significant head start on weeds. Furthermore, water in the ponded fields (paddies) itself acts as herbicide, leading to minimal competition in the initial crop growth phase. Crop-weed competition intensifies as growth progresses. The crop-weed competition is comparatively more in upland dry direct-seeded rice (DSR), as both crop and weeds emerge at the same time and compete for nutrients, light, and space with the rice seedlings. Weeds caused 10%–15%, 35%–45%, and 20%–25% reduction in yield in transplanted rice (TPR), DSR, and wet-seeded rice, respectively (Rao et al. 2007).

DSR is gaining popularity as a potential alternative to conventional transplanting with reduced input requirements, reduced methane and CO_2 emissions, increased adaptability to climate change, and increased economic returns. Weed infestation is the major setback in DSR, which is held accountable for the yield loss ranging from 15% to 20%, but in severe cases, it may exceed 50% or even cause complete crop failure (Jabran and Chauhan 2015). Further, dry-DSR rice has more diverse weed floristic composition than TPR. Aerobic weeds such as *Dactyloctenium aegyptium* (L.) Willd. and *C. rotundus* are common in unpuddled DSR but are not found in TPR due to hypoxic edaphic environment (ponded water) that inhibits weed germination. Some weeds, especially red/weedy rice, are a major limitation in dry-DSR > wet-DSR (Johnson Sunder Singh et al. 2024). Recently, there has been a shift toward herbicide-tolerant crops (for example, non-genetically modified basmati crops such as Pusa Basmati 1979 and Pusa Basmati 1985) for reducing area under TPR in favor of DSR. These herbicide-tolerant rice varieties, if cultivated following the stewardship guidelines, could provide an effective solution for weed management (weedy rice in particular) in DSR.

Wheat Wheat is the important staple food and the second cereal crop in terms of area and production after rice in India. Wheat is the most important cereal crop of temperate regions, and it requires a cool climate with moderate rainfall. Yield losses in wheat due to *Phalaris minor* Retz. alone may be up to 50%, and heavy infestations may lead to total crop failure (Malik and Singh 1995; Singh 2007). The crop biomass was significantly reduced at 200 plants m^{-2} of *P. minor* (Bhan and Froud-Williams 2006). The biology of *P. minor* is already discussed in detail by Chaudhary et al. (2021). Narrowing the crop row width, increasing the crop density, and sowing competitive cultivars can also result in a dense canopy soon after emergence and may considerably reduce the quantity of light reaching the soil surface.

In the rice-wheat cropping system, production systems have changed significantly for wheat and rice. It was demonstrated that though seeds of *P. minor* get distributed in upper 10 cm soil depth during the intensive tillage and puddling performed during the preceding rice crop, there was 50% reduction in the emergence of the first cohort of *P. minor* in zero-till wheat compared to conventional-till sown wheat, irrespective of the texture of soil (Singh et al. 1999; Franke et al. 2007). Further, TPR-wheat cropping system has *P. minor* as predominant weed in wheat. However, DSR/cotton/corn-wheat cropping systems have *Avena* spp. along with *P. minor* (Kaur et al. 2022). Recently, adoption of zero/conservation tillage with or without residue mulching and dry-direct seeding (without puddling) in wheat and rice, respectively, has significantly modified the weed flora and management strategies. In conservation agriculture systems of rice-wheat, where residue of the previous crop is retained on the soil surface as mulch (e.g., Happy Seeder, Surface Seeder, and Smart Seeder-sown wheat), light exposure of *P. minor* seeds is prohibited, which leads to reduction in germination of *P. minor*. The presence of residue as mulch moderates soil temperature and moisture along with its effect on weed germination and emergence that ultimately enhances the crop vigor. Thus, sustainable weed control is achieved with Integrated Weed Management (IWM), combining conservation tillage practices with residue as mulch on the soil surface (Chauhan et al. 2012). Crop residues may help in weed control through allelopathic effects and/or by blocking light transmission. However, the amount of crop residue has a differential effect on weed germination, and a smaller residue load may not be sufficient for covering the soil surface in even and uniform way, and may rather trigger weed germination and pose challenges for weed management under zero tillage. The efficacy of pre-emergence herbicide in residue-retained fields may be lowered.

Corn Corn, being a C_4 plant, is one of the most vibrant food grain crops having wider adaptability under diverse soil and climatic conditions. In India, corn is the third most important cereal crop after rice and wheat, being used both as grain and fodder crop. Along with the main use of corn as food crop and animal feed, it is a major raw material for diverse industrial uses. Corn production steadily increased to 33.7 million tons during 2021–2022, with a targeted production of 35.5 million tons projected for 2023–2024.

Weeds pose a significant economic challenge to corn cultivation, jeopardizing crop quality by contaminating seeds and vying for essential resources such as light, water, and nutrients (Table 2.1). Studies by Mukhtar et al. (2007) have demonstrated that uncontrolled weed growth can lead to staggering yield reductions of 58%–62% in winter and 67%–79% in summer, accompanied by a substantial 65% decrease in plant height under similar weedy conditions. The first 20–60 days in the rainy season and 30–45 DAS in winter corn are the most critical periods of weed competition for the crop. The presence of weeds reduces the corn yields by 27%–60%, depending upon the season, floristic composition, and weed growth.

Pulses There are eleven types of pulses as identified by the United Nations Food and Agriculture Organization, including *Phaseolus vulgaris* L. (common bean), *Vicia faba* L. (broad bean), *Pisum sativum* L. (field pea), *Cicer arietinum* L. (chickpea), *Vigna unguiculata* (L.) Walp. (cowpea), *Cajanus cajan* (L.) Millsp. (pigeon pea), *Lens culinaris* Medik. (lentil), *Vigna subterranea* (L.) Verdc. (Bambara bean), *Vicia sativa* L. (common vetch), and *Lupinus albus* L. (lupin). India is the largest producer (25% of global production), consumer (27% of world consumption), and importer (14%) of pulses in the world. Pulses production surged to a peak of 27.3 million tonnes in India during 2021–2022. India leads the charge as the world's top producer of pigeon pea. Chickpea, the world's third-most cultivated legume, holds a position of immense significance in the semi-arid tropics, as it exhibits remarkable resilience in harsh environmental conditions, including heat and drought.

Table 2.1 Distribution of corn-associated weeds in India

Predominant weeds	Distribution	References
Echinochloa colona, Trianthema portulacastrum, Cyperus rotundus, Eleusine indica	Banswara, Rajasthan	Porwal (2000)
Echinochloa colona, Brachiaria ramosa, Digitaria sanguinalis, Dactyloctenium aegyptium, Eleusine indica, Setaria glauca, Sorghum halepense, Panicum spp., *Ageratum conyzoides, Galinsoga parviflora, Commelina benghalensis, Lindernia cilata, Polygonum hydropiper, Euphorbia geniculata, Oxalis latifolia, Cyperus rotundus*	Pantnagar, Uttaranchal	Pandey et al. (2001)
Cynodon dactylon, Cyperus rotundus, Echinochloa crus-galli, Echinochloa colona, Agropyron repens, Parthenium hysterophorus, Digitaria sanguinalis, Eclipta alba, Euphorbia hirta, Commelina benghalensis	Pantnagar	Sarma and Gautam (2006)
Cyperus rotundus, Digera arvensis, Commelina benghalensis, Euphorbia hirta, Parthenium hysterophorus, Cleome viscosa	Bihar	Jat et al. (2012)
Commelina benghalensis, Ageratum conyzoides, Echinochloa colona, Panicum dichotomiflorum, Cyperus iria, Digitaria sanguinalis, Polygonum alatum, Aeschynomene indica	Palampur, Himachal Pradesh	Kumar et al. (2012)
Echinochloa colona, Digitaria sanguinalis, Cyperus rotundus, Commelina communis, Phyllanthus niruri, Eclipta alba	Hyderabad	Sanodiya et al. (2013)
Celosia argentea, Commelina benghalensis, Dactyloctenium aegyptium, Digera arvensis, Eleusine indica, Echinochloa colona, Corchorus trilocularis, Leptochloa chinensis, Rumex acetosella	New Delhi	Singh et al. (2015a)
Parthenium hysterophorus, Melilotus alba, Cyperus rotundus, Trianthema portulacastrum, Dactyloctenium aegyptium, Cynodon dactylon, Digera muricata, Amaranthus viridis, Commelina benghalensis, Eragrostis cilianensis, Chenopodium album, Trichodesma indicum, Digitaria sanguinalis, Euphorbia geniculata, Echinochloa colona	Hyderabad	Ram et al. (2017)

Table 2.2 Weed flora associated with pulse crops

Season	Weed type	Common weed species
Winter	Broadleaved weeds	*Chenopodium album, Fumaria parviflora, lathyrus* spp., *Melilotus alba, Vicia sativa, Cirsium arvense, Rumex dentatus, Rumex spinosus, Anagallis arvensis, Veronica agrestis, Medicago denticulata, Ageratum conyzoides, Argemone mexicana, Asphodelus tenuifolius, Carthamus oxyacantha, Coronopus didymus, Gnaphalium indicum, Lathyrus aphaca, Launaea nudicaulis, Solanum nigrum, Spergula arvensis, Vicia hirsuta, Amaranthus viridis, Physalis minima*
	Grasses	*Phalaris minor, Avena ludoviciana, Poa annua, Polypogon monspliensis, Lolium temulentum, Echinochloa colonum, Cynodon dactylon, Eleusine indica, Dactyloctenium aegyptium, Digitaria sanguinalis, Echinochloa crus-galli, Setaria glauca, Saccharum spontaneum, Sorghum halepense, Eragrostis tenella*
Summer	Annual grasses	*Echinochloa colona, Dactyloctenium aegyptium, Eleusine indica, Digitaria sanguinalis, Commelina benghalensis, Eragrostis* spp.
	Broadleaved weeds	*Trianthema portulacastrum, Digera arvensis, Phyllanthus niruri, Celosia argentea, Cleome viscosa, Cucumis trigonus, Eclipta alba, Euphorbia hirta, Trianthema monogyna, Convolvulus arvensis*
	Perennials	*Cyperus rotundus, Cyperus difformis, Cyperus iria*

Adapted from Kaur and Kaur (2021)

Weed infestation poses a significant threat to pulse cultivation in India, leading to substantial yield losses and economic challenges for farmers (Table 2.2). In India, pulses are mainly cultivated (84%) under rainfed conditions alongside non-legume crops, exacerbating the competition between crops and weeds. Significant losses were reported in *Vigna radiata* L. (green gram), common bean, and *Vigna aconitifolia* (Jacq.) Maréchal (moth bean) crops by Pramanick et al. (2014) and Kumar et al. (2016).

Cotton Cotton, known as "White Gold," is a key commercial crop in India and globally, with India being the largest producer, cultivating 13 million ha. India is the only country in the world that cultivates all four species of cotton. These species are *Gossypium arboreum* (Asian Cotton), *Gossypium herbaceum* (Asian Cotton), *Gossypium barbadense* (Egyptian Cotton), and *Gossypium hirsutum* L. (American Upland Cotton). Cotton is a crucial fiber crop, and its seeds are used in the vegetable oil industry and as fodder for milch cattle. Cotton crop is a long-duration crop, with its life cycle spanning 150–180 days depending on the variety, soil type, and climatic conditions.

Cotton crop faces significant challenges from abiotic and biotic stresses, particularly weed competition during the early stages of growth. The pervasive weed infestation in cotton cultivation is a critical challenge in all major cotton-growing regions in India, including the north, central, and south zones. Despite regional variations in cotton cultivation, these zones exhibit a comparable pattern of yield reduction due to the relentless competition from weeds. Weeds compete for moisture, nutrients, light, and space, leading to substantial yield losses of cotton up to 45%–85%. For instance, Pagar et al. (1995) reported from Nagpur that weed-free plots produced a seed cotton yield of 987 kg/ha, while unweeded plots yielded only 226 kg/ha, resulting in a yield reduction of 77%. The first 15–60 DAS are crucial for cotton growth, as weeds compete intensely during this period, making the crop particularly vulnerable to weed competition. If timely weed control is not implemented, grass weeds can aggressively rob essential nutrients and moisture from the soil, significantly reducing the resources available for cotton plants. This nutrient competition can hinder the growth and development of cotton, leading to reduced crop yield and quality. Effective weed management is crucial to prevent these invasive weeds from outcompeting cotton, ensuring that the crop receives adequate nutrients for optimal growth (Deshpande et al. 2006). Manual weeding is typically carried out 15–20 DAS, during the early leaf stage of the cotton crop. It is then repeated after 35–40 days, just before square formation, and finally after 55–60 days, prior to flowering. It helps in keeping the plot weed free and maintains soil moisture.

In the central zone, Raju and Tayade (2013) observed that cotton fields in gravelly soils are predominantly dominated by annual grasses and perennial weeds, particularly *Cyperus rotundus*, *Cynodon dactylon*, *Digitaria ciliaris* Retz. Koeler, and *Dinebra retroflexa* Vahl. Panzer. Furthermore, broadleaved weeds such as *Celosia argentea* (L.) and *Euphorbia hirta* (L.) are also prevalent in these cotton fields. In the North zone, Nayak et al. (2024) observed the predominance of several weed species in cotton fields. The most prevalent weed was *Cyperus rotundus* (57%), followed by *E. colona* (13%), *Parthenium hysterophorus* L. (11%), *T. portulacastrum* (7%), *Cynodon dactylon* (7%), and *Phyllanthus niruri* L. (6%). In southern Indian conditions at Madurai, Tamil Nadu, Chandrasekaran (2019) observed the dominance of *T. portulacastrum* (42%), *Cyperus rotundus* (34%), *Cynodon dactylon* (14%), and *Phyllanthus niruri* L. (3%), along with minor weed flora in cotton.

Sugarcane Sugarcane, a native of India, is widely cultivated in tropical and subtropical agroecosystems of the world as main and ratoon crop to produce sugar as well as bioenergy. In India, it is the largest crop in terms of production value amongst the important commercial crops. One-third of the sugarcane production is processed in sugar factory while two-third is used for making jaggery and *khandsari*. Weeds are considered major constraints to higher yields in sugarcane production. Weeds can reduce potential sugarcane yield by 12%–72% depending on the severity of infestation by competing for moisture, nutrients, and light during the growing season (Kanwar et al. 1992). Weeds remove 4 times more nitrogen and 2.5 times of potassium than that of sugarcane crops during the first 7 weeks' period. The duration of crop-weed competition in sugarcane is much greater than in other short season row crops due to long duration, wider row spacing, slow initial growth phase, and abundant supply of irrigation water and fertilizers (Shyam and Singh 2015). Weed problem is also very serious in the ratoon crop of sugarcane due to its little preparatory cultivation. Weed-free environment during initial stage after planting or ratooning is essential for realizing higher cane yield. Effective weed management should provide weed-free environment for the sugarcane from planting or ratoon onward as the case may be, up to at least 90–120 days (Singh et al. 2011). Owing to the initial slow growth of sugarcane, the inter-row spaces that is not covered by sugarcane canopy provide a promising microclimate for profuse growth of weeds under wide row planting.

More than 200 weed species comprising annuals, perennials, and parasitic weeds have been reported to invade the sugarcane fields. A mixed population of grasses, broadleaved weeds, and sedges grows with sugarcane under different agroclimatic conditions, seasons, edaphic conditions, crop rotations, weed management practices, and agronomic practices followed for cultivation of the crop. Weeds such as *Cynodon dactylon*, *Cyperus rotundus*, *Rottboellia cochinchinensis* (Lour.) Clayton, *Ipomoea* spp., and *Striga* sp. are very difficult to control in sugarcane. The infestation of binding weeds like *Ipomoea* spp., is a serious menace in many sugarcane-growing areas after final earthing up and causes a 15%–25% reduction in cane yield (Sarala et al. 2011). These binding weeds take support from cane stalk, spread, and wrap on the top plane, and exerts their weight resulting in lodging of cane and reduced cane and sugar yield (Marimuthu et al. 2002).

Striga spp., an obligate parasitic weed, is adversely affecting cane cultivation in some sugarcane-growing districts of tropical zone like Belgaum, Bagalkot, and Vijayapura (Manjunatha et al. 2018). *Striga* (three species, *Striga asiatica*, *S. gesnerioides*, and *S. hermonthica*) becomes a serious problem when sugarcane cultivation is done in the poor fertile soils along with the moisture deficit conditions. *Striga* spp. are annual, photosynthetically active, obligate root parasites that depend on host plant for their growth. A single *Striga* plant produces numerous miniature seeds, which remain viable in soil for several years and will not germinate until host plant like sugarcane grows near to them and produces chemical signals called strigolactones. *Striga* is completely dependent on sugarcane for survival for nutrients and water and invades sugarcane host through root system. This eventually results in chlorosis, stunting growth and death of sugarcane crop. Infestation of *Striga* spp. is more common in low input, dry, infertile, and marginal soils with poor management practices (Dafaallah 2019). *Striga* management practices involve improvement of soil fertility, which has to be improved, and moisture stress should be avoided.

2.4 Evolution of Herbicide Resistance in Weeds

In India, *P. minor* was the first weed to evolve herbicide resistance against isoproturon during 1991 (Malik and Singh 1995). To date, 7 unique cases of 4 weeds, namely *Phalaris minor* (4 cases), *Rumex dentatus* (1 case), *Cyperus difformis* (1 case) and *Echinochloa crus-galli var crus-galli* (1 case), have evolved herbicide resistance in India (Table 2.3). There are 11 and 4 unique cases of herbicide resistance in *P. minor* that are confirmed globally and in India, respectively (Heap 2025). Globally, there is only one unique case of *Rumex dentatus* that has evolved herbicide resistance (Dhanda et al. 2022). Globally, there are 12 and 54 confirmed unique cases of *Cyperus difformis* and *Echinochloa crus-galli var crus-galli*, respectively.

Table 2.3 Confirmed unique cases of weeds resistant to different herbicide modes of actions in India

Year	Situation	Sites of action	Candidate herbicide(s)
Phalaris minor			
1991	Wheat	PSII inhibitors—serine 264 binders, HRAC group 5 (legacy C1 C2)	Isoproturon
1994		Inhibition of acetyl CoA carboxylase, HRAC group 1 (legacy A)	Diclofop-methyl, pinoxaden
2006		Multiple resistance: Three sites of action Inhibition of acetolactate synthase, HRAC group 2 (legacy B); Inhibition of acetyl CoA carboxylase, HRAC group 1 (legacy A); PSII inhibitors—serine 264 binders, HRAC group 5 (legacy C1 C2)	Clodinafop-propargyl, fenoxaprop-ethyl, sulfosulfuron, isoproturon, mesosulfuron-methyl, pinoxaden, pyroxsulam
2013		Inhibition of acetolactate synthase, HRAC group 2 (legacy B)	Iodosulfuron-methyl-Na, mesosulfuron-methyl
Rumex dentatus			
2014	Wheat	Inhibition of acetolactate synthase, HRAC group 2 (legacy B)	Florasulam, iodosulfuron-methyl-Na, mesosulfuron-methyl, pyroxsulam
Cyperus difformis			
2017	Rice	Inhibition of acetolactate synthase, HRAC group 2 (legacy B)	Bispyribac-sodium
Echinochloa crus-galli var. crus-galli			
2017	Rice	Inhibition of acetolactate synthase, HRAC group 2 (legacy B)	Bispyribac-sodium

Adapted from Heap (2025)
Note: PSII inhibitors: herbicides belonging to the Photosystem II inhibitors group; ACCase inhibitors: herbicides belonging to the acetyl-CoA Carboxylase inhibitors group; ALS inhibitors: herbicides belonging to the acetolactate synthase inhibitors group

Recently, herbicide resistance was also characterized in *Commelina communis* against imazethapyr in soybean. Amongst these unique cases, evolution of herbicide resistance against isoproturon in *Phalaris minor* was devastating, which affected about 0.8–1.0 million ha of wheat-growing regions (Malik and Singh 1995; Franke et al. 2007). Due to resistance evolution in *P. minor*, crop yield decreased by 25%–50% and farmers were forced to plow their fields under heavy infestations, during late 1990s (Kaur et al. 2022). In isoproturon-resistant *P. minor*, metabolic detoxification of the herbicide was the reason for evolution of resistance (Singh et al. 1998; Kumar and Guru 2016).

2.5 Strategies for Weed Management

The extent of crop-weed competition is influenced by crop, cultivar being grown, agronomic practices, and climatic factors. "Prevention is better than cure," and in this context, maintaining clean cultivation practices is crucial for abating weed menace in crops. Prevention is a cornerstone of integrated pest management, considered the most cost-effective strategy for growers (Norris et al. 2003). Preventive management is highly efficient regardless of property size, applicable from small vegetable crop seedbeds to extensive areas dedicated to major field crops (Christoffoleti et al. 2007). This accentuates the importance of keeping the field, its boundaries, water channels, farm roads, and the entire farm area free from weeds. It also highlights the use of clean (weed-seed free) seed, water, organic manure, and chemical fertilizers as prominent preventive weed control techniques. Farmyard manure should be well-decomposed to ensure that weed seeds are destroyed, and farm machinery must be thoroughly cleaned before use.

Cultural weed management techniques play a crucial role in weed control by completely removing weeds and providing competitive advantages to crops, as they reduce weed establishment. Cultural methods of weed control include planting the crop in right time with optimum planting density, raised bed sowing, residue mulching, earthing up, and applying plant nutrients in bands to boost the early growth of the crop for early canopy cover. Angiras et al. (2010) found that raised seedbeds had the lowest seed numbers of certain weeds at shallow depths (e.g., *D. sanguinalis*, *E. colona*, *Panicum dichotomiflorum* Michx.), while zero tillage had the lowest numbers at deeper depths. Growing trap crops like sorghum, corn, and cowpea to stimulate suicidal germination and reducing soil seed bank of *Striga*. Application of adequate farmyard manure or organic manure improves soil fertility, enhances soil moisture retention, and increases the competitive ability of sugarcane over *Striga*. Application of extra doses of nitrogen, split application of fertilizers, and light and frequent irrigations are also recommended in the Striga-infested areas (Singh et al. 2018).

Cultural weed management in pulses encompasses a range of strategies vital for mitigating weed competition and optimizing crop productivity. Timing of sowing is a critical factor, as evidenced by the research of Malik and Yadav (2014), who found that sowing pigeon peas in May resulted in reduced weed density and biomass, ultimately leading to higher yields compared to later sowings in June. Additionally, selecting appropriate planting methods can further suppress weed growth, as demonstrated by Muhammad et al. (2007), who showed that flat planting methods were more effective than ridge planting in reducing weed density and biomass.

An effective weed management program necessitates an in-depth understanding of weed species identification, their biology, associated weed communities, and their seed dispersal capabilities. Greater understanding of the factors influencing the germination of weed seeds could facilitate the development of more effective cultural weed management practices (Chauhan and Johnson 2010). Since weed communities are shaped by climatic conditions and soil characteristics, it is imperative to comprehend the location-specific critical periods of crop-weed competition. Moreover, understanding the traits of crop varieties, including their cross-resistance and multiple resistances to herbicides, is essential for optimizing weed control strategies and ensuring sustainable crop production. Some weed management options are discussed below.

Land Preparation Proper seedbed preparation significantly reduces weed problems, while neglected fields can lead to persistent issues that last for years until the weed seed bank is depleted. Conservation tillage affects weed dynamics through its effect on seed germination and emergence behavior (Chauhan et al. 2012). Deep plowing and collection and destruction of vegetative/asexual parts of weeds while preparing the field for planting crops would reduce their infestation in the crop. Deep summer plowing every 2–3 years immediately after harvest using subsoiler or moldboard plow, followed by two crosswise harrowings before the pre-monsoon showers eliminates the perennial weed seed propagules of *Saccharum sponatneum* L., *Ischaemum pilosum* Klein ex Willd. Wight, *C. rotundus*, and *C. dactylon*. Deep summer plowing effectively exposes the rhizomes and vegetative propagules of these perennial weeds to intense summer heat, thereby mitigating weed proliferation in succeeding crops.

Soil solarization with a 0.05 mm transparent polyethylene cover in April–May after irrigation effectively suppressed weeds and boosted baby corn yields compared to non-solarized soil. Proper land preparation and irrigation before solarization maximize its weed control efficacy and increase yields of baby corn and peanut (Thimmegowda et al. 2007). Further, puddling in rice fields is helpful in reducing the infestation of *C. dactylon* and *C. rotundus*. Most terrestrial cropped field weeds cannot survive in ponded fields; thus. water management is an excellent method to control diverse weed flora.

Blind tillage, a non-selective tillage technique usually followed in sugarcane, can be employed prior to crop sowing to manage early-season weeds. This practice disrupts the soil surface indiscriminately, without targeting specific crop rows, uprooting, and burying the newly emerged weed seedlings. By diminishing the competition for vital resources such as nutrients, water, and light, blind tillage significantly reduces weed pressure, thereby fostering an environment conducive to robust crop growth.

Stale Seedbed Technique The stale seedbed technique involves the removal of initial flushes of weeds before the sowing of rice, cotton, corn, and pulses. This method permits the germination of weed seeds situated just beneath the soil surface, which are then eliminated prior to crop sowing. By depleting the weed seed bank before sowing the crop, this technique mitigates competition between the crop and weeds during critical growth phases. The technique entails thoroughly moistening a meticulously prepared field through irrigation or rainfall, encouraging weed seeds to germinate. Subsequently, a shallow tillage operation or the application of non-residual herbicides can be utilized to destroy the dense flush of weed seedlings.

The stale/false seedbed technique aims to diminish the weed seed bank by leveraging seed germination biology. Its effectiveness hinges on various factors influencing weed seed germination and seedling emergence, such as soil temperature, diurnal temperature fluctuations, soil moisture, light exposure, soil nitrate levels, and soil gaseous conditions (Merfield 2017). Leguminous cover crops such as cowpea or *Crotalaria juncea* L. (sunn hemp) are highly effective in suppressing weed growth and can be judiciously combined with the stale seedbed technique to provide a sustainable weed management strategy. Further, weedy rice and volunteer rice plants can easily be controlled with this eco-friendly technique.

Crop Rotation Cultivation of diverse crops in rotation affects weed floristic composition present in a field, primarily due to different agronomic practices. Crop rotation involves the sequential cultivation of diverse crops within the same field. Effective rotations include crops that mitigate the prevalence of troublesome weeds in subsequent seasons. Crop-bound (*Striga*) and crop-associated weeds (*Echinochloa* spp. in rice, *P. minor* in wheat) can easily be controlled following the crop rotation. The strategic integration of crop and herbicide rotations has proven highly effective in weed management. By incorporating diverse crops into rotations, one can utilize herbicides with different modes of action, thereby mitigating the risk of the evolution of herbicide resistance (Norsworthy et al. 2012).

Monocultures significantly influence floristic composition, potentially exacerbating the spread of resistant and detrimental weed species. Crop rotation exerts a profound influence on the soil weed seed bank, facilitating the management of both weed species and their seeds and seedlings (Cardina et al. 2009). Such diversity in rotational crops offers multiple advantages for soil health and subsequent crops in terms of pest management. For example, the fibrous root systems of cereals help alleviate soil compaction, nitrogen-fixing legumes bolster soil fertility, and rotating crops with vegetable crops ensures a high level of fertilizer residual carryover. Consequently, crop rotations emerge as highly efficacious strategies, providing incremental yield benefits compared to monoculture systems.

Rotating cotton with alternative crops can effectively diminish the prevalence of established cotton-associated weeds. Implementing a cotton-cereal rotation, for instance, is instrumental in controlling various broadleaf weeds. The practice of rotating cotton with corn not only enhances soil organic carbon content, primarily due to the substantial biomass produced by corn (Mitchell and Entry 1998), but also benefits from corn's rapid growth and extensive canopy coverage, which effectively suppresses weeds that could otherwise impact cotton in subsequent crops. Cotton-based rotation systems can significantly impact both the cotton crop and its environment, potentially resulting in a 2.6%–4.5% increase in seed cotton yield when legumes are incorporated into the rotation, as opposed to conventional cotton monoculture. This yield enhancement is attributed to improved macroaggregate formation, stemming from deeper root systems, leaf drop, augmented rhizosphere activity, and root exudates. In central India, a biennial rotation of cotton and soybean has been shown to enhance both cotton productivity and nitrogen use efficiency relative to continuous cotton cultivation. Conversely, in the irrigated regions of northern

India, rotations involving cotton with wheat, chickpea, or barley have proven to be the most advantageous.

Intercropping Intercropping, a practice that involves growing complementary crops in the same field, can significantly suppress weed growth. This method reduces reliance on herbicides, offering an efficient, cost-effective, and environmentally friendly alternative (Gnanavel and Natrajan 2014). Intercropping not only lessens the emergence of secondary weeds that may appear after the intercrops have fully covered the ground, but it also necessitates manual intervention, such as hand weeding, to manage weeds that emerge prior to full canopy cover. In the realm of conservation agriculture, cover crops or mulches are recommended as intercrops or companion crops to aid in weed suppression. Intercropping with legumes and cover crops reduces *C. dactylon* infestation since this weed does not tolerate shade (Singh et al. 2018).

Cotton, a long-duration crop characterized by its wide spacing and slow early growth, often leaves substantial interspaces that are prone to weed infestation, resulting in approximately 45% crop loss. These expansive interspaces present prime opportunities for weed proliferation. In India, short-duration legumes such as clusterbean (*Cyamopsis tetragonoloba* (L.) Taub.), soybean, cowpea, pigeon pea, black gram (*Vigna mungo* (L.) Hepper), and green gram are commonly grown between widely spaced cotton plants to enhance weed suppression and overall system productivity. A 2-year study conducted by Chandrasekaran (2019) demonstrated that intercropping cotton with black gram effectively reduced the density of grass weeds compared to sole cotton. Moreover, black gram was more successful in controlling sedge populations and achieved a notable reduction in broadleaved weeds compared to other systems.

Mulching Mulching involves applying a protective layer of material to the soil surface. The efficacy of weed suppression is intricately linked to the mulch area index, which limits light penetration to newly germinated weed seeds. Mulch inhibits weed seedling growth by preventing light penetration or excluding specific wavelengths crucial for photosynthesis (Kamara et al. 2000). Singh et al. (2015b) observed that mulching significantly decreased mean seasonal soil evaporation compared to areas without mulch. Additionally, mulching enhanced grain yield, yield attributes, and water use efficiency. The literature suggests that mulching improves soil moisture retention, leading to enhanced water-holding capacity (Jabran 2019).

For optimal results, organic mulches should be applied at a thickness of 10–15 cm to ensure sufficient light exclusion. The principle behind mulching is to obstruct sunlight from reaching the photosynthetic portions of weeds, thereby inhibiting their development. Many weed seeds require light to trigger germination; thus, mulches diminish the germination rates of these seeds. Furthermore, emerging weeds may succumb to starvation due to the lack of photosynthesis, or, if allelopathic mulches are used, may die from chemical growth inhibition. In corn cultivation, Pandey et al. (2000) observed that mulching with pine needles improved yield attributes and water use efficiency compared to areas without mulch. Dry leaves of sugarcane or any other crop can be applied as mulch in the interspaces of sugarcane to check weeds. In addition to weed control, this trash will add nutrients to the soil upon decomposition (Singh et al. 2018).

In cotton cultivation, various types of mulch, such as organic crop residues, dry shredded cotton stalks, or polyethylene film employed to effectively suppress weed growth. According to Nalayini et al. (2009), polyethylene mulch with a thickness of 30 μm achieved complete weed control in cotton, although it was less effective against *C. rotundus*. Wheat straw, when used as mulch at a rate of 10 t/ha, has been shown to significantly diminish weed biomass, enhance growth parameters, and boost seed cotton yield by 38.4% compared to untreated controls (Singh et al. 2021). Ramzan et al. (2012) assessed various organic mulch materials for their efficacy in weed management and revealed that different mulching materials, except plastic mulch, were less effective than manual weeding. Plastic mulch exhibited the lowest weed density, surpassing other materials such as wheat straw, banana leaves, sugarcane trash, and no mulch. Nalayini et al. (2023) reported a marked reduction in weed density in cotton fields under tropical conditions in Coimbatore, Tamil Nadu, when stale seedbed techniques were integrated with leguminous cover crops such as sunn hemp and cowpea. This integrative approach outperformed the conventional practice of pre-emergence application of pendimethalin followed by two manual weedings. The synergy of cover crops and stale seedbed techniques proved highly efficacious in depleting the weed seed bank in cotton cultivation, offering a superior alternative for sustainable weed management.

In addition to traditional mulching, cover crops such as sunn hemp and *Sesbania bispinosa* (Jacq.) W.Wight (spiny sesbania) represent a viable alternative for weed suppression. These fast-growing legumes contribute to nutrient capture, nitrogen fixation, the alleviation of plow pans, water retention, and the provision of organic matter when incorporated into the soil prior to cotton sowing (Sharma and Ghosh 2000). The effectiveness of cover crops is contingent upon timely planting and robust biomass production. Cover crops also offer soil erosion protection and can be cultivated between wide row spaces of cotton or in the field's side rows. Before flowering, typically 45–60 days post-sowing, the cover crops can be cut and applied as mulch. This biomass serves as a physical barrier to weed growth and may also release allelopathic chemicals to further inhibit weed proliferation.

Mechanical Methods Mechanical and physical methods of weed control have been employed since the advent of agriculture and remain integral to weed management strategies in India. These methods encompass a range of practices, including tillage, hoeing, hand weeding, digging, intercultural operations, and mulching. Mechanized weeding entails physically eliminating weeds from the cropping system to diminish competition with crops. Beyond weed management, it can also aerate the soil and enhance its quality, occasionally exerting a greater influence on crop yield than solely focusing on weed control.

Mechanical weeders span from manual weeding to advanced implements propelled by tractors. Tools for mechanical cultivation include hoes, harrows, tines, brush weeders, and cutting apparatus such as mowers and sickles. The selection of these tools depends on factors like the type of crop, weed species, and the economic circumstances of farmers. Manual weeding with hand tools has long been practiced as a conventional means of weed management. Brush weeding involves uprooting and burying weeds between crop rows, proving effective in wet conditions. Harrowing, a traditional method, is commonly employed before and after corn seedling emergence, especially in hilly regions, to foster corn seedling competitiveness and prepare the land for subsequent weeding. Additionally, heavy machinery like mechanical hoes and mowing machines is utilized for weed control. Manual hand pulling of weeds, removal of weeds by using small hand tools like hand hoes, and control of weeds by using inter-cultivation implements drawn by animals or tractors all constitute mechanical methods of weed control. Hand weeding remains a practical and effective approach for managing both annual and perennial weeds, though it is labor-intensive and particularly challenging during heavy rainfall. These techniques are particularly advantageous in areas where labor is abundant and affordable. The timing of hoeing is critical; it should be neither too early nor too late to ensure effective weed control during the crop's most vulnerable stages of competition. Research indicates that physical methods, such as burying weeds to a depth of 1 cm, are particularly effective, followed by cutting at the soil surface to manage weed populations.

Research indicates that two hand weedings at 21 and 42 DAS in irrigated corn are more effective in reducing weed dry matter and nutrient depletion by weeds (Subramanyam et al. 2001). Various yield parameters were significantly enhanced in weed-free plots managed through hand weeding compared to control plots, contributing to crop development and increased corn grain yield (Shrestha et al. 2018). In corn, Sharma et al. (2010) found that hand weeding at 25 and 45 DAS resulted in higher yields with minimal weed density and dry weight. Similarly, Prasad et al. (2008) reported that manual weeding at 15 and 30 DAS achieved the highest weed control efficiency (70.9%) along with higher grain yield. Though effective, it is labor-intensive and costly, remaining a viable method for weed control in both cropped and uncropped soils (Sharma et al. 2000). Hand hoeing at 25 and 40 DAS resulted in the lowest weed density and dry weight, contributing to 44.22% of yield enhancement over the control in green gram (Kaur et al. 2010).

In cotton fields, weed removal using tools such as the hand hoe or wheel hoe is generally more efficient than hand pulling. Commonly, manual weeding is the major and most effective method of weed control, which demands more labor and is cumbersome. Many times, this operation may not be possible if it rains. In central India, mechanical weed management in cotton typically begins around 30 DAS. This approach includes three to four rounds of hoeing (inter-cultivation) at 15-day intervals, supplemented by two to three rounds of manual hand weeding. Specifically, hoeing and hand weeding conducted at 20, 40, and 60 DAS have been shown to be highly effective in managing weeds and are widely practiced in the cotton-growing regions of Vidarbha, Maharashtra (Patil et al. 1997). Similarly, Kakade et al. (1999) observed that employing manual weed control methods, specifically three rounds of hoeing combined with three rounds of hand weeding, resulted in a significantly higher number of bolls and increased boll weight, culminating in a 50.8% increase in seed cotton yield compared to the untreated weedy check. *Cyperus rotundus*, one of the world's most problematic perennial weeds, is difficult to control using conventional methods like hand weeding and hoeing.

Timely removal of weeds, particularly before these reach the flowering stage, is crucial; otherwise, they can pose a persistent challenge in subsequent cropping seasons. Incorporating weeds and farm wastes into on-site composting practices can recycle biomass effectively back into the field. Efforts to mechanize the weeding process in cotton have been pursued globally to enhance weed management efficiency and economic viability. In this context, two self-propelled rotary weeders and a tractor-operated sweep weeder have proven to be effective for inter-row weeding in cotton. Dixit et al. (2011) reported a weed control efficiency of 94%–95%, along with cost savings of 50%–60% and labor savings of 70% compared to manual weeding.

Mechanical weeding is a popular method of weed control in sugarcane. Weeds are generally removed by blind hoeing operations within 20–35 days after planting, and weeding/interculture is done from rows of standing crop with the help of power tillers or bullock–/tractor-drawn tillers. Earthing up operation after fertilizer application, either manually or by using ridge plows, incidentally controls weeds to some extent. Mechanical weeding at 30, 60, and 90 days after planting is an effective practice for weed management in sugarcane (Patel et al. 2024). Though interculture is not efficient for controlling weeds (as weeds within crop row are not removed), it is less costly.

In recent advancements, robotic weeding has shown significant potential in enhancing the efficiency of mechanical weed management in corn. Recent advancements in robotic systems have significantly improved mechanical weed management in corn. Robotic weeding, guided by precise navigation data, has been shown to enhance weed control efficiency by targeting specific areas with minimal soil disturbance (Bakker et al. 2010). Furthermore, innovative approaches to weed detection, such as the use of unmanned aerial vehicles (UAVs) combined with improved machine learning models like the enhanced YOLO v5n, offer promising results in mapping and identifying weed infestations with high accuracy. Chen et al. (2024) demonstrated that their novel weed detection strategy achieved a high R^2 value of 0.96 and a root mean square error of 3.08% in detecting weeds across various coverage levels. Building on this, Zhu et al. (2022) introduced a YOLOX-based weeding robot with a blue light laser, reducing weed dry weight by 85% and maintaining low seedling injury. These innovations collectively enhance weeding efficiency and contribute to sustainable agricultural practices by reducing herbicide usage.

Chemical Method Herbicides have been pivotal in weed management strategies for the past half-century, with their utilization increasing notably in recent years. Prolonged rainfall often renders soils unmanageable, obstructing timely weed control measures and coinciding with acute labor shortages during critical early growth stages. Under these challenging conditions, precise chemical weed control stands out as an efficient and economically viable solution. Herbicides contribute significantly to weed control, environmental preservation, and the conservation of labor required for weed management practices.

In this, herbicides are applied either as pre-emergence or post-emergence. Amongst herbicide application timings, pre-plant or pre-emergence applications offer initial advantages in weed suppression. The effectiveness of pre-emergence herbicides is contingent upon sufficient soil moisture to activate weed suppression. Meanwhile, advanced formulations of post-emergence herbicides offer broad-spectrum control, proving invaluable during labor-deficient periods for targeted weed management. The application rate and timing of herbicides are intricately tied to specific environmental and crop conditions, necessitating precise, site-specific application methodologies to maximize their efficacy. According to the Central Insecticide Board & Registration Committee, there are many herbicides recommended for weed control in rice and wheat crops, as given in Tables 2.4 and 2.5, respectively (Anonymous 2025b).

Earlier, there were recommendations of pre-emergence herbicides such as atrazine, alachlor and simazine in corn. Further, post-emergence herbicides are included in the package of this crop for weed control (Table 2.6). The highest stover and grain yield, as well as other parameters such as grain weight and grains per cob, were achieved with the sequential application of atrazine at 750 g a.i./ha with tembotrione at 100 g a.i./ha at 15–20 DAS (Bhagat et al. 2019; Lukangila et al. 2024). Chhokar et al. (2019) investigated combinations of mesotrione and atrazine for managing diverse weed flora and observed higher weed control efficiency with the application of mesotrione + atrazine at 91 + 909 g a.i./ha, leading to significant grain yield. Maximum weed control efficiency (98%) was attained with the pre-emergence application of atrazine at 500 g a.i./ha along with pendimethalin at 250 g a.i./ha in corn grown on sandy loam soils (Patel et al. 2006). Brankov et al. (2024) investigated post-emergence herbicides for controlling weedy sunflower in corn, highlighting the effectiveness of tembotrione, bentazone, mesotrione, and *ALS* inhibitors such as rimsulfuron and nicosulfuron.

Chemical weed control is crucial for enhancing yield in pulse crops like green gram, black gram, soybean, and clusterbean (Table 2.7). Pratap et al. (2024) demonstrated that post-emergence application of fomesafen 11.1% w/w + fluazifop-p-butyl 11.1% w/w (25% SL) at 250–312.5 g a.i./ha effectively controlled diverse weeds in green gram, achieving the highest grain yield of 1220 kg/ha with no phytotoxic effects. Similarly, Ciontu et al. (2024) found that combining pre-emergence herbicides, such as aclonifen and s-metolachlor, with post-emergence treatments in chickpea effectively managed weeds, though challenges remained with perennial species like *Convolvulus arvensis*. These findings underscore the importance of strategic herbicide application in pulse crop management.

In cotton, herbicides like fluchloralin and pendimethalin are effective against both grass and broadleaf weeds. Table 2.8 outlines the optimal timing and dosages of these herbicides to manage annual grass weeds effectively when they reach 3–4 inches in height or are 20–25 days old. Notably, pendimethalin is a potent

Table 2.4 Herbicides approved for use in rice in India

Herbicide name	Dosage (g a.i./ha)	Waiting period before harvest (days)
Rice nursery		
Bispyribac sodium 10% SC	20–25	–
Penoxsulam 21.7% SC	22.5	60
Pyribenzoxim 5% EC w/w	30	–
Direct seeded rice		
Azimsulfuron 50% DF	35	59
Bispyribac sodium 10% SC	20–25	78
Carfentrazone ethyl 40% DF	25	102
Cyhalofop butyl 10% EC	75–80	90
Fenoxaprop-p-ethyl 6.7% w/w EC	56.6–60.38	61
Florpyrauxifen-benzyl 2.7% w/w EC	31.25–37.5	85
Imazethapyr	100	75
Metamifop 10% EC	100	87
Oxyfluorfen 0.35% GR	100–150	–
Oxyfluorfen 23.5% EC	150–240	–
Paraquat dichloride 24% SL	300–800	–
Pendimethalin 30% EC	1000–1500	–
Pendimethalin 5% G	1000–1500	–
Penoxsulam 2.67% OD	25	85
Pretilachlor 30.7% EC	450–600	110
Propanil 80% w/w DF	2000–3000	84
Pyribenzoxim 5% EC w/w	30	89
Bispyribac sodium 2% + 2,4-D sodium salt 54.3% w/w SP	25.0 + 678.75	79
Bispyribac sodium 38% + Chlorimuron ethyl 2.5% + Metsulfuron methyl 2.5% (w/w) WG	43	64
Florpyrauxifen-benzyl 1.31% w/w + Penoxsulam 2.1% w/w OD	15.63 + 25	83
Florpyrauxifen-benzyl 2.13% w/w + Cyhalofop-butyl 10.64% w/w EC	150–180	92
Penoxsulam 1% + Pendimethalin 24% w/w SE	22.5–25.0 + 540–600	100
Penoxsulam 1.02% + Cyhalofop-butyl 5.1% OD	120–135	60
Pretilachlor 30.0% + Pyrazosulfuron ethyl 0.75% WG	600 + 15	113
Triafamone 20% + Ethoxysulfuron 10% WG	44 + 22.5	83
Pretilachlor 27.4% w/w + Florpyrauxifen-benzyl 0.91% w/w EC	600 + 20	95
Pyribenzoxim 5% EC w/w	30	89
Direct seeded imazethapyr-herbicide tolerant rice		
Imazethapyr 10% SL	100	75

(continued)

Table 2.4 (continued)

Herbicide name	Dosage (g a.i./ha)	Waiting period before harvest (days)
Puddled transplanted rice		
Anilophos 2% G	400–500	30
Azimsulfuron 50% DF	35	59
Bensulfuron methyl 60% DF (pre-emergence)	60	88
Bensulfuron methyl 60% DF (post emergence)	60	71
Bentazone 480 g/l SL	960	71
Bispyribac sodium 10% SC	20–25	78
Butachlor 5% GR	125–187	90–105
Butachlor 50% EW	125–150	–
Chlorimuron ethyl 25% WP + surfactant	6	60
Cinmethylin 10% EC	75–100	110
Clomazone 50% EC	400–500	90
2,4-D Ethyl ester 4.5% GR (having 2,4-D acid 4% w/w)	1000	–
Ethoxysulfuron 15% WDG	12.5–15	110
Fenoxaprop-p-ethyl 9% w/w DF	56.25	65
Fenoxaprop-p-ethyl 9.3% w/w EC (9% w/v)	56.25	70
Fenoxaprop-p-ethyl 6.7% w/w EC	56.6–60.38	61
Florpyrauxifen-benzyl 2.7% w/w EC	25–31.25	73
Flucetosulfuron 10% WG	25	90
Ipfencarbazone 25% SC (22.81% w/w)	125	91
MCPA, amine salt 40% WSC	800–2000	–
Metsulfuron methyl 20% WP	4	60
Metsulfuron methyl 20% WG	4	71
Orthosulfamuron 50% WG	60–75	65
Oxadiargyl 80% WP	100	97
Oxadiargyl 6% EC	100	97
Oxadiazon 25% EC	500	–
Oxyfluorfen 0.35% GR	100–150	–
Paraquat dichloride 24% SL	300–800	–
Pendimethalin 30% EC	1000–1500	–
Pendimethalin 5% G	1000–1500	–
Penoxsulam 21.7% SC	20–22.5 to 22.5–25	60
Penoxsulam 2.67% OD	22.5–25	60
Pretilachlor 50% EC	500–750	75–90
Propanil 35% EC	2000–3000	90
Pyribenzoxim 5% EC w/w	30	79
Pyrazosulfuron ethyl 10% WP	10–15	95
Pyrazosulfuron ethyl 70% WDG	21	43

(continued)

Table 2.4 (continued)

Herbicide name	Dosage (g a.i./ha)	Waiting period before harvest (days)
Quinclorac 250 g/l SC	187.5–250	78
Anilofos 24% + 2,4-D ethyl Ester 32% EC	240 + 320 to 360 + 480	90
Bensulfuron methyl 0.6% + Pretilachlor 6% GR	60 + 600	88
Bensulfuron-methyl 4.8% + Pretilachlor 48% OD	60 + 600	102
Bispyribac sodium 0.25% + Penoxsulam 0.25% + Pyrazosulfuron ethyl 0.20% GR	25 + 25 + 20	73
Bispyribac sodium 2% + 2,4-D sodium salt 54.3% w/w SP	25.0 + 678.75	79
Bispyribac sodium 7.6% + Fenoxaprop-p-ethyl 11.4% w/w WG	37.8 + 56.8	54
Bispyribac sodium 9.5% + Penoxsulam 7.8% w/v SC	23.75 + 19.5	78
Bispyribac sodium 20% + Pyrazosulfuron ethyl 15% WDG	20 + 15	130
Bispyribac-sodium 15% + Metsulfuron-methyl 2.4% + chlorimuron ethyl 3.6% OD	26.25 + 4.2 + 6.3	81
Bispyribac sodium 38% + chlorimuron ethyl 2.5% + Metsulfuron methyl 2.5% (w/w) WG	43	64
Clomazone 20% + 2,4-D Ethyl Ester 30% EC	250–375	100–110
Florpyrauxifen-benzyl 1.31% w/w + Penoxsulam 2.1% w/w OD	15.63 + 25	83
Florpyrauxifen-benzyl 2.13% w/w + Cyhalofop-butyl 10.64% w/w EC	150	67
Metsulfuron methyl 10% + Chlorimuron ethyl 10% WP	4	90
Pendimethalin 38.4% + Pyrazosulfuron ethyl 0.85% ZC	900 + 20	86
Penoxsulam 0.97% w/w + Butachlor 38.8% w/w SE	820	60
Penoxsulam 1.02% + Cyhalofop-butyl 5.1% OD	120–135	60
Penoxsulam 2% + oxadiazon 21% w/w SC	20 + 210	78
Pyriftalid 31.0% w/w + Bensulfuron methyl 15.7% w/w SC	116.1 + 58.9	99
Triafamone 20% + Ethoxysulfuron l0% WG	44 + 22.5	83
Bispyribac sodium 0.25% + Penoxsulam 0.25% + Pyrazosulfuron ethyl 0.20% GR	25 + 25 + 20	73
Bispyribac sodium 9.5% + Penoxsulam 7.8% w/v SC	23.75 + 19.5	78
Pretilachlor 27.4% w/w + Florpyrauxifen-benzyl 0.91% w/w EC	600 + 20	72
Pyribenzoxim 5% EC w/w	30	79
Butachlor 50% EC	1250–2000	90–120

(continued)

Table 2.4 (continued)

Herbicide name	Dosage (g a.i./ha)	Waiting period before harvest (days)
2,4-D Ethyl Ester 38% EC (having 2,4-D acid 34% w/w)	850	–
2,4-D Ethyl Ester 20% WP	900	–
Fenoxaprop-p-ethyl 9% w/w DF	56.25	65
Flufenacet 60% DF	120	90–110
Fenoxaprop-p-ethyl 5% + chlorimuron ethyl 0.6% + pretilachlor 50% ME	50 + 6 + 500	76
Oxadiargyl 1% + Pretilachlor 6% GR	100 + 600	85
Pretilachlor 6% + Pyrazosulfuron ethyl 0.15% (H)	600 + 15	83
Pretilachlor 6.0% + Pyrazosulfuron ethyl 0.15% GR	600 + 15	83

Adapted from Anonymous (2025b)

Table 2.5 Herbicides approved for use in wheat in India

Herbicide name	Dosage (g a.i./ha)	Waiting period before harvest (days)
Carfentrazone ethyl 40% DF	20	80
Clodinafop-propargyl 15%WP	60	110
Clodinafop-propargyl 15% w/w DF	60	70
2,4-D Dimethyl amine salt 58% SL	500–750	–
2,4-D Sodium salt technical (having 2,4-D acid 80% w/w) (earlier registered as 80%WP)	500–840	90
2,4-D Ethyl Ester 38% EC (having 2,4-D acid 34% w/w)	450–750	–
2,4-D Ethyl Ester 20% WP	900	–
2,4-D Sodium salt 50% (w/w) + Metribuzin 15% (w/w) WP	400 + 120	120
Dicamba 48% w/v SL	240	91
Diclofop methyl 28% EC	700–1000	90
Fenoxaprop-p-ethyl 10% EC	100–120	110
Flumioxazin 50% SC	125	137
Isoproturon 50% WP	1000	–
Isoproturon 75% WP	1000	60
MCPA, amine salt 40% WSC	1000	–
Methabenzthiazuron 70% WP (PE –2DAS)	1050–1400	100
Methabenzthiazuron 70% WP (post –EM 30 DAS)	1050–1750	100
Methabenzthiazuron 70% WP (early POE.16–18 DAS)	700–870	100
Metribuzin 70% WP	350–525	30
Metsulfuron methyl 20% WP	4	80
Metsulfuron methyl 20% WG	4	76

(continued)

Table 2.5 (continued)

Herbicide name	Dosage (g a.i./ha)	Waiting period before harvest (days)
Paraquat dichloride 24% SL (pre-plant (minimum tillage) before sowing)	1000	120–150
Pendimethalin 30% EC	1000–1500	–
Pinoxaden 5.1% EC	40–45	90
Pyroxasulfone 85% w/w WG	127.5	131
Sulfosulfuron 75% WG	25	110
Triallate 50% EC	1250	150
Triasulfuron 20% WG	20	81
Bixlozone 50% + Metribuzin 10% WG	625 + 125 to 750 + 150	120
Carfentrazone ethyl 20% + Sulfosulfuron 25% WG	20 + 25	110
Clodinafop propargyl 15% + Metsulfuron methyl 1% WP	60 + 4	100
Clodinafop propargyl 9% + Metribuzin 20% WP (w/w)	54 + 120	120
Fenoxaprop-p-ethyl 7.77% w/w + Metribuzin 13.6% w/w EC	100 + 175	110
Halauxifen-methyl 1.21% w/w + Fluroxypyr-meptyl 38.9% w/w EC	6.05 + 194.5	87
Halauxifen-methyl 20.8% + Florasulam 20.0% w/w WG (with surfactant polyglycol 26–2 N)	12.76	71
Mesosulfuron methyl 3% + Iodosulfuron methyl sodium 0.6% WG	12 + 2.4	96
Metribuzin 42% + Clodinafop propargyl 12% WG	210 + 60	92
Metsulfuron methyl 10% + Carfentrazone ethyl 40% DF	5 + 20	100
Pendimethalin 35% + Metribuzin 3.5% w/w SE	875 + 87.5 to 1050 + 105	123
Pendimethalin 40% + Metribuzin 8% EC	1000 + 200	107
Sulfosulfuran 75% + Metsulfuron methyl 5%WG	30 + 2	110

Adapted from Anonymous (2025b)

pre-emergence herbicide that effectively controls up to 75% of weed species, including annual grasses (Raju and Tayade 2013). In India, the prevalent practice for weed management involves the application of pre-emergence herbicides, notably pendimethalin, followed by two to three rounds of inter-cultivation. The efficacy of pre-emergence herbicides, especially fluchloralin is optimized when they are incorporated 1–2 inches into moist soil using a blade harrow (*Bakhar*) subsequent to application. Weed management methodologies utilizing these herbicides typically require additional measures, including 2–3 rounds of inter-cultivation for soil aeration at 21-day intervals, along with a hand weeding operation at 30–35 DAS to eliminate any uncontrolled or resistant weed populations. The efficacy of pre-emergence herbicides is substantially enhanced when integrated into a

Table 2.6 Herbicides approved for use in corn in India

Herbicide name	Dosage (g a.i./ha)	Waiting period before harvest (days)
Atrazine 50% WP	500–1000	–
2,4-D Dimethyl amine salt 58% SL	500	50–60
2,4-D Sodium salt technical (having 2,4-D acid 80% w/w) (earlier registered as 80%WP)	1000	120 (pre-em) 90 (post-em)
2,4-D Ethyl Ester 38% EC (having 2,4-D acid 34% w/w)	900	50–60
2,4-D Ethyl Ester 20% WP	900	–
Dicamba 48% w/v SL	360	69
Diuron 80% WP	800	–
Halosulfuron methyl 75% WG	67.5	45
Paraquat dichloride 24% SL [pre-plant (minimum tillage) before sowing]	200–500	90–120
Paraquat dichloride 24% SL (post emergence directed inter row application at 2–3 leaf stage of weeds)	200–500	90–120
Pyroxasulfone 85% w/w WG	127.5	103
Tembotrione 34.4% SC	120	55
Topramezone 336 g/l w/v SC	25.2 to 33.6	83
Tolpyralate 40% (w/v) SC	50	61
Halosulfuron methyl 5% + Atrazine 48% WG	56.25 + 540	37
Mesotrione 2.27% w/w + Atrazine 22.7% w/w SC	875	42
Saflufenacil 68 g/l + Dimethenamid- p 600 g/l EC	68 + 600 to 85 + 750	98
Topramezone 10 g/l + Atrazine 300 g/l SC	775	90
Topramezone 32 g/l + Dimethenamid-p 538 g/l EC	32 + 538	85
Tembotrione 9% + Atrazine 45% WG	135 + 675	81

Adapted from Anonymous (2025b)

comprehensive weed management strategy. For instance, in cotton cultivation, the pre-emergence application of pendimethalin at 1000 g a.i./ha, followed by quizalofop-ethyl at 50 g a.i./ha at the 2–4 leaf stage of weeds and one subsequent hoeing, resulted in a weed control efficiency of 71%. However, the weed-free check exhibited a superior efficiency of 90.3% (Singh et al. 2016).

Generally, pre-emergence herbicides offer superior weed control during the critical initial 30–40 days of cotton growth. However, as the weed seed bank in the soil continues to contribute to subsequent weed flushes, these emerging weeds begin to compete with cotton crop for essential nutrients and soil moisture. In such scenarios, application of post-emergence herbicides not only curtails hand weeding costs but also mitigates potential crop losses. Post-emergence herbicides are predominantly foliage-active and require a minimum of 1–4 h of rain-free weather to ensure optimal efficacy. These herbicides have increasingly gained the trust of farmers, particularly in recent normal and wet seasons. Commonly used herbicides in cotton include graminicides such as quizalofop ethyl, fenoxaprop methyl, and

Table 2.7 Herbicides recommended for various pulses

Herbicide common name	Pulse crop	Dose (g a.i./ha)	Waiting period before harvest (days)
Pendimethalin 30% EC	Soybean	750–1000	110
	Pigeon pea	700–1000	133
Pendimethalin 38.7% CS	Soybean	580.5–677.25	40
Pendimethalin 30% + Imazethapyr 2% EC	Soybean	750 + 50 to 900 + 60	90
Metribuzin 70% WP	Soybean	350–525	30
Imazethapyr 10% SL	Soybean	75–100	72
	Black gram	75	56
	Green gram	75	46
	Pigeon pea	75	125
Imazethapyr 70% WG + Surfactant	Soybean	70	56
Clodinafop-propargyl 12.5% EC	Green gram	125	75
Metolachlor 50% EC	Soybean	1000	–
Pyroxasulfone 85% w/w WG	Soybean	127.5	94
Quizalofop ethyl 5% EC	Soybean	37.5–50.0	95
	Black gram		52
Quizalofop ethyl 10% EC	Soybean	37.5–45.0	69–103
Quizalofop-p-tefuryl 4.41% EC	Soybean	30–40	30
Quizalofop ethyl 10% EC + Chlorimuron ethyl 25% WP + Surfactant (twin pack)	Soybean	37.5 + 9	75
Quizalofop ethyl 7.5% + Imazethapyr 15% w/w EC	Soybean	32.81 + 65.625	75
Sulfentrazone 39.6% w/w SC	Soybean	360	88
Fenoxaprop-p-ethyl 6% + Chlorimuron ethyl 0.9% + Imazethapyr 10% SC	Soybean	60 + 9 + 100	64
Fluthiacet-methyl 2.5% + Quizalofop-ethyl 10% EC	Soybean	12.5 + 50	67
Fluazifop-p-butyl 11.1% w/w + Fomesafen 11.1% w/w SL	Soybean	250	71
Fomesafen 12.5% + Fenoxaprop-p-ethyl 10% + Chlorimuron ethyl 0.9% ME	Soybean	125 + 100 + 9	78
Fomesafen 12% + Quizalofop ethyl 3% w/w SC	Soybean	180 + 45	71
Fomesafen 12.5% + Quizalofop ethyl 4.68% EC	Soybean	125 + 46.8	77
Fomesafen 17.5% + Clodinafop-propargyl 12.5% w/w ME	Soybean	175 + 125	101
	Green gram		83
Fomesafen 16.8% w/w + Propaquizafop 5.2% w/w ME	Black gram	210 + 65	51
	Green gram		46

(continued)

Table 2.7 (continued)

Herbicide common name	Pulse crop	Dose (g a.i./ha)	Waiting period before harvest (days)
Haloxyfop-R methyl 12.8% + Imazethapyr 10% (w/w) ME	Soybean	105.6 + 82.5	88
Propaquizafop 10% EC	Soybean	50–75	21
	Black gram	75–100	21
Propaquizafop 2.5% + imazethapyr 3.75% w/w ME	Soybean	50 + 75	80
	Black gram		56
	Clusterbean		69
Imazethapyr 35% + imazamox 35% WG	Soybean	70	56
	Clusterbean		64
	Pigeon pea		125
Sodium aciflourofen 16.5% + clodinafop propargyl 8% EC	Soybean	165 + 80	61
	Black gram		54
	Green gram		54
Diclosulam 0.9% + pendimethalin 35% SE	Soybean	22.5 + 875	107
Bentazone 480 g/l SL	Soybean	960	62
Chlorimuron ethyl 25% WP + Surfactant	Soybean	9	45
Clethodim 25% w/w (240 g/L) EC	Soybean	120–180	62
Clomazone 50% EC	Soybean	750–1000	90
Diclosulam 84% WDG	Soybean	22–26	60
Ethalfluralin 35.65% w/w EC (36% EC)	Soybean	720	107
Fenoxaprop-p-ethyl 9.3% w/w EC (9% w/v)	Soybean	100	100
	Black gram	56.25–67.5 g	43
Fluazifop-p-butyl 13.4% EC	Soybean	125–250	90
Fluchloralin 45% EC	Soybean	1000–1500	120–150
Flumioxazin 50% SC	Soybean	125	110
Fluthiacet methyl 10.3% EC	Soybean	13.6	73
Haloxyfop R methyl 10.5% w/w EC	Soybean	108–135	60
	Black gram	108	60
Sulfentrazone 28% + Clomazone 30% WP	Soybean	350 + 375	84

Adapted from Anonymous (2025b)

propaquizafop (Singh et al. 2016; Rao 2018). Notably, pyrithiobac sodium, an early post-emergence herbicide with both soil and foliar activity, demonstrates significant efficacy in controlling *E. hirta*, *C. argentea*, and various other grasses, achieving over 75% weed control even under challenging wet monsoon conditions.

In sugarcane, many pre-emergence herbicides such as atrazine, metribuzin, ametryn, diuron, oxyfluorfen, oxadiazon, and pendimethalin have been found to be useful for controlling weeds up to 45–90 days after planting (Patel et al. 2024, Table 2.9). Pre-emergence herbicides must be applied on the soil surface as a blanket spray within 3–4 days of planting. In intercropped sugarcane, chemical weed control is

Table 2.8 Herbicides approved for use in in cotton in India

Herbicide name	Dosage (g a.i./ha)	Waiting period before harvest (days)
Clethodim 13% w/w EC (clethodim 12% w/v EC)	120	135
Diuron 80% WP	750–1500	–
Ethalfluralin 35.65% w/w EC (36% EC)	720	107
Fenoxaprop-p-ethyl 9.3% w/w EC (9% w/v)	67.5	87
Fluazifop-p-butyl 13.4% EC	167	104
Fluchloralin 45% EC	900–1200	180
Glufosinate ammonium 13.5% SL (15% w/v)	375–450	96
Haloxyfop R methyl 10.5% w/w EC	108	86
Paraquat dichloride 24% SL	300–500	150–180
Pendimethalin 30% EC	750–1250	150
Pendimethalin 38.7% CS	580.5–677.25	101
Propaquizafop 10% EC	62.5	115
Pyrithiobac sodium 10% EC	62.5–75	160
Quizalofop-ethyl 5% EC	50.5	94
Pyrithiobac sodium 3.1% w /w + pendimethalin 34.0% w/w ZC	649.25–742	152
Pyrithiobac sodium 6% w/w + Quizalofop-ethyl 4% w/w EC	60 + 40 to 75 + 50	160

Adapted from Anonymous (2025b)

not employed, as most of the intercrops are sensitive to pre-emergence herbicides. In such conditions, pendimethalin (1500 g a.i./ha) can be used as a pre-emergence herbicide (Singh and Lal 2008). However, weeds need to be controlled either through interculture or use of post-emergence herbicides in sugarcane as it is a long duration crop. Post-emergence herbicides such as halosulfuron and 2,4-D are commonly used in sugarcane. Nonselective herbicides such as paraquat, glyphosate, and glufosinate are used for controlling weeds in the inter-row spaces as directed spray in sugarcane and cotton, and the protective hood is used to stop herbicide fluid from falling on the crop stems and leaves (Singh and Kaur 2004). Alternatively, use of topramezone 33.6 g a.i./ha in combination with atrazine 900 g a.i./ha provided excellent control of *C. dactylon* and other broadleaved weeds (Ducca et al. 2020). Application of ethoxysulfuron 15% WG at 60 g a.i./ha effectively controlled *C. rotundus* and some broad-leaved weeds such as *Trianthema* spp., *Digera arvensis*, *Cleome viscosa* L. and *Ipomoea* spp. in sugarcane (Singh et al. 2014). Application of halosulfuron methyl 75% WG at 90 g a.i./ha as foliar spray at 45 DAS was effective for the control of *C. rotundus* in sugarcane without any adverse effects on succeeding black gram crop (Chand et al. 2014). Pre-emergence application of sulfentrazone at 1200 g a.i./ha efficiently killed *C. rotundus* tubers with subsequent reductions in the number of viable tubers (Singh et al. 2018).

In long-duration crops such as cotton and sugarcane, sequential application of herbicides is a requirement for sustainable weed control. Sequential application of atrazine or metribuzin or diuron followed by post emergence application of 2,4-D

Table 2.9 Herbicides approved for use in sugarcane in India

Herbicide name	Dosage (g a.i./ha)	Waiting period before harvest (days)
Ametryn 80% WDG	2000	311
Atrazine 50% WP	500–2000	–
Metribuzin 70% WP	1050–2000	60
Metribuzin 70%WG	1400–2000	309
Clomazone 50% EC	750–1000	296
Halosulfuron methyl 75% WG	60–67.5	294
Sulfentrazone 39.6% w/w SC	720	306
2,4-D Dimethyl amine salt 58% SL	3500	–
2,4-D Na salt technical (with 2,4-D acid 80% w/w)	2000–2600	300
2,4-D Ethyl Ester 38% EC (with 2,4-D acid 34% w/w)	1200–1800	300–330
Diuron 80% WP	1600–3200	–
Metsulfuron methyl 20% WP	6	346
2,4-D Sodium salt 48% + metribuzin 32% + Chlorimuron ethyl 0.85% WDG	2020	315
2,4-D sodium salt 44% + metribuzin 35% + Pyrazosulfuron ethyl 1.0% WDG	1320 + 1050 + 30	249
Ametryn 73.1% w/w + Trifloxysulfuron sodium 1.8% w/w WG	937.5–1125	221
Clomazone 22.5% w/w + metribuzin 21% w/w WP	563 + 525	307
Hexazinone 13.2% + Diuron 46.8% WP	264 + 936	282–306
Mesotrione 2.27% w/w + Atrazine 22.7% w/w SC	875	190
Sulfentrazone 28% + clomazone 30% WP	700 + 750	302
Topramezone 10 g/l + atrazine 300 g/l SC	930	268
Halosulfuron methyl 6% + metribuzin 50%WG	67.5 + 562.5	247
Halosulfuron methyl 12% w/w + Metribuzin 55% WG	54 + 247.5	294
Mesotrione 2.27% w/w + Atrazine 22.7% w/w SC	875	190
Metribuzin 16.8% + 2,4-D Sodium salt 56% WP	525 + 1750	308
Sulfentrazone 28% + Clomazone 30% WP	700 + 750	302
Topramezone 10 g/l + Atrazine 300 g/l SC	930	268

Adapted from Anonymous (2025b)

amine provided the most effective control of *Ipomoea* spp. in Punjab, India (Bhullar et al. 2012). Application of metribuzin at 1250 g a.i./ha + 2,4-D at 1500 g a.i./ha, effectively controlled *Ipomoea* when it is at 45 cm in Lucknow, India (Singh et al. 2018). To control *Striga* after emergence but before it flowers, 2,4-D Na salt can be applied at 1000 g a.i./ha as a directed spray and spray should be repeated two or three times with 15 days intervals. Pendimethalin at 2000 g a.i./ha when applied as a directed spray just before the second irrigation within sugarcane rows after full earthing up may control *Striga* at the emergence stage. Chemical control of *R.*

cochinchinensis is difficult as it is resistant to many herbicide groups like triazines (atrazine) and triazinones (metribuzin). The directed post-emergence application of paraquat and ametryn reduced infestation of *R. cochinchinensis* in sugarcane (Anusha et al. 2023). Clomazone, a triketone, is recommended to control *R. cochinchinensis* infestation in sugarcane both as pre-planting and pre-emergence after planting (Tropaldi et al. 2021).

Integrated Weed Management Traditional weed management methods can be both costly and labor-intensive and herbicides applied at sowing typically do not provide enduring weed control. IWM involves combination of different weed management methods along with agronomic practices (Harker and O'Donovan 2013). The reliance on one weed control technique for controlling weeds is reduced by following stewardship guidelines, and thus its effectiveness is being maintained. The objective of IWM is to maintain weed within manageable limits while taking care of weed shifts and evolution of herbicide resistance. Effective IWM strategies include crop rotation, stale/false seedbed, mulching, cover cropping, competitive cultivars, cropping density, geometry, and time of sowing along with judicious use of herbicides (Kaur et al. 2022).

The periodic weed survey/surveillance showed that the predominance of weeds keeps on shifting from one weed to another weed mainly due to adoption of improved agronomic practices like changes in cropping systems, better irrigation facilities, fertilizer application, sowing time, crop cultivars, and particularly weed management/herbicide use (Hamill et al. 2004). Therefore, weed survey/surveillance in any crop gives an insight into the extent of weed problems and knowledge of associated weed composition, which helps in devising weed management practices before it takes alarming proportions. Continuous application of a particular cropping practice can contribute inadvertently to a shift in dominance and distribution of weeds (Harker and O'Donovan 2013). Weeds readily adapt to practices that are continuously practiced due to selection pressure and weed hardiness. For example, continuous use of herbicides with the same mode of action imposes selection pressure, and there is faster evolution of herbicide resistance in weeds. Therefore, there should be integration of weed management tactics in conservation agriculture systems to prevent over-reliance on herbicides (Chauhan et al. 2012; Singh et al. 2024b). In formulating a weed management program, it is important to systematically manipulate recommended production practices in synchronization with the current location-specific farming activities. A better understanding of the mechanisms of interaction between the different species in an agricultural field population can lead to more effective weed management practices.

Effective weed control is crucial for achieving profitability in cotton cultivation, particularly during the early stages of growth when weeds pose the greatest competition. Pendimethalin application can maintain a weed-free status for 30–40 days; however, if weather conditions allow, hoeing can be performed subsequently, or alternative chemical methods should be utilized in its absence. This holistic strategy not only enhances crop productivity but also ensures effective weed management. Reddy et al. (2023) demonstrated that applying pendimethalin 38.7% CS at 750 g a.i./ha as a pre-emergence treatment, followed by pyrithiobac sodium 10 EC at 100 g a.i./ha and quizalofop ethyl 5 EC at 50 g a.i./ha as post-emergence treatments at 25 DAS, coupled with inter-cultivation at 55–60 DAS, proved highly effective in controlling weeds and resulted in increased seed cotton yield. For efficient and cost-effective weed management in irrigated cotton, it is recommended to apply pendimethalin at 1000 g a.i./ha as a pre-emergence treatment, followed by one hand weeding at 35–40 DAS and a mixture of pyrithiobac sodium 50 g a.i./ha with quizalofop ethyl 50 g a.i./ha at 60 DAS (Nalayini et al. 2012).

It is very important and necessary to use a combination of different methods, including different herbicides at different times along with mechanical weeding, to effectively control weeds and maximize corn yield. Applying half of the total recommended dose of atrazine as a pre-emergence herbicide, followed by hand weeding at 28 DAS, led to a significant improvement in corn yield by effectively reducing weed infestation compared to the sole application of herbicide at its full recommended dose (Pandey et al. 2002). Ramachandran et al. (2012) observed that a combination of alachlor at 1000 g a.i./ha applied as pre-emergence along with brown manuring resulted in superior yield attributes and the highest grain and stover yield. Further, pre-emergence application of atrazine 1000 g a.i./ha on the third day after planting + hand weeding on 45th day after planting and earthing up on the 60 days after planting combined with post-emergence spraying of 0.5% 2,4-D sodium salt (5 g/l) + 2% urea (20 g/l) on the 90th day after planting is recommended for complete control of *Striga asiatica* in sugarcane.

Seed predation by granivore fauna, including ants and other insects, presents a promising tool for weed management, particularly in zero-till systems where weed seeds accumulate on the soil surface (Chauhan et al. 2012). Encouraging weed seed predators by retaining crop residues in fields can be an effective and cost-efficient approach. These environmentally friendly methods can complement existing weed control practices as part of an integrated weed management (IWM) strategy. However, the abundance of natural enemies needed for effective weed control may be limited in specific agricultural situations.

Unmanned aerial vehicles (UAVs) equipped with various sensors such as optical, multispectral, and hyperspectral play a crucial role in precision agriculture by enabling accurate weed mapping and targeted herbicide application. In addition to UAVs, autonomous laser weeding represents a cutting-edge approach for in-field weed control. Andreasen et al. (2022) describe how these systems utilize artificial intelligence and deep learning to target and destroy weeds with laser beams, achieving significant reductions in weed populations. Laser robots, such as the Laser Weeder, can eliminate up to 200,000 weeds per h, making them highly effective in large-scale operations (Papadopoulos 2022). These technologies not only improve weed management efficiency but also minimize herbicide use, thus reducing environmental impact and preserving biodiversity. Overall, the integration of UAVs and laser weeding into mechanical weed management strategies offers promising advancements for pulse crop cultivation, enhancing both precision and sustainability.

2.6 Conclusions

This chapter elaborates the critical importance of addressing weed challenges to sustain agricultural productivity and ensure food security in India. Weeds, with their diverse species and adaptive strategies, pose significant economic threats, leading to substantial yield losses if left uncontrolled. Therefore, a broader understanding of the influence of chemical and non-chemical weed control treatments on weed population dynamics enables the better prediction of ecological changes. Inter-cultivation has been employed as a weed control strategy, but weather variability and heavy rains often hinder timely weed control. Further, herbicide use in India is only 16% of the total pesticide use compared to 52% use of herbicide on a global scale. Still, there are reports of seven unique cases of herbicide-resistant weeds in rice and wheat crops in India. Weeds must be exposed to constantly changing management methods by exploiting IWM practices to keep them "off balance." Through a comprehensive exploration of preventive, cultural, mechanical, chemical, and biological management strategies tailored to different crops, this chapter emphasizes the necessity of an integrated approach to weed management. By empowering farmers with knowledge and tools for effective weed control, we can mitigate yield losses, optimize crop productivity, reduce production costs, and promote sustainable agricultural practices that safeguard the future of Indian agriculture.

References

Ahmad R, Khuroo AA, Hamid M, Charles B, Rashid I (2019) Predicting invasion potential and niche dynamics of *Parthenium hysterophorus* (congress grass) in India under projected climate change. Biodivers Conserv 28:2319–2344

Andreasen C, Scholle K, Saberi M (2022) Laser weeding with small autonomous vehicles: friends or foes? Front Agron 4:841086. https://doi.org/10.3389/fagro.2022.841086

Angiras NN, Chopra P, Kumar S (2010) Weed seed bank and dynamics of weed flora as influenced by tillage and weed control methods in maize (*Zea mays* L.). Agric Sci Dig 30(1):6–10

Annapurna C, Singh JS (2003) Variation of *Parthenium hysterophorus* in response to soil quality: implications for invasiveness. Weed Res 43:190–198

Anonymous (2025a) Statistical report on value of output from agriculture and allied sectors (2011–12 to 2023–24). National Statistics Office, National Accounts Division, Ministry of Statistics and Programme Implementation, Government of India. 365pp. Available on www.mospi.gov.in

Anonymous (2025b) Major uses of pesticides (registered under the Insecticides Act, 1968). Government of India, Ministry of Agriculture & Farmers Welfare, Central Insecticides Board & Registration Committee, Faridabad. 62pp. Available on https://ppqs.gov.in/en/divisions/cib-rc/major-uses-of-pesticides

Anusha S, Krishnapriya V, Vinu V, Geetha P, Kumar RA (2023) *Rottboellia cochinchinensis*: a noxious grass weed. Krishi Sci 4(10):38–41

Bakker T, van Asselt K, Bontsema J, van Henten EJ (2010) Robotic weeding of a maize field based on navigation data of the tractor that performed the seeding. IFAC Proc 43(26):157–159. https://doi.org/10.3182/20101206-3-JP-3009.00027

Bhagat S, Kumar A, Puniya R (2019) Effect of herbicides and their combinations on weeds, productivity and profitability of maize in rainfed sub-tropics of Jammu. Indian J Weed Sci 51(4):358–361

Bhan VM, Choudary DBB (1976) Germination, growth and reproductive behaviour of *Phalaris minor* Retz. as affected by date of planting. Indian J Weed Sci 8:126–130

Bhan R, Froud-Williams R (2006) *Phalaris* spp. competition with wheat using an additive design series. In: Preston C, Watts J, Crossman N (eds) Proceedings of 15th Australian weeds conference Adelaide. Weed Management Society of South Australia Inc., South Australia, pp 417–419

Bhullar MS, Walia US, Singh S, Singh M, Jhala AJ (2012) Control of morning glories (*Ipomoea* spp.) in sugarcane (*Saccharum* spp.). Weed Technol 26(1):77–82. https://doi.org/10.1614/WT-D-11-00112.1

Brankov M, Simić M, Piskackova T, Zarić M, Rajković M, Pavlović N, Dragičević V (2024) A post-emergence herbicide program for weedy sunflower (*Helianthus annuus* L.) control in maize. Phytoparasitica 52(1):12. https://doi.org/10.1007/s12600-024-01126-w

Cardina J, Herms CP, Doohan D (2009) Crop rotation and tillage system effects on weed seedbanks. Weed Sci 50:448–460

Chand M, Singh S, Bir D, Singh N, Kumar V (2014) Halosulfuron methyl: a new post emergence herbicide in India for effective control of *Cyperus rotundus* in sugarcane and its residual effects on the succeeding crops. Sugar Tech 16(1):67–74. https://doi.org/10.1007/s12355-013-0263-4

Chandrasekaran H (2019) Evaluation of suitable intercrop and nutrient management on weed control and seed cotton yield. J Appl Sci 19:447–452

Chaudhary A, Kaur S, Chauhan BS (2021) *Phalaris minor* and *Phalaris paradoxa*. In: Chauhan BS (ed) Biology and management of problematic crop weed species. Academic Press, Elsevier, pp 335–356

Chauhan BS (2013) Growth response of itch grass (*Rottboellia cochinchinensis*) to water stress. Weed Sci 61(1):98–103. https://doi.org/10.1614/WS-D-12-00060.1

Chauhan BS, Johnson DE (2010) The role of seed ecology in improving weed management strategies in the tropics. Adv Agron 105:221–262

Chauhan BS, Singh RG, Mahajan G (2012) Ecology and management of weeds under conservation agriculture: a review. Crop Prot 38:57–65

Chen P, Xia T, Yang G (2024) A new strategy for weed detection in maize fields. Eur J Agron 159:127289. https://doi.org/10.1016/j.eja.2024.127289

Chhokar RS, Sharma RK, Gill SC, Singh RK (2019) Mesotrione and atrazine combination to control diverse weed flora in maize. Indian J Weed Sci 51(2):145–150

Christoffoleti PJ, Carvalho SJP, Nicolai M, Doohan D, VanGessel M (2007) Prevention strategies in weed management. In: Upadhyaya MK, Blackshaw RE (eds) Non-chemical weed management: principles, concepts and technology. CABI, Oxfordshire, pp 1–15

Ciontu VM, Grădilă M, Jalobă D, Palade AC, Gîdea M (2024) Impact of some pre-and post-emergent herbicides applied to the chickpea crop (*Cicer arietinum* L.). Rom Agric Res 41:287–294

Correia NM (2016) Biology and management of *Rottboellia cochinchinensis*. Rev Bras Herb 15(1):89–96

Dafaallah AB (2019) Biology and physiology of witchweed (*Striga* spp.): a review. Int J Acad Multidiscip Res 3(10):42–51

Dalley CD, Viator RP, Richard EP (2013) Integrated management of Bermuda grass (*Cynodon dactylon*) in sugarcane. Weed Sci 61(3):482–490

Deshpande WS, Pawar WS, Mankar PN, Bobae PN, Chimote AN (2006) Integrated weed management in rainfed cotton. Indian J Agron 51(1):68–69

Dhanda S, Kaur S, Chaudhary A, Jugulam M, Hunjan MS, Sangha MK, Bhullar MS (2022) Characterization and management of metsulfuron-resistant *Rumex dentatus* biotypes in Northwest India. Agron J 114:366–378

Dixit A, Manes GS, Chandel R (2011) Comparative performance evaluation of weeders for cotton crop. J Inst Eng (India) 92:10–13

Ducca AS, Vargas PD, Sabate S, Romero ER (2020) Topramezone: new herbicide registered in sugarcane for post-emergent management of *Cynodon dactylon* in Tucuman, Argentina. Sugar Tech 22(4):738–740

Franke AC, Singh S, McRoberts N, Nehra AS, Godara S, Malik RK, Marshall G (2007) *Phalaris minor* seed bank studies: longevity, seedling emergence and seed production as affected by tillage regime. Weed Res 47:73–83

Gharde Y, Singha PK, Dubey RP, Gupta PK (2018) Assessment of yield and economic losses in agriculture due to weeds in India. Crop Prot 107:12–18

Giraldeli AL, Silva AFM, Bassin LC, Araujo LDS, Silva GSD, Pagenotto ACV, Filho RV (2021) Chemical management of *Cyperus rotundus* L. and other weeds at sugarcane in PSS system. Biosci J 237:e37056. https://doi.org/10.14393/BJ-v37n0a2021-51527

Gnanavel I, Natrajan SK (2014) Ecofriendly weed control option for sustainable agriculture—a review. Agric Rev 35:172–183

Hamill AS, Holt JS, Mallory-Smith CA (2004) Contributions of weed science to weed control and management. Weed Technol 18:1563–1565

Harker KN, O'Donovan JT (2013) Recent weed control, weed management, and integrated weed management. Weed Technol 27:1–11

Heap I (2025) The International Herbicide-Resistant Weed Database. www.weedscience.org. Accessed: 21 Jan 2025

Hofmeijer MA, Melander B, Salonen J, Lundkvist A, Zarina L, Gerowitt B (2021) Crop diversification affects weed communities and densities in organic spring cereal fields in northern Europe. Agric Ecosyst Environ 308:107251. https://doi.org/10.1016/j.agee.2020.107251

Holm L, Pancho JV, Herberger JP, Plucknett DL (1979) A geographical atlas of worst weeds. Wiley, Toronto

Holm FA, Kirkland KJ, Stevenson FC (2000) Defining optimum herbicide rates and timing for wild oat (*Avena fatua*) control in spring wheat (*Triticum aestivum*). Weed Technol 14:167–175

Horowitz M (1972) Development of *Cynodon dactylon* (L.) Pers. Weed Res 12:207–220

Jabran K (2019) Role of mulching in pest management and agricultural sustainability. Springer International Publishing, Cham, p 62. https://doi.org/10.1007/978-3-030-22301-4

Jabran K, Chauhan BS (2015) Weed management in aerobic rice systems. Crop Prot 78:151–163

Jat RK, Gopar R, Gupta R (2012) Conservation agricultural in maize-wheat cropping systems of eastern India: weed dynamics and system productivity. In: Extended summaries vol 3: 3rd international agronomy congress. Indian Society of Agronomy, New Delhi, pp 893–894

Johnson Sunder Singh JS, Kandasamy T, Swaminathan M, Muthurajan R, Palanisamy MA, Dhashnamurthi V, Manickam S (2024) Role of herbicide-tolerant (HT) rice in the weed management of direct seeded crop: challenges and opportunities. Czech J Genet Plant Breed 60(4):159–180. https://doi.org/10.17221/26/2024-CJGPB

Jones EAL, Contreras DJ, Everman WJ (2021) *Digitaria ciliaris*, *Digitaria ischaemum*, and *Digitaria sanguinalis*. In: Chauhan BS (ed) Biology and management of problematic crop weed species. Academic Press, Elsevier, pp 173–196

Kakade SV, Patil BM, Karunakar AP, Sethi HN, Thakare PD (1999) Effect of sequential application of pre-emergence and post-emergence herbicides on weeds and yield of cotton. Crop Res 17(2):175–178

Kamara AY, Akobundu IO, Chikoye D, Jutzi SC (2000) Selective control of weeds in an arable crop by mulches from some multipurpose trees in southwestern Nigeria. Agrofor Syst 50:17–26

Kanwar RS, Singh S, Sodhi RS, Garcha AIS (1992) Comparative performance of different herbicide combination for weed control in sugarcane. Indian Sugar 42(8):621–625

Kareemulla K, Krishnan P, Ravichandran S, Kumar BG, Sharma S, Bhatta R (2021) Spatiotemporal analysis of size and equity in ownership dynamics of agricultural landholdings in India vis-à-vis the world. Sustainability 13:10225. https://doi.org/10.3390/su131810225

Kathiresan G, Avudaithai S, Kannappan K (2004) Control of twining weed (*Ipomoea sepiaria*) in sugarcane. Sugar Tech 6(1&2):53–58

Kaur M, Aggarwal NK (2017) *Trianthema portulacastrum* L. The noxious weed and its control. Adv Plants Agric Res 6(3):213–218

Kaur S, Chauhan BS (2023) Challenges and opportunities to sustainable crop production. In: Guleria P, Kumar V, Mo B (eds) Plant small RNA in food crops. Elsevier Inc., pp 25–43. ISBN: 978-0-323-91722-3

Kaur R, Kaur S (2021) Strategies for integrated weed management in major pulse crops. Indian Farming 71(8):14–20

Kaur G, Brar HS, Singh G (2010) Effect of weed management on weeds, nutrient uptake, nodulation, growth and yield of summer mungbean (*Vigna radiata*). Indian J Weed Sci 42(1&2):114–119

Kaur A, Batish DR, Chauhan BS, Kaur S, Singh HP, Kohli RK (2021) Parthenium hysterophorus. In: Chauhan BS (ed) Biology and management of problematic crop weed species. Academic Press, Elsevier, pp 311–333

Kaur S, Dhanda S, Yadav A, Sagwal P, Yadav DB, Chauhan BS (2022) Current status of herbicide-resistant weeds and their management in the rice-wheat cropping system of South Asia. Adv Agron 172:307–354

Kubiak A, Wolna-Maruwka A, Niewiadomska A, Pilarska AA (2022) The problem of weed infestation of agricultural plantations vs. the assumptions of the European biodiversity strategy. Agronomy 12:1808. https://doi.org/10.3390/agronomy12081808

Kumar N, Guru SK (2016) Herbicide resistance mechanism of *Phalaris minor* in Uttarakhand. J Crop Weed 12:129–133

Kumar S, Rana SS, Chander N, Angiras NN (2012) Management of hardy weeds in maize under mid-hill conditions of Himachal Pradesh. Indian J Weed Sci 44(1):11–17

Kumar N, Nath CP, Hazra KK, Sharma AR (2016) Efficient weed management in pulses for higher productivity and profitability. Indian J Agron 61(4):5199–5213

Lukangila MAB, Tabu HI, Mugisho DB, Lubobo AK, Mbombo AKMW (2024) Evaluating the effects of manual hoeing and selective herbicides on maize (*Zea mays* L.) productivity and profitability. Heliyon 10(13):e33294. https://doi.org/10.1016/j.heliyon.2024.e33294

Maheshwari P, Singh B (1934) A preliminary note on the morphology of the aerial and underground flowers of *Commelina benghalensis* Linn. Curr Sci 3:158–160

Malik RK, Singh S (1995) Little seed canary grass (*Phalaris minor*) resistance to isoproturon in India. Weed Tech 9:419–419

Malik RS, Yadav A (2014) Effect of sowing time and weed management on performance of pigeon pea. Indian J Weed Sci 46(2):132–134

Mandal AK, Dheebakaran G, Banik M, Kumar A, Kumar V (2017) Impact of climate change on adaptation capacity of Horse purslane (*Trianthema portulacastrum*). Pharma Innov 6(11):414–418

Manjunatha HP, Nirmalnath PJ, Chandranath HT, Ammanna S, Jagadeesh KS (2018) Field evaluation of native arbuscular mycorrhizal fungi in the management of Striga in sugarcane (*Saccharum officinarum* L.). J Pharmacogn Phytochem 7(2):2496–2500

Marimuthu R, Ahmed SN, Giridharan S (2002) Control of ipomoea in sugarcane. Indian Sugar 51(11):807–809

Matus-Cádiz M, Hucl P (2002) Morphological variation within and among five annual *Phalaris* species. Can J Plant Sci 82:85–88

Merfield CN (2017) The BHU Future Farming Centre. Lincoln, Canterbury. https://www.bhu.org.nz/future-farming-centre. Accessed: 21 Jan 2025

Millholon RW (1992) Effect of itchgrass (*Rottboellia cochinchinensis*) interference on growth and yield of sugarcane (*Saccharum* spp. hybrids). Weed Sci 40(1):48–53

Mitchell CC, Entry JA (1998) Soil C, N, and crop yields in Alabama's long-term 'old rotation' cotton experiment. Soil Tillage Res 47:331–338

Muhammad A, Ahmad HK, Ejaz A, Khan MA, Sagoo AG (2007) Effect of sowing methods and weed control techniques on yield and yield components of chickpea. Pak J Weed Sci Res 13(1/2):49–61

Mukhtar AM, Eltahir SA, Siraj OM, Hamada AA (2007) Effect of weeds on growth and yield of maize (*Zea mays* L.) in Northern State, Sudan. Sudan J Agric Res 8:1–7

Nalayini P, Anandham R, Sankaranarayanan K, Rajendran TP (2009) Polyethylene mulching for enhancing crop productivity and water use efficiency in cotton (*Gossypium hirsutum*) and maize (*Zea mays*) cropping system. Indian J Agron 54(4):409–419

Nalayini P, Sanakaranarayanan K, Velmourougane K (2012) Herbicide mixtures and herbicide rotation for efficient, economical and broad spectrum weed control in winter irrigated Bt Cotton of southern India. In: Proceedings of third international agronomy congress. Indian Society of Agronomy, New Delhi, pp 709–711

Nalayini P, Blaise D, Mundafale HR (2023) Stale bed techniques and leguminous cover crops as component of integrated weed management in irrigated cotton. Indian J Weed Sci 55(1):46–49

Nayak S, Singh JK, Sen M, Singh MK, Sahoo S (2024) Evaluating bio-efficacy of early post-emergence herbicides on weed control, yield, and economic outcomes in Bt Cotton (*Gossypium hirsutum* L.). Afr J Biol Sci 6(14):9737–9749

NITI Aayog (2018) Demand and supply projections towards 2033: crops, livestock, fisheries and agricultural inputs—the working group report. https://www.niti.gov.in/sites/default/files/2023-02/Working-Group-Report-Demand-Supply-30-07-21.pdf. Accessed: 21 Jan 2025

Norris RF, Caswell-Chen EP, Kogan M (2003) Concepts in integrated pest management. https://cir.nii.ac.jp/crid/1130000793655843072. Accessed: 21 Jan 2025

Norsworthy J, Ward S, Shaw D, Llewellyn R, Nichols R, Webster T, Barrett M (2012) Reducing the risks of herbicide resistance: best management practices and recommendations. Weed Sci 60(Spl):31–62

Pagar PC, Rajgire HJ, Darange SO, Nikam VJ (1995) Impact of integrated weed management on yield of rainfed cotton (*Gossypium hirsutum* L.). J Soils Crops 5(2):160–162

Pandey AK, Prakesh V, Mani VP, Singh RD (2000) Effect of rate of nitrogen and time of application on yield and economics of babycorn (*Zea mays*). Indian J Agron 45(2):338–343

Pandey AK, Prakash V, Singh RD, Mani VP (2001) Integrated weed management in maize (*Zea mays*). Indian J Agron 46(2):260–265

Pandey AK, Prakash V, Gupta HS (2002) Effect of integrated weed-management practices on yield and economics of babycorn (*Zea mays*). Indian J Agric Sci 72(4):206–209

Papadopoulos L (2022) This new farming robot uses lasers to kill 200,000 weeds per hour. Interesting Engineering. https://interestingengineering.com/innovation/farming-robot-lasers-200000-weeds-per-hour. Accessed 15 Aug 2024

Patel VJ, Upadhyay PN, Patel JB, Meisuriya MI (2006) Effect of herbicide mixtures on weeds in kharif maize (*Zea mays* L.) under middle Gujarat conditions. Indian J Weed Sci 38(1&2):54–57

Patel T, Dudhat MS, Patel D, Thanki JD (2024) Comparative evaluation of agronomical, mechanical and chemical management of weeds and their impact on sugarcane productivity. Indian J Weed Sci 56(2):167–175

Patil BM, Satao RN, Lahariya GS (1997) Integrated weed management in cotton. PKV-Res J 21(2):220–221

Porwal MK (2000) Economics of weed-control measures in winter maize (*Zea mays*). Indian J Agron 45(2):344–347

Pramanick B, Bera PS, Kundu CK, Bandopadhyay P, Brahmachari K (2014) Effect of different herbicides used in transplanted rice on weed management in rice-lathyrus cropping system. J Crop Weed 10(2):433–436

Prasad A, Singh G, Upadhyay RK (2008) Integrated weed management in maize (*Zea mays* L.) and maize+ blackgram. Indian. J Weed Sci 40(3&4):191–192

Pratap T, Singh VV, Tripathi N (2024) Evaluation of post-emergence herbicides for weed management in greengram. Indian J Weed Sci 56(1):46–51

Raju AR, Tayade AS (2013) Integrated weed management. In: Nagrare VS, Kranthi S, Kranthi KR, Naik VCB, Kumar R, Dharajothi B, Udikeri SS, Mukherjee AK, Mukherjee PK, Monga D, Kumar AS, Narkhedkar NG, Banu G, Raju AR, Tayade AS, Prakash AH, Khader SESA (eds) Handbook of cotton plant health. Central Institute for Cotton Research, Nagpur, pp 88–93

Ram P, Sreenivas G, Rani LP (2017) Impact of sustainable weed management practices on growth, phenology and yield of rabi grain maize (*Zea mays* L.). Int J Curr Microbiol Appl Sci 6(7):701–710

Ramachandran A, Veeramani A, Prema P (2012) Effect of brown manuring on weed growth, yield and economics of irrigated maize. Indian J Weed Sci 44(3):204–206

Ramakrishnan PSR, Kumar S (1971) Productivity and plasticity of wheat and *Cynodon dactylon* (L.) Pers. in pure and mixed stands. J Appl Ecol 8(1):85–98

Ramzan M, Waquar-ul-Haq MA, Jamro GH, Jaffar MA (2012) Effect of mulch techniques on weed control and yield of cotton. Lasbela Univ J Sci Technol 1:5–12

Rao AS (2018) Flauazifop-p-butyl against grasses in cotton and its residual effect on succeeding blackgram. Indian J Weed Sci 50(4):402–404

Rao AN (2021) *Echinochloa colona* and *Echinochloa crus-galli*. In: Chauhan BS (ed) Biology and management of problematic crop weed species. Academic Press, Elsevier, pp 197–239

Rao AN, Johnson DE, Sivaprasad B, Ladha JK, Mortimer AM (2007) Weed management in direct-seeded rice. Adv Agron 93:153–255

Reddy V, Kulkarni S, Chavan M, Ajayakumar MY, Rudramurthy. (2023) Effects of integrated weed management practices on weed parameters and yield of Bt cotton (*Gossypium hirsutum* L.). Int J Environ Clim Chang 13(11):324–331

Sanodiya P, Jha AK, Shrivastava A (2013) Effect of integrated weed management on seed yield of fodder maize. Indian J Weed Sci 45:214–216

Sarala NV, Rao KS, Nagamadhuri KV, Kumar MH (2011) Evaluation of post emergence herbicides for control of creeper weeds in sugarcane. J Sugarcane Res 1(2):78–81

Sarma CK, Gautam RC (2006) Effect of tillage, seed rate and weed control methods on weeds and maize (*Zea mays* L.). Indian J Weed Sci 38(1&2):58–61

Satao RN, Tayade AS, Murarkar SR (1995) Control of *Cynodon dactylon* (L.) pers. Crop Res 10(2):216–220

Sharma A, Ghosh A (2000) Effect of green manuring with *Sesbania aculeata* and nitrogen fertilization on the performance of direct-seeded flood-prone lowland rice. Nutr Cycl Agroecosyst 57:141–153. https://doi.org/10.1023/A:1009863100224

Sharma AR, Toor AS, Sur HS (2000) Effect of interculture operations and scheduling of atrazine application on weed control and productivity of rainfed maize (*Zea mays*) in Shiwalik foothills of Punjab. Indian J Agric Sci 70(11):757–761

Sharma AR, Singh R, Dhyani SK, Dube RK (2010) Moisture conservation and nitrogen recycling through legume mulching in rainfed maize (*Zea mays*)–wheat (*Triticum aestivum*) cropping system. Nutr Cycl Agroecosyst 87:187–197

Shetty SVR, Sivakumar MVK, Ram SA (1982) Effect of shading on the growth of some common weeds of the semi-arid tropics. Agron J 74:1023–1029

Shrestha A, Thapa B, Devkota M, Subedi R (2018) Comparative efficiency of different weed management practices on yield and economic in summer maize in dang. Adv Crop Sci Technol 6(2):1–4

Shyam R, Singh R (2015) Control of nut sedge and other weeds in sugarcane with ethoxysulfuron. Indian J Weed Sci 47(1):43–45

Singh S (2007) Role of management practices on control of isoproturon-resistant littleseed canarygrass (*Phalaris minor*) in India. Weed Technol 21:339–346

Singh A, Kaur C (2004) Weed control with pre and post emergence herbicides application in spring planted sugarcane. Sugar Tech 6(1&2):93–94

Singh AK, Lal M (2008) Weed management in spring planted sugarcane (*saccharum* spp hybrid) based intercropping systems. Indian J Agric Sci 78(1):35–39

Singh S, Kirkwood RC, Marshall G (1998) Effect of monooxygenase inhibitor piperonylbutoxide on herbicidal activity and metabolism of isoproturon in resistant biotypes of *Phalaris minor*. Pestic Biochem Physiol 59:143–153

Singh S, Kirkwood RC, Marshall G (1999) Biology and control of *Phalaris minor* Retz. (littleseed canarygrass) in wheat. Crop Prot 18:1–16

Singh W, Singh R, Malik RP, Mehta R (2011) Effect of planting density and weed management options on weed dry weight and cane yield of spaced transplanted sugarcane (*Saccharum officinarum* L.) after wheat harvest in subtropical India. Indian J Weed Sci 43(1&2):97–100

Singh R, Pratap T, Pal R, Singh VP, Rekha SJ (2014) Management of nut sedge in sugarcane by ethoxysulfuron. Indian J Weed Sci 46(4):342–345

Singh AK, Parihar CM, Jat SL, Singh B, Sharma S (2015a) Weed management strategies in maize (*Zea mays*): effect on weed dynamics, productivity and economics of maize-wheat (*Triticum aestivum*) cropping system in Indo-Gangetic plains. Indian J Agric Sci 85(1):87–92

Singh KB, Jalota SK, Gupta RK (2015b) Soil water balance and response of spring maize (*Zea mays*) to mulching and differential irrigation in Punjab. Indian J Agron 60(2):279–284

Singh K, Singh HP, Singh K (2016) Weed management in cotton with pre- and post-emergence herbicides. Indian J Weed Sci 48(3):348–350

Singh VP, Dileep K, Singh KK (2018) Management of problematic weeds in sugarcane. Indian Farming 68(11):58–63

Singh K, Dhillon BS, Limba V (2021) Effect of mulching on growth and productivity of Bt cotton (*Gossypium hirsutum* L.). Int J Curr Microbiol Appl Sci 10(1):2610–2615

Singh A, Gharde Y, Singh PK (2024a) Explaining distributional patterns of *Trianthema portulacastrum* and *Ageratum conyzoides* in India under future climatic scenarios using ensemble modelling approach. Indian J Weed Sci 56(1):63–72

Singh T, Choudhary A, Dhanda S, Kaur S (2024b) Integrated weed management is the key to delay the evolution of herbicide resistance in weeds under conservation tillage—insights. Indian J Weed Sci 56(3):223–229

Singh R, Kaur S, Shergill LS, Kaur T, Singh G (2025) Seed-shattering phenology of *Phalaris minor* and *Avena ludoviciana* at wheat harvest in North-Western India. Indian J Weed Sci 57(2):172–176

Stoller EW, Sweet RD (1987) Biology and life cycle of purple and yellow nut sedge (*Cyperus rotundus* and *C. esculentus*). Weed Technol 1(1):66–73

Subramanyam D, Reddy MP, Reddy BR (2001) Economics of weed management practices in irrigated maize (*Zea mays* L.). Andhra Agric J 48(3&4):318–320

Thimmegowda MN, Nanjappa HV, Ramachandrappa BK (2007) Influence of tillage and moisture regimes with soil solarization on weed dynamics and yield of baby corn-groundnut crop sequence. Indian J Weed Sci 39(1&2):13–16

Tidemann BD, Geddes CM, Beckie HJ (2021) *Avena fatua* and *Avena sterilis*. In: Chauhan BS (ed) Biology and management of problematic crop weed species. Academic Press, Elsevier, pp 43–66

Tropaldi L, Carbonari CA, de Brito IPFS, de Matos AKA, de Moraes CP, Velini ED (2021) Dynamics of clomazone formulations combined with sulfentrazone in sugarcane (*Saccharum* spp.) straw. Agriculture 11(9):854. https://doi.org/10.3390/agriculture11090854

Walker SR, Evenson JP (1985) Biology of *Commelina benghalensis* L. in South-Eastern Queensland. 1. Growth, development and seed production. Weed Res 25:239–244

Yaduraju NT (2012) Weed management perspectives for India in the changing agriculture scenario in the country. Pak J Weed Sci Res 18:703–710

Zhang WM, Watson AK (1997) Effect of dew period and temperature on the ability of *Exserohilum monoceras* to cause seedling mortality of *Echinochloa* species. Plant Dis 81:629–634

Zhu H, Zhang Y, Mu D, Bai L, Zhuang H, Li H (2022) YOLOX-based blue laser weeding robot in corn field. Front Plant Sci 13:1017803. https://doi.org/10.3389/fpls.2022.1017803

Chapter 3
Biology and Management of Weeds in Iran

Alireza Hasanfard and Seyed Vahid Eslami

Abstract Weeds significantly reduce agricultural productivity and pose a serious threat to food production and economic livelihoods. Weeds aggressively compete with crops for essential resources such as water, nutrients, and sunlight, thereby adversely affecting crop performance. Weeds represent a critical and persistent challenge in Iranian agriculture, causing an estimated 25% to 30% reduction in overall crop yields. Although traditional methods such as plowing and weeding, combined with modern technologies including drones, biological control, and the development of herbicide-resistant crop varieties, are employed in Iran, substantial yield losses persist. In Iran, weeds are recognized not only as a major threat to crop productivity and food security but also as disruptors of natural ecosystems, contributing to biodiversity decline. The increasing demand for agricultural products, driven by population growth, has led to the expansion of cultivated areas, which in turn has created additional habitats conducive to weed proliferation. Consequently, both the diversity and density of weeds have increased, complicating their management, particularly for vital crops such as wheat. This complexity is further exacerbated by changes in weed species composition, the evolution of herbicide resistance, and variability in climatic conditions. A comprehensive understanding of weed biology enables identification of critical growth stages and reproductive periods, facilitating targeted control measures to prevent weed establishment and seed production. This chapter explores the biology of key weed species affecting major crops in Iran, including wheat, rice, corn, chickpea, and rapeseed, and presents successful research findings that can guide the future of weed management in the country.

A. Hasanfard (✉)
Department of Agrotechnology, Faculty of Agriculture, Ferdowsi University of Mashhad, Mashhad, Iran
e-mail: alireza.hasanfard@mail.um.ac.ir

S. V. Eslami
Department of Plant Production and Genetics, Faculty of Agriculture, University of Birjand, Birjand, Iran

B. S. Chauhan, K. Jabran (eds.), *Biology and Management of Weeds in Asia and Africa*, Sustainability Sciences in Asia and Africa,
https://doi.org/10.1007/978-981-95-7314-1_3

Keywords Biodiversity · Competition · Food security · Livelihoods · Weed composition

3.1 Introduction

The presence of weeds in agricultural systems poses serious challenges to sustainable crop production. Human activities have significantly contributed to the distribution and propagation of weed species, which ultimately impact crop yields and environmental health (Bullock and Pufal 2020; Ansong 2024). This issue of weeds has evolved over time due to anthropogenic disturbances in natural ecosystems and the selection pressures resulting from various agricultural practices. Indeed, human actions serve as the primary medium for weed dispersal and establishment in agricultural production systems. Historically, human intervention has been a driving force in shaping plant communities. With the advent of agriculture, selective pressures have led to the proliferation of weed species that are well adapted to agricultural environments. For instance, the intense selection pressure resulting from the application of herbicides, such as glyphosate, can lead to the evolution of herbicide-resistant weed biotypes or shifts in the relative abundance of specific weed species within the community (Owen 2008). With the clearing of land for crop cultivation, new weed species have often been introduced, facilitating their spread through soil disturbance and the creation of suitable habitats. This process has accelerated in recent years due to urbanization, increased international trade, and globalization. Agricultural practices such as tillage, irrigation, and the use of fertilizers have further contributed to the establishment of weeds in crop fields (Owen 2008; Sweeney et al. 2008; Armengot et al. 2016). Tillage, for instance, creates a favorable environment for weed seeds to germinate and grow by exposing them to sunlight and oxygen. Similarly, irrigation systems provide a reliable water supply that can sustain weed growth. The application of chemical fertilizers also enhances nutrient availability in the soil, potentially promoting weed proliferation. Sweeney et al. (2008) reported that spring applications of nitrogen fertilizer increase the total inorganic nitrogen content in the soil and promote weed growth. However, the impact of nitrogen on weed emergence varies depending on the weed species, seed source, and environmental conditions.

The presence of weeds in Iran, a country characterized by diverse geographical features and climatic conditions, has long posed a significant challenge to agricultural production. These species exhibit remarkable adaptations that enable them to survive and thrive in various environmental conditions, creating unique challenges for farmers and land managers throughout the country (Hasanfard et al. 2021). The introduction and spread of weeds in Iran can be attributed to several factors. Human activities, including agriculture, trade, and transportation, have played a crucial role in the dispersal of weed seeds and propagules (Nosratti et al. 2020; Bullock and Pufal 2020). Furthermore, Iran's strategic geographical location at the crossroads of major trade routes has facilitated the introduction and establishment of invasive weed species from neighboring regions.

Weeds in Iran exhibit a diverse array of life histories and growth forms. Both annual and perennial species are prevalent, with some demonstrating unique adaptations to environmental stressors such as drought, salinity, and extreme temperatures. Notable examples of economically important weed species include wild oats (*Avena* spp.), annual/rigid ryegrass (*Lolium rigidum* Gaudin), and redroot pigweed (*Amaranthus retroflexus* L.) (Hasanfard et al. 2022; Amini et al. 2014; Esmaeilzadeh et al. 2012). The prevalence of weeds in Iran's agricultural systems adversely affects crop yields, jeopardizing food security and rural livelihoods. Weeds compete with crops for limited resources, including water, nutrients, and light, and can serve as hosts for pests and diseases that threaten crop health (Anwar et al. 2021). Furthermore, managing weeds can be labor-intensive and costly, particularly in regions with limited access to advanced agricultural technologies and inputs.

Numerous studies have highlighted the detrimental effects of weed competition on major crops, including wheat (*Triticum aestivum* L.), rice (*Oryza sativa* L.), corn (*Zea mays* L.), chickpeas (*Cicer arietinum* L.), and rapeseed (*Brassica napus* L.) (Nabati et al. 2023; Nosratti et al. 2020; Cherati et al. 2011). Weed competition can result in substantial yield losses across various crops in Iran. On average, yield reductions can reach as high as 23% for wheat, 21% for barley (*Hordeum vulgare* L.), 35% for rice, 20% for corn, and 50% for chickpeas (Nosratti et al. 2020).

In addition to reducing crop yields, weeds can also adversely affect crop quality in several ways. Weed seeds can contaminate harvested crops, leading to reduced quality and potential rejection by buyers (Buddenhagen et al. 2023). Some weed species can act as hosts for pests and disease pathogens, increasing the risk of crop infestation and reducing overall crop health (Kumar et al. 2021). Certain weeds release allelochemicals that can inhibit the growth and development of crops, potentially diminishing their quality (Weston and Duke 2003). A study conducted in Iran demonstrated that extracts from various parts (root, stem, and leaf) of field bindweed (*Convolvulus arvensis* L.) and common mallow (*Malva sylvestris* L.) reduced the proline and protein content in sesame (*Sesamum indicum* L.) seeds (Alipour et al. 2019). To minimize the adverse effects of weeds on crop yield and quality in Iran, it is crucial to implement integrated weed management strategies.

In Iran, yield losses due to weed competition can range from 20% to 50%, depending on the crop species. This leads to considerable financial losses to farmers and a reduction in income from crop sales. For instance, it is estimated that the presence of weeds in Khuzestan Province, Iran, can reduce rice yields by 62% (Anafjeh et al. 2024), resulting in an annual loss of $13.5 million. Managing weeds can be labor-intensive and costly, particularly in regions with limited access to advanced agricultural technologies and inputs. Farmers may need to invest in additional labor, herbicides, or equipment to control weeds effectively. These increased production costs can erode farmers' profits and discourage investment in agriculture. Weed infestations can also compromise crop quality, making it challenging for Iranian farmers to meet international quality standards and access lucrative export markets. This can result in lost revenue and a diminished reputation for Iran's agricultural exports. Furthermore, reduced crop yields and quality due to weed infestations can negatively impact Iran's food security, particularly in rural areas where subsistence farming is prevalent. This situation can lead to increased poverty, malnutrition, and social inequality, especially among smallholder farmers. Additionally, weeds can harbor pests and diseases, contaminate water resources, and reduce biodiversity. These environmental impacts can lead to further economic costs to society, such as increased healthcare expenses, water treatment, and habitat restoration efforts.

To mitigate the economic impact of weed infestations in Iran, it is essential to invest in research, extension services, and policy initiatives that promote sustainable and integrated weed management practices. By addressing the challenges posed by weeds, Iran can enhance agricultural productivity, improve food security, and foster sustainable economic growth. Biological studies play a crucial role in the development of effective and sustainable weed management strategies. Understanding weed biology and ecology enables the formulation of targeted control methods that minimize environmental impacts and reduce the risk of evolution of herbicide resistance in weeds. Understanding the biology of weeds, including their growth patterns, seed dormancy, and reproductive strategies, is essential for developing effective management techniques. By targeting specific stages of weed development, we can significantly reduce their proliferation. Ultimately, biological studies contribute to a more comprehensive and sustainable approach to weed management by considering the ecological context in which weeds thrive. This knowledge is crucial for developing practices that mitigate the impacts of weeds while preserving environmental health and promoting sustainable agricultural production.

3.2 Biology

3.2.1 Weed Biology in Iranian Crops

Weeds pose a significant threat to the yields of major crops in Iran. The decline in yields of crops is primarily attributed to weeds' ability to compete for essential resources such as light, water, nutrients, and space. Understanding weed biology enables the development of targeted control measures that exploit weeds' vulnerabilities and life cycles. By possessing knowledge of weed biology, it becomes possible to identify the most vulnerable stages in a weed's life cycle and apply control measures at the optimal time. Additionally, understanding weed biology helps in selecting the most effective methods for controlling specific weed species while minimizing environmental harm and reducing crop yield losses.

Iran's weed biology is complex and diverse, reflecting the country's varied climate, geography, and long history of agriculture. A notable example of this complexity is the weed species swallow-wort (*Cynanchum acutum* L.). This plant is recognized in Iran as an invasive species with significant potential to thrive in agricultural fields, particularly those cultivated with wheat and barley. This perennial, broad-leaved weed is widely distributed across Iran and exhibits considerable morphological variation in response to the country's diverse environmental conditions. *Cynanchum acutum* grows in a range of habitats, from arid and semiarid regions to humid and subtropical zones, demonstrating its ability to adapt to different climatic conditions. Its morphology can vary significantly across different populations, with plants exhibiting distinct growth habits and reproductive strategies depending on environmental factors such as temperature, rainfall, and soil type. This adaptability of *C. acutum* reflects the complex and diverse weed biology found in Iran, shaped by the interplay of the country's varied climate, geography, and agricultural history.

Weed species present in Iran exhibit a wide range of life cycles, growth habits, and reproduction strategies, allowing them to adapt and persist in different environments. Understanding the biology of Iran's weeds, including their reproduction, dispersal, and competitive abilities, is crucial for designing effective and sustainable weed management strategies. In conclusion, the biology of Iran's weeds is multifaceted, reflecting the diverse ecological and agricultural contexts in which they exist.

Although extensive research has been conducted on the biology of weeds that impact important crops in Iran, ongoing studies on weed biology and ecology are essential to address the challenges posed by these unwanted plants in the country's agricultural sector. Therefore, in the following sections, we will discuss the biology and ecology of weeds that affect several major crops in Iran, including wheat, rice, corn, chickpeas, and rapeseed.

The diversity of weeds in wheat, rice, corn, chickpeas, and rapeseed fields in Iran is remarkably high, presenting a significant challenge to crop production. In these crop fields, a multitude of weed species can coexist, each exhibiting unique adaptive traits that enable them to exploit available resources and thrive under diverse environmental conditions. For instance, more than 400 species of weeds from 44 different plant families have been documented in the wheat fields of Iran. It has been reported that threehorn bedstraw (*Galium tricornutum* Dandy) and hedge parsley [*Torilis arvensis* (Huds.) Link] were the most abundant species in chickpea fields in Kurdistan province of Iran during the autumn planting, while *C. arvensis* was the most abundant species during the spring planting (Mousavi et al. 2007). One of the primary reasons for the diversity of weed species in Iran is its varied geography. The country features arid deserts, humid coastal regions, and towering mountains, which create a wide range of ecological niches that support numerous weed species. This topic will be further explored in the section titled "Weed Biology and Adaptation to Iran's Climate."

As previously mentioned, the diversity of weeds in the specified crops is quite significant. Table 3.1 presents a list of the most important weeds associated with Iran's top five crops, which have been extensively documented in the literature.

Table 3.1 List of some of the most significant weeds affecting major crops in Iran (Hasanfard et al. 2022; Nosratti et al. 2020; Hasanfard and Izadi-Darbandi 2020; Esmaeilzadeh et al. 2012)

Crops	Weed species	Family	Growth form	Growing season[a]
Wheat	*Capsella bursa-pastoris*	Brassicaceae	Annual broadleaf	Winter/summer
	Chenopodium album	Amaranthaceae	″	Summer
	Descurainia sophia	Brassicaceae	″	Winter/summer
	Galium tricornutum	Rubiaceae	″	″
	Lamium amplexicaule	Lamiaceae	″	Winter
	Malva parviflora	Malvaceae	″	Winter/summer
	Polygonum aviculare	Polygonaceae	″	″
	Rapistrum rugosum	Brassicaceae	″	″
	Sinapis arvensis	″	″	″
	Alopecurus myosuroides	Poaceae	Annual grass	Winter
	Avena spp.[c]	″	″	Winter/summer
	Hordeum spontaneum	″	″	Winter
	Lolium rigidum	″	″	″
	Phalaris minor	″	″	″
	Secale cereale	″	″	″
	Cirsium arvense	Asteraceae	Perennial broadleaf	Perennial
	Convolvulus arvensis	Convolvulaceae	″	″
Rice	*Cyperus difformis*	Cyperaceae	Sedge	Perennial[b]
	Echinochloa crus-galli	Poaceae	Annual grass	Summer
	Azolla spp.	Azollaceae	Perennial broadleaf	Perennial aquatic fern
Corn	*Amaranthus* spp.	Amaranthaceae	Annual broadleaf	Summer
	Chenopodium album	″	″	″
	Portulaca oleracea	Portulacaceae	″	″
	Solanum nigrum	Solanaceae	″	″
	Echinochloa crus-galli	Poaceae	Annual grass	″
	Convolvulus arvensis	Convolvulaceae	Perennial broadleaf	Perennial
	Cynodon dactylon	Poaceae	Perennial grass	″
	Sorghum halepense	″	″	″
Chickpeas	*Centaurea* spp.	Asteraceae	Annual broadleaf	Summer
	Cuscuta campestris	Convolvulaceae	″	Annual parasitic/ summer
	Galium tricornutum	Rubiaceae	″	Winter/summer
	Sinapis arvensis	Brassicaceae	″	″
	Solanum nigrum	Solanaceae	″	Summer
	Polygonum aviculare	Polygonaceae	″	Winter/summer

(continued)

Table 3.1 (continued)

Crops	Weed species	Family	Growth form	Growing season[a]
	Avena spp.	Poaceae	Annual grass	〃
	Cirsium arvense	Asteraceae	Perennial broadleaf	Perennial
	Convolvulus arvensis	Convolvulaceae	〃	〃
	Cyperus rotundus	Cyperaceae	Sedge	〃
Rapeseed	*Descurainia sophia*	Brassicaceae	Annual broadleaf	Winter
	Lepyrodiclis holosteoides	Caryophyllaceae	〃	Summer
	Malcolmia africana	Brassicaceae	〃	Winter
	Rapistrum rugosum	〃	〃	Winter/summer
	Sinapis arvensis	〃	〃	〃
	Avena spp.	Poaceae	Annual grass	Winter/summer
	Hordeum spontaneum	〃	〃	Winter
	Lolium rigidum	〃	〃	〃
	Phalaris minor	〃	〃	〃
	Cardaria draba	Brassicaceae	Perennial broadleaf	Perennial
	Cirsium arvense	Asteraceae	〃	〃

[a]These classifications may vary slightly depending on local climate conditions; however, the categorizations mentioned above are generally accepted
[b]Certain regions may perceive it as a summer annual
[c]It is either winter or summer, depending on the species

3.3 Seed Ecology

The ecology of weed seeds encompasses various aspects related to their production, dispersal, dormancy, germination, and survival in the environment (Benvenuti 2007). Understanding weed seed ecology is essential for developing effective weed management strategies, as seeds play a crucial role in the dynamics and persistence of weed populations. Key factors in weed seed ecology include the following.

3.3.1 Seed Production

Weed species frequently produce substantial quantities of seeds, which enhance their invasiveness and persistence in both agricultural and natural ecosystems (Benvenuti 2007; Nasim and Shabbir 2012). Factors influencing seed production include plant health, environmental conditions, and the presence of pollinators and seed predators. For instance, the majority of *Amaranthus* species are prevalent in Iran's major agricultural farms, where they significantly reduce the yield of crops such as corn. *Amaranthus retroflexus*, spiny amaranth (*Amaranthus spinosus* L.), and smooth amaranth (*Amaranthus hybridus* L.) are monoecious plants, possessing both male and female flowers on the same individual. On average, each plant can produce between 100,000 and 300,000 seeds (Sellers et al. 2003). A study conducted in Iran revealed that competition between corn and *A. retroflexus* led to a decrease in total dry matter, leaf area index, and the growth rate of corn (Sheibany et al. 2009). Furthermore, as weed density increased to 65 plants per meter of corn row, both the grain yield and biological yield of corn were adversely affected. Another example, wild melon (*Cucumis melo* L.), has been observed in several major crops in Iran, including corn. It has been reported that *C. melo*, recognized as an invasive species in Iran, can produce a substantial number of fruits and seeds—up to 5000—within a growth cycle averaging 75 days. Furthermore, effective weed management is crucial during May and June, as this period coincides with the peak seed production of *C. melo* (Sohrabi et al. 2016).

In conclusion, the prolific seed production of weed species, such as *Amaranthus* spp., plays a critical role in their invasiveness and persistence within agricultural and natural ecosystems. Factors such as plant health and environmental conditions significantly impact seed production, while the presence of pollinators and seed predators also influences seed dispersal. Understanding the factors that affect weed seed production and addressing crop-weed competition through effective management strategies are essential for maintaining crop productivity and controlling the spread of invasive weed species.

3.3.2 *Seed Dispersal*

Weed seeds can be dispersed through various mechanisms, including wind, water, animals, and human activities (Davies and Sheley 2007). Seed dispersal plays a crucial role in the survival, spread, and establishment of weed species in new environments. Several mechanisms contribute to this process:

3.3.2.1 Anemochory (Wind Dispersal)

One method of horizontal seed distribution is wind dispersal, which occurs when seeds are transported through the surrounding environment by wind currents (Ren et al. 2021). During this process, seeds may land on the soil surface or on plant debris. Wind dispersal can occur either in conjunction with the parent plant or through the independent dispersal of seeds or fruits. Certain species, such as those in the daisy family, possess fruits with specialized evolutionary adaptations for movement, featuring wing-like or feather-like structures that enhance their ability to travel through the air. These fruits function like parachutes, enabling them to be carried over long distances (Zhang et al. 2014). Considering the diversity and prevalence of weed families in Iran, it can be inferred that one of the primary factors influencing the dispersal of weed seeds across the country is wind-mediated dispersal. Jafari et al. (2018) reported that in countries such as Iran, among various spatial processes, long-distance seed dispersal predominantly facilitated by wind has played a significant role in the expansion of dandelion (*Taraxacum officinale* Weber ex F.H.Wigg.) populations. This observation aligns with the lightweight seed morphology characteristic of *Taraxacum* species, which enables efficient wind dispersal over considerable distances. Furthermore, *T. officinale* is frequently documented in Iranian literature as a common species in urban green spaces and agricultural fields (Zabihollahi et al. 2008), indicating its widespread distribution across most regions of Iran.

3.3.2.2 Hydrochory (Water Dispersal)

In this mechanism, seeds often have buoyant outer layers that facilitate their transportation by streams, rivers, and even ocean currents (Jones et al. 2020). The flow of water enables seeds to reach new locations along riverbanks and coastal regions, thereby promoting the spread of invasive weeds into previously uncolonized areas.

3.3.2.3 Zoochory (Animal Dispersal)

Animals play a crucial role in the dispersal of seeds away from parent plants, thereby facilitating the invasion of weeds into new ecosystems. Many weed seeds possess hooks, barbs, or sticky coatings that adhere to animals, enabling long-distance transport. Additionally, seeds that pass through an animal's digestive system may benefit from being deposited in nutrient-rich manure, which enhances their germination potential. Weeds such as cocklebur (*Xanthium strumarium* L.) produce seeds that cling to animal fur, feathers, or wool. For instance, Nabati et al. (2023) documented the presence of *X. strumarium* in chickpea fields in Iran. This distribution mechanism may be one of the primary factors contributing to the widespread presence of this species on major agricultural farms in Iran.

Iranian farmers continue to practice transhumance, a traditional method involving the seasonal movement of livestock between pastures and farms, particularly following crop harvests. This approach offers several benefits for weed control, as grazing on weed species helps manage their growth. However, a significant issue arises when livestock are not confined for a sufficient duration to allow for the complete passage of seeds through their digestive systems before being moved to non-infested areas. While digestion by livestock can destroy many weed seeds when they graze on mature weeds or consume screenings and chaff, not all seeds are eliminated (Frick and Johnson 2006). Although digestion substantially reduces the number of viable seeds, those that survive are often deposited in nutrient-rich manure, which enhances their chances of germination. For instance, studies have shown that up to 22% of spotted knapweed (*Centaurea stoebe* L.) seeds remain viable after passing through the digestive tracts of sheep and mule deer. Another study reported initial viability rates of 40% for leafy spurge (*Euphorbia esula* L.)

seeds ingested by sheep and 60% for those ingested by goats; however, viability dropped to 0% by the fifth day post-ingestion, indicating rapid seed degradation within the digestive system (Frost and Launchbaugh 2003).

3.3.2.4 Anthropochory (Human-Mediated Dispersal)

Human activities significantly accelerate the spread of invasive species by transporting seeds beyond their natural habitats, sometimes even across continents. Many weed seeds have adaptations that enable them to survive harsh conditions, such as being carried in contaminated soil, on equipment, or within crops. Common activities, including agriculture, gardening, construction, and transportation, unintentionally facilitate the spread of weed seeds, with seeds often hitching a ride on machinery, vehicles, or clothing. The presence of invasive plant species in Iran has significantly increased over the past decade, primarily due to human-mediated dispersal. For example, mesquite [*Prosopis juliflora* (Sw.) DC.] has rapidly spread across the arid regions of Iran (Amiri et al. 2022). This expansion has been facilitated by human activities, including the transport of contaminated soil and the use of machinery in construction and agricultural projects. The introduction of this invasive species, initially intended for purposes such as windbreaks and reforestation, has led to its unintended spread, which crowds out native vegetation and causes ecological imbalances. Anecdotal evidence also indicates the presence of certain species of this invasive plant in wheat and barley fields.

3.3.2.5 Ballochory (Explosive Mechanisms)

Explosive mechanisms, referred to as ballochory, involve plants that utilize self-propelled seed dispersal to scatter their seeds. Examples include touch-me-nots (*Impatiens noli-tangere* L.) and jewelweed (*Impatiens capensis* Meerb.), which possess seedpods that burst open upon maturation, forcefully ejecting seeds away from the parent plant. Although the dispersal range is relatively short, this method effectively promotes local colonization and reduces competition between the parent plant and its offspring, thereby enhancing the survival chances of new seedlings. Redstem filaree [*Erodium cicutarium* (L.) L'Hér.] is a species commonly found in wheat and barley fields in Iran. By utilizing stored elastic energy, similar to that of a coiled awn, this species can project its seeds up to half a meter away from the parent plant. The seeds then penetrate the soil with their spines, resulting in self-burial (Evangelista et al. 2011).

In general, long-distance dispersal can introduce weed species to new habitats, enabling them to establish in regions where they may lack natural predators or competition. This phenomenon can lead to the proliferation of invasive species, which disrupt ecosystems, outcompete native plants, and threaten biodiversity.

3.3.3 Seed Dormancy

Many weed seeds exhibit dormancy, a state in which they remain viable but do not germinate, even under favorable conditions (Batlla et al. 2020). The mechanisms of dormancy include physical factors (e.g., hard seed coat), physiological factors (e.g., embryo immaturity), and chemical factors (e.g., the presence of germination inhibitors). Common Iranian weeds, such as wild mustard (*Sinapis arvensis* L.), *A. retroflexus*, and lambsquarters (*Chenopodium album* L.), are recognized for their capacity to enter dormancy, which enables them to endure periods of drought or extreme heat. For example, *S. arvensis* can enter a state of secondary dormancy when subjected to harsh conditions, such as low moisture or elevated temperatures.

Dormancy allows the weed seeds to establish a persistent soil seed bank, remaining viable for many years and ensuring successful germination when conditions become favorable. Soltani et al. (2016) reported that *S. arvensis* seeds can establish a persistent soil seed bank, with secondary dormancy induced by water stress or increased burial depth. Nosratti et al. (2019) reported that freshly harvested seeds of soldier thistle [*Picnomon acarna* (L.) Cass.], an invasive weed that affects rainfed crops in Iran, exhibited dormancy and required a period of post-ripening to overcome this dormancy. The germination rate of the seeds increased over time after harvest (Fig. 3.1). Immediately following seed harvest, the germination rate was 20%. Germination increased significantly from the 4th week onward. By the 8th week, the germination rate reached 80%, and by the 12th week, 100% germination was achieved. In general, seed dormancy is a biological survival strategy that prevents seeds from germinating until conditions are optimal for growth, thereby increasing the likelihood of successful establishment. This process is regulated by internal seed factors, including hormones and physical barriers, in response to environmental cues.

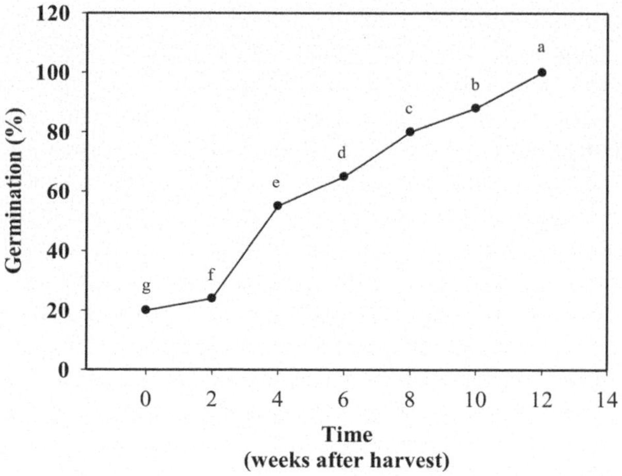

Fig. 3.1 The germination of *Picnomon acarna* seeds after harvest was observed when incubated at a temperature of 20/10 °C (16 hours high/8 hours low) under a photoperiod of 16 hours of light and 8 hours of darkness. Points on the chart labeled with different letters indicate significant differences at the $P = 0.05$ level, as determined by Fisher's Least Significant Difference test. (Adapted from Nosratti et al. 2019)

3.3.4 Seed Germination

Germination ecology refers to the ways in which environmental factors and biological interactions influence the germination of weed seeds (Garcia et al. 2020). Factors such as temperature, light, moisture, and soil characteristics play crucial roles in determining when and how weed seeds germinate. For instance, some seeds require specific temperature ranges or light exposure to initiate germination, while moisture availability is essential for activating metabolic processes within the seed. Additionally, interactions with other organisms—such as soil microbes, which can either promote or inhibit seed growth, and insects, which may damage or disperse seeds—significantly impact germination success. Understanding these dynamics is vital for predicting and managing weed infestations in ecosystems. For instance, cogongrass [*Imperata cylindrica* (L.) P. Beauv.] poses a significant threat to date palm (*Phoenix dactylifera* L.) and sugarcane (*Saccharum officinarum* L.) fields in Iran, particularly in the southern regions, primarily due to its invasive rhizomes (Hamidavi et al. 2021). Research indicates that its germination is favored by light/dark cycles and that it is sensitive to salinity and osmotic stress, with viability decreasing at elevated temperatures. The plant exhibits remarkable drought tolerance, allowing it to withstand extended periods without irrigation. Given Iran's arid and semiarid climate (Rastgoo and Hasanfard 2022), its presence in cereal fields, such as wheat, is expected to increase in the coming years.

Muskweed (*Myagrum perfoliatum* L.) is a winter weed that poses a significant challenge in Iran, particularly in Kermanshah Province, where it adversely affects winter and autumn crops, including wheat and barley. Honarmand et al. (2016) found that seeds of this species encapsulated in siliques had a low germination rate of 2.8%, while naked seeds achieved a germination rate of 50%. Treatments such as sulfuric acid and potassium nitrate significantly enhanced germination, with optimal conditions observed at 20/10 °C. This species exhibits physiological dormancy and is capable of forming a persistent seed bank due to its protective fruit wall.

Chicory (*Cichorium glandulosum* Boiss. et Huet) is a species that has recently proliferated in the autumn crops of northwestern Iran, particularly in wheat fields. Amini et al. (2024) reported that temperature regimes significantly influence the germination percentage of *C. glandulosum*. The greatest germination percentages for the Tabriz (93%) and Marand (94%) populations were observed at day/night temperatures of 20/10 °C (Fig. 3.2). In contrast, the germination percentages of seeds at day/night temperatures of 5/2 °C, 10/5 °C, and 30/20 °C were the lowest for both populations. These researchers also reported that in both studied populations, seed germination remained unaffected at an osmotic potential of −0.2 MPa (Fig. 3.3). However, as the osmotic potential decreased to −0.6 and −0.8 MPa, seed germination declined by 44% and 70% in the Tabriz population, and by 49% and 69% in the Marand population, respectively (Fig. 3.3). The osmotic potentials required to inhibit 50% of germination in *C. glandulosum* for the Tabriz and Marand populations were −0.68 MPa and −0.62 MPa, respectively.

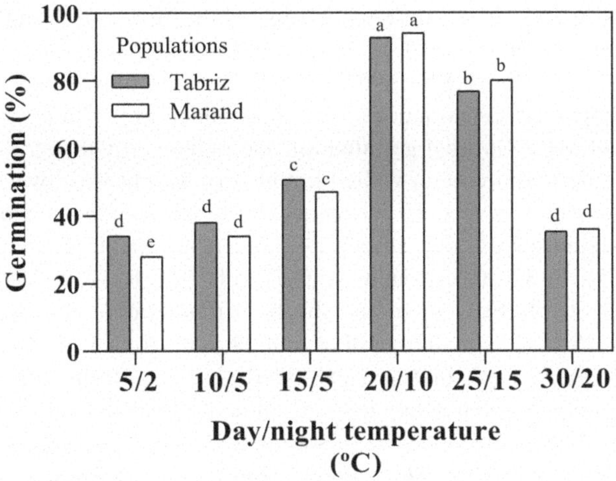

Fig. 3.2 Effect of altering day/night temperatures on the germination percentage of *Cichorium glandulosum* populations from Tabriz and Marand. Columns with the same letters indicate that there are no significant differences in the mean values by Fisher's protected Least Significant Difference test ($P \leq 0.05$). (Modified from Amini et al. 2024)

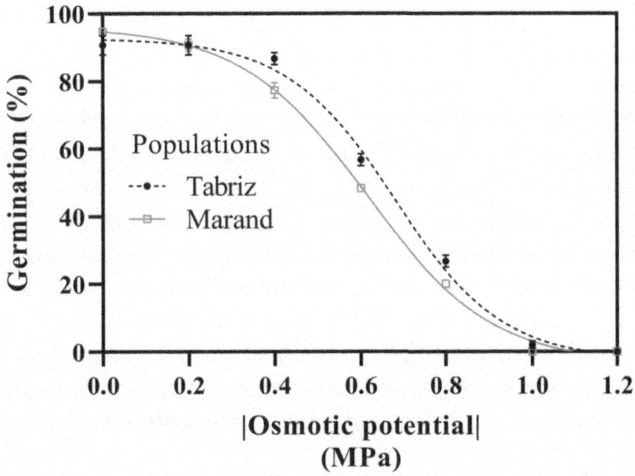

Fig. 3.3 Effect of osmotic potential on the germination percentage of *Cichorium glandulosum* populations from Tabriz and Marand. The vertical bars represent the standard error. (Modified from Amini et al. 2024)

Wild oat (*Avena sterilis* subsp. *ludoviciana*) is a significant weed in Iran, particularly in wheat fields, where it competes with crops for resources and substantially impacts agricultural productivity (Hasanfard et al. 2022). The germination ecology of *A. ludoviciana* is influenced by several environmental factors, including moisture, temperature, light conditions, and soil characteristics. Seeds require adequate soil moisture for optimal germination, typically thriving at temperatures between 15 °C and 25 °C (Rastgoo et al. 2023a). While light can promote germination, the seeds are also capable of germinating in darkness, demonstrating their adaptability. Soil type, pH, and nutrient availability further affect germination success, particularly in disturbed soils that are common in agricultural settings. Additionally, the seeds exhibit physiological dormancy, necessitating specific environmental cues to break this dormancy. Understanding these factors is essential for effective management in agricultural fields. An experiment conducted at the Shahrood Agriculture

Research Center in Iran investigated the effects of drought stress on the germination and seedling growth of *A. ludoviciana* under varying irrigation levels (Nezami et al. 2017). The results indicated that both drought stress and irrigation significantly impacted germination, with rates declining sharply at water potentials below −1 MPa. Optimal germination and radicle growth were observed at 60% of wheat's water requirement, while plumule dry weight was lowest at both the 60% and 80% irrigation levels. Experiments on *A. ludoviciana* evaluated the seed germination responses of both susceptible and resistant biotypes under varying temperatures (5–30 °C) and drought stress levels (0.0 to −1.2 MPa) (Rastgoo et al. 2023a). Drought stress increased the base temperature for germination from 3.24 °C to 7.15 °C for the susceptible biotype and from 3.12 °C to 7.43 °C for the resistant biotype. Both biotypes exhibited similar germination responses, with maximum emergence occurring at 15 °C under optimal moisture conditions.

High concentrations of sodium chloride (NaCl) create osmotic stress, which can hinder water uptake by seeds. This leads to reduced germination rates and delayed germination. Some species exhibit greater salt tolerance than others, affecting their success in germinating under saline conditions. In a study, Nosratti et al. (2019) reported that the germination of *P. acarna* seeds was significantly inhibited by NaCl. The percentage of seed germination decreased from 95.4% to 4.1% as the NaCl concentration increased from 0 to 256 mM, with complete inhibition of germination observed at higher concentrations. The effects of sodium chloride on weed germination highlight the complex interplay between abiotic factors and plant ecology. By understanding how different weed species respond to salinity, ecologists and agronomists can better predict species distribution and develop more effective management strategies in agricultural and natural ecosystems.

In general, understanding the germination biology and ecology of weed seeds is essential for developing effective weed management strategies. This knowledge helps identify the optimal conditions for seed germination and emergence, enabling targeted interventions that reduce competition with crops.

3.3.5 Seed Bank Dynamics

Weed seeds can remain viable in the soil seed bank for several years, creating a reservoir for potential future infestations. The dynamics of these seed banks are complex and involve various factors, such as seed longevity, dormancy, and predation, which collectively shape the size and composition of the seed bank over time.

One of the primary determinants of a weed seed bank's resilience is seed longevity—the length of time seeds remain viable in the soil (Humphries and Florentine 2022). Certain weed species produce seeds that can endure in the soil for years, and in some cases, even decades. This longevity allows these species to persist through adverse conditions, ensuring that they can re-emerge when the environment becomes suitable, such as after disturbances like tillage or changes in land use. For example, *A. retroflexus* and *C. album* pose significant challenges to Iran's major crops, particularly corn (Hamidzadeh Moghadam et al. 2023). A key factor contributing to the widespread presence of these species in Iranian farms is their seed longevity (Bajwa et al. 2019). This characteristic allows them to survive in the soil for extended periods, facilitating adaptation and enabling them to thrive in various environments. Consequently, they can rapidly expand their presence and persist in diverse agricultural settings, further enhancing their prevalence in the region.

Dormancy is a critical factor influencing seed bank dynamics. As previously mentioned, seeds can enter a dormant state due to various environmental conditions, including light, temperature, and moisture availability. This dormancy serves as a strategic adaptation, allowing seeds to avoid germination under unfavorable conditions (Klupczyńska and Pawłowski 2021). Some species exhibit physiological dormancy, which requires specific cues—such as cold stratification or exposure to light—to break dormancy and trigger germination. Others may display morphological dormancy, where seeds must undergo physical changes to become viable. This diversity of dormancy strategies results in a staggered germination pattern, enabling weed populations to exploit resources over time rather than all at once.

Predation and environmental interactions significantly influence the composition of seed banks. Seed predation by insects, rodents, and birds can reduce the number of viable seeds in the soil. The intensity of predation may vary depending on habitat type and the presence of natural enemies, which, in turn, affects the persistence of various weed species. Practices such as cover cropping, crop rotation, and appropriate tillage methods can help deplete the seed bank over time, thereby reducing the likelihood of future infestations (Cechin et al. 2022). Furthermore, employing integrated pest management techniques that consider the ecological roles of seed predation and dormancy can further enhance the effectiveness of weed control strategies.

3.4 Weed Biology and Adaptation to Iran's Climate

The adaptation of weed biology to Iran's diverse climate is evident in several characteristics, including the following:

Drought Tolerance Some weeds have developed adaptive mechanisms to tolerate water scarcity in the arid and semiarid regions of Iran. An example of a weed species that tolerates water scarcity in arid and semiarid regions of Iran, particularly in wheat fields, is littleseed canary grass (*Phalaris minor* Retz.) (Bayat and Zargar 2021). This grass weed poses a significant challenge to wheat cultivation in the country, especially under water-limited conditions. *Phalaris minor* has developed several adaptations to survive and thrive in environments with low water availability. It has an extensive and fibrous root system that enables it to access moisture from deeper soil layers, allowing it to compete effectively with wheat plants for limited water resources. Additionally, this weed species exhibits a high water-use efficiency, allowing it to produce more biomass per unit of water used compared to wheat. The adaptation of this species for germination in arid conditions is noteworthy.

A study conducted in Iran on the drought tolerance of three *Phalaris* species—*P. minor*, hood canarygrass (*Phalaris paradoxa* L.), and shortspike canarygrass [*Phalaris brachystachys* Link]—revealed that *P. minor* demonstrated a superior ability to germinate under elevated levels of drought stress (Rezvani et al. 2021). Consequently, its widespread prevalence in Iran, particularly in wheat and rapeseed crops, can be attributed to its exceptional drought tolerance. *Phalaris minor* can tolerate high temperatures and elevated levels of salinity, two common abiotic stresses in arid and semiarid regions of Iran (Rezvani et al. 2021). These adaptive traits have made it a successful invader of wheat fields and a significant challenge for farmers in areas affected by water scarcity.

Temperature Resilience Some weed species, such as annual ground cherry (*Physalis divaricata* L.), which is commonly found in corn, have evolved to withstand extreme temperature fluctuations across various regions of the country. In Iran, weed species exhibit differing levels of tolerance to high temperatures, shaped by their unique adaptations and the specific climatic conditions of each area. Common Iranian weed species, including *Avena* spp., have been observed to demonstrate enhanced resilience to temperature extremes, encompassing both freezing and elevated temperatures.

Freezing stress poses a significant environmental challenge that can restrict the survival and distribution of plant species, including weeds. However, certain weed species have evolved a variety of physiological and molecular mechanisms that allow them to tolerate freezing stress and thrive in cold environments (Østrem et al. 2018; Hasanfard et al. 2021). These adaptations include the accumulation of protective solutes, such as sugars, amino acids, and antifreeze proteins, which help prevent ice formation and cellular damage. In addition, some weeds possess enhanced antioxidant defense systems that can mitigate oxidative stress induced by freezing temperatures. Morphological adaptations, such as a reduced leaf surface area or the development of dense, insulating leaf hairs, also contribute to the tolerance of freezing stress in certain weed species. These adaptations not only allow weeds to thrive in cold climates but also confer a competitive advantage over less tolerant species, thereby enhancing their invasiveness in agricultural and natural ecosystems. In Table 3.2, the tolerance of various weeds to freezing stress in significant crops in Iran is presented.

The broad spectrum of germination temperatures has significantly contributed to the proliferation of weeds. For instance, *Avena* spp. can germinate within a temperature range of 5 to 30 °C, with an optimal range of 15 to 20 °C (Rastgoo et al. 2023a). However, mature plants exhibit greater tolerance to extreme temperatures, with some studies indicating that *Avena* spp. can withstand temperatures as low as −10 °C and as high as 40 °C for brief periods.

Salinity Tolerance Salinity tolerance is a crucial factor in the survival and distribution of weed species in Iran, especially in areas where soil salinity poses a significant challenge. Numerous Iranian weed species have developed a range of physiological, morphological, and molecular mechanisms to withstand and adapt to

Table 3.2 Threshold of tolerance to freezing stress in major weeds of wheat and chickpea

Crops	Crop tolerance (survival % at −12 °C)	Weeds	Growth stage	Weed tolerance (survival % at −12 °C)	Reference
Wheat (cv. Mihan)	85	*Avena sterilis* ssp. *ludoviciana*	Three- to six-leaf stage	100	Hasanfard et al. (2021)
		Rapistrum rugosum (L.) All.	"	80	
Wheat (*cv.* Pishgan and Ghezel khoushe)	100	*Hordeum spontaneum* Koch.	Two- leaf stage	64	Izadi-Darbandi et al. (2020)
		Secale cereale L.	"	100	
Wheat (*cv.* Pishgan and Ghezel khoushe)	100	*Hordeum spontaneum* Koch.	Four- leaf stage	60	Hasanfard and Izadi-Darbandi et al. (2020)
		Secale cereale L.	"	100	
Chickpea (*cv.* Saral)	100	*Sinapsis arvensis* L.	Two-to four- leaf stage	100	Rastgoo et al. (2023b)

elevated salinity levels in their environment. Other adaptations to salinity in Iranian weeds include the production of antioxidants that counteract the harmful effects of reactive oxygen species induced by salt stress, as well as the development of specialized salt-excreting glands on leaf surfaces. These adaptations not only enhance the survival of these weed species in saline environments but also confer a competitive advantage over less tolerant species, facilitating their proliferation and persistence in disturbed habitats. Certain weeds, such as wild barley (*Hordeum spontaneum* Koch.), can thrive in saline soils, which are common in various agricultural regions of Iran. The study conducted by Rezvani et al. (2021) indicates that *P. minor* exhibits a higher level of tolerance to salt stress during germination compared to *P. paradoxa* and *P. brachystachys*. This finding suggests that *P. minor* has the ability to germinate under challenging environmental conditions characterized by elevated salt concentrations. The capacity of *P. minor* to thrive in adverse conditions may have significant implications for its invasiveness and persistence in habitats affected by drought and salinity.

Elevation Adaptation Iran's diverse terrain, particularly its mountainous regions, has fostered the development of distinct weed populations adapted to varying altitudes. For example, ecotypes of *A. ludoviciana* have evolved specific traits that enable them to thrive at different elevations. These adaptations may include alterations in growth patterns, reproductive strategies, and mechanisms for stress tolerance. For instance, weed populations at higher altitudes may exhibit enhanced cold tolerance, more efficient water utilization, or modified photosynthetic processes to cope with the environmental conditions associated with increased elevation.

The adaptations discussed highlight the remarkable ability of weed species in Iran to survive and reproduce under diverse environmental conditions. These adaptations enhance the resilience and versatility of weed populations, making them highly competitive in various habitats. For instance, the ability to tolerate a wide range of temperatures, salinity levels, and altitudes allows weed species to thrive in diverse regions across Iran, from coastal areas to high-elevation mountain ranges. Furthermore, their capacity to adapt to human-induced disturbances, such as crop cultivation and herbicide applications, significantly enhances their persistence in agroecosystems. These adaptive traits present considerable challenges for weed management, as they can reduce the effectiveness of control measures. For example, herbicide-resistant weeds may require alternative control methods, while the presence of diverse weed ecotypes adapted to different environmental conditions necessitates tailored management strategies (Norsworthy et al. 2012; Vencill et al. 2012). Consequently, understanding the biology and ecology of local weed populations is essential for developing effective control strategies. This knowledge can inform the selection of appropriate cultural, mechanical, and chemical methods to target specific weed species and their adaptations, ultimately improving weed management outcomes and minimizing their impact on crop production and natural ecosystems.

3.5 Phenology

Weed phenology refers to the study of the timing of life cycle events in weed species, including germination, flowering, seed set, and senescence. Understanding the phenology of weeds is crucial for developing effective management strategies in agriculture. Some important aspects of weed phenology include the following.

3.5.1 Germination and Growth Stages

Most weeds require specific temperature and moisture conditions for germination, which can vary significantly among species. Certain weed seeds possess dormancy mechanisms that inhibit germination until environmental conditions are optimal. After germination, weeds enter a phase of vegetative growth, during which they develop leaves and roots. This stage is crucial for competing for resources. Weeds then transition to reproductive stages, during which they produce flowers and seeds. The timing of these transitions can be influenced by environmental factors such as photoperiod and temperature.

3.5.2 Flowering and Seed Set

Many weeds bloom at specific times of the year, often coinciding with agricultural cycles. Understanding these patterns is essential for planning effective interventions. Some weeds are self-pollinating, while others rely on pollinators, which can significantly influence seed production and dispersal. Seed set and dispersal are critical components of phenology, significantly influencing the timing of reproductive events in plants and their subsequent success across various environments. Phenological patterns determine when plants flower, set seeds, and disperse them, often in response to environmental cues such as temperature, daylight, and moisture availability (Alm et al. 1991). For many species, optimal seed set occurs under favorable conditions, ensuring that seeds mature and are ready for dispersal when environmental factors are most conducive to germination and survival. This timing can also be synchronized with the activities of pollinators and seed dispersers, thereby enhancing the likelihood of seed dispersal through mechanisms such as wind, water, or animal movement. Understanding these phenological relationships is essential for predicting how plants will respond to changing climates and ecosystems.

Cucumis melo is recognized as an invasive species in Golestan Province, Iran, where it has been reported to adversely affect a variety of crops, including corn and soybeans [*Glycine max* (L.) Merr.]. In a study examining the phenology and seed production of *C. melo* in relation to weather conditions and weed management, the impact of temperature on development was analyzed through the accumulation of thermal time across seven planting dates from April to August (Sohrabi et al. 2016). The results revealed slight variations in developmental rates, particularly for sowings in August, with thermal time values ranging from 251- to 733-degree days. Weed emergence occurred 5 to 20 days after planting, and *C. melo* produced up to 5000 seeds during an average growth cycle of 75 days. This research confirms the importance of managing *C. melo* in May and June due to its high seed production and suggests that temperature alone may not fully account for development, as additional factors, such as day length, also play a significant role.

A study examined the growth patterns of key weed species in Qaemshahr, Iran, to enhance herbicide application and agronomic planning (Salehian et al. 2013). Seeds of *Avena* spp., *S. arvensis*, *P. minor*, *P. paradoxa*, and *L. rigidum* were planted on 14 different dates in a greenhouse. The results indicated significant effects of planting dates on leaf number, particularly for *S. arvensis* and *Avena* spp. Wild oat was influenced by temperature and photoperiod, while *S. arvensis* responded positively to extended daylight hours. *Lolium rigidum* exhibited the longest life cycle at 295 days, whereas wild oat had the shortest at 104 days. These findings suggest that weed emergence models can improve control measures.

3.5.3 Senescence and Management Implications

After seed set, weeds undergo senescence, during which they die off; however, their seeds can remain viable in the soil for several years. Understanding when weeds complete their life cycle can help optimize the timing of herbicide applications and other control methods. By synchronizing cultural practices, mechanical control methods, and chemical applications with the specific growth phases of weeds, farmers and land managers can enhance their effectiveness (Buhler et al. 2000; Bajwa et al. 2015). For example, timing actions such as herbicide application or mechanical weeding to coincide with critical moments in the weed lifecycle can significantly reduce seed production and spread, thereby minimizing competition with crops. Additionally, analyzing the growth patterns of weeds enables proactive strategies, allowing for targeted management efforts that exploit vulnerable stages in the weed lifecycle. This informed approach not only protects crop yields but also promotes healthier ecosystems by decreasing reliance on chemical treatments and encouraging more sustainable farming practices.

3.5.4 Management of Weeds in Important Crops of Iran

While there are no specific databases or resources available to provide the exact percentage of studies on weed management in Iran, particularly concerning major crops, it can be estimated based on trends in Iranian literature that chemical weed management, specifically the use of herbicides, constitutes approximately 50% to 60% of the research conducted. Following this, studies on cultural practices, such as crop rotation and intercropping, are gaining traction and likely represent around 20% to 30% of the research. Other methods, including biological, physical, and integrated weed management strategies, occupy subsequent positions in terms of research focus. A notable gap in the literature on weed management in Iran is the insufficient research dedicated to promoting preventive measures against weed presence in agricultural fields. The following sections discuss various successful experiences in weed management utilizing different methods.

3.5.4.1 Prevention

One of the most fundamental, straightforward, and effective methods of weed management is prevention. Preventing the growth of weeds is more cost-effective and easier than removing them once they have become established. Additionally, preventive measures are safer and more sustainable. Most weeds in agricultural fields originate from seeds present in the soil seed bank; however, there are other pathways through which new weed seeds can be introduced. Among the various sources of weed seeds are fertilizers, mulches, and contaminated seeds from agricultural crops. It is important to note that preventing the contamination of a field with weeds, while simultaneously reducing control costs, facilitates the management of these plants within the agricultural system. This can be achieved by delaying the establishment of existing weeds or preventing the establishment of weeds that are not currently present in the area.

Managers must enhance the implementation of cultural and biological weed control methods to address the increasing demand for reduced herbicide use and tillage practices. However, these methods often prove ineffective against high-density infestations and can result in significant weed seed production. Consequently, effective preventive weed management is essential. The study by Anderson (2007) emphasizes the importance of integrating weed prevention and control strategies to effectively limit the spread of weeds in the semiarid steppe of the United States. By reducing reliance on herbicides by 50%, this approach highlights proactive measures, such as crop rotation and no-till practices, that disrupt weed life cycles and minimize seed production. This preventative focus not only curtails weed populations but also supports sustainable farming practices, ultimately leading to healthier ecosystems and reduced long-term challenges in weed management. In Iran, the implementation of cultural practices such as crop rotation, intercropping, and cover cropping serves as an effective preventive measure for weed management. These strategies are designed to disrupt the growth cycles of weeds and reduce weed seed banks, thereby alleviating weed pressure in agricultural systems.

A study conducted in Iran investigated the impact of crop rotations on soil seed banks in comparison to non-crop lands (Hosseini et al. 2014). The findings indicate that the weed seed bank, particularly in wheat fields, is significantly influenced by crop rotation practices and levels of disturbance, both of which are crucial for effective weed management. Recently, Iran has made significant strides in weed prevention by implementing stringent regulations regarding the quality of fertilizers and seeds, which have notably reduced contamination. Educational programs aimed at informing farmers about the dangers of using unverified products have led to the widespread adoption of certified seeds and fertilizers.

3.5.4.2 Chemical Control

Chemical weed control is an effective strategy for controlling unwanted vegetation, particularly when targeting specific weed species. Herbicides quickly reduce weed populations, thereby enhancing crop yields. However, the overall effectiveness of this approach depends on several factors, including timing, weather conditions, and proper application techniques to prevent the development of resistance. For instance, Ebrahimpour et al. (2022) investigated the effects of herbicide combinations on weed control in irrigated wheat in Abyek County, Iran. They found that specific treatments, such as bromoxynil plus MCPA from BASF tank-mixed with clodinafop propargyl from Golsam Gorgan, increased grain yield by 88% compared to untreated plots. The study concluded that increased rainfall and favorable temperatures significantly enhanced the efficacy of the herbicides. Zand et al. (2010) reported that experiments were conducted during 2006–2007 at six research stations of the Iranian Plant Protection Research Institute to evaluate the efficacy of various herbicides for weed control in wheat. The results indicated that treatments with sulfosulfuron plus metsulfuron-methyl, as well as bromoxynil plus MCPA in combination with clodinafop propargyl, provided effective weed control and improved wheat grain yield at most locations. In contrast, the efficacy of different formulations of mesosulfuron-methyl plus iodosulfuron-methyl-sodium varied across locations. Notably, the oil dispersion formulation performed better than the water-dispersible granules formulation at the Shiraz and Gorgan stations, which was partly attributed to the more uniform rainfall distribution during the growing season and the wheat cultivar used, respectively. Overall, the findings demonstrated that the oil dispersion formulation was superior to the water-dispersible granules formulation in enhancing grain yield at most experimental sites.

The efficacy of herbicides can be negatively impacted by freezing conditions due to several factors. Low temperatures can hinder the absorption of herbicides by plants, reduce their translocation within the plant, and even cause herbicide molecules to become trapped in ice crystals, resulting in diminished weed control. Therefore, it is essential to consider weather conditions and optimal temperature ranges for effective herbicide application. Since the application of mesosulfuron-methyl plus iodosulfuron-methyl-sodium plus mefenpyr-diethyl is prevalent in wheat fields in Iran, a study was conducted to evaluate the efficacy of this herbicide on sterile oat and turnipweed [*Rapistrum rugosum* (L.) All.]—two significant winter weeds—under two conditions: freezing before spray (FBS) and freezing after spray (FAS) (Hasanfard et al. 2023). The survival percentages for sterile oat under FBS and FAS at the twofold dose recommended in Iran (36 g ai ha^{-1}) were 40% and 13%, respectively. For *R. rugosum*, the survival percentages were 20% and 7% under the same conditions. In contrast, both weeds were completely controlled under no freezing (NF) conditions with the twofold recommended dose. The dose required to control 50% of individuals (LD_{50}) for sterile oat in NF, FBS, and FAS were 11.8, 31.5, and 14.2 g ai ha^{-1}, respectively, while for *R. rugosum*, the LD_{50} values were 10.6, 18.4, and 10.7 g ai ha^{-1}, respectively (Table 3.3). Therefore, adjusting the timing of herbicide application, especially for winter weeds, can significantly enhance their control effectiveness.

In Iran, corn is sown from mid-April to early September, which influences herbicide efficacy due to varying environmental conditions. Pourmorad Kaleibar et al. (2021) developed a model to optimize herbicide dosages using a tank mix of nicosulfuron (40 OD, 40 g a.i. L^{-1}, Bisterfeld) and bromoxynil + MCPA (Bromicide MA®, 400 EC, 400 g a.i. L^{-1}, Nufarm) based on thermal time after sowing. Trials conducted in 2016 and 2017 in Karaj, Iran, demonstrated that weed weight increased linearly in the absence of treatment, while effective herbicide dosages decreased with thermal time. The model accurately predicted weed weight and yield loss, achieving an R^2 value of $\geq$0.90. The authors recommend applying herbicides early, no later than 600 thermal units after corn sowing, to facilitate reduced dosages, as they observed a consistent decrease in herbicide efficacy with delayed application.

Table 3.3 The dose of mesosulfuron-methyl plus iodosulfuron-methyl-sodium plus mefenpyr-diethyl required for 50% mortality (LD_{50}) of *Avena ludoviciana* and *Rapistrum rugosum* under different freezing treatments

Plant species	Freezing treatment	$b \pm$ SE	$d \pm$ SE (%)	$LD_{50} \pm$ SE (g ai ha^{-1})
	No freezing	3.27 ± 0.40	98.7 ± 2.04	11.8 ± 0.56
Avena ludoviciana	Freezing before spray	1.84 ± 0.41	96.5 ± 2.38	31.5 ± 2.30
	Freezing after spray	2.70 ± 0.37	99.2 ± 2.08	14.2 ± 0.72
	No freezing	4.52 ± 0.70	97.4 ± 1.71	10.6 ± 0.36
Rapistrum rugosum	Freezing before spray	1.56 ± 0.19	98.3 ± 2.38	18.4 ± 1.30
	Freezing after spray	2.95 ± 0.33	100.4 ± 1.87	10.7 ± 0.45

Modified from Hasanfard et al. (2023)
Abbreviation: b is the slope of the curve around e (LD_{50}), d is the upper limit for Y (survival %). LD_{50} were separated using their 95% confidence intervals. The data shows the parameter's mean ± Standard Error (SE)

Field studies were conducted in Shiraz, Iran, during 2000 and 2001 to evaluate the effects of individual and combined herbicide treatments on weed control and corn yield (Bijanzadeh and Ghadiri 2006). In this study, 14 different herbicide combinations were applied at two doses. The results indicated that all herbicide treatments reduced weed biomass compared to the weedy control plots. The greatest reduction in weed biomass was observed with the atrazine + alachlor combination, while the lowest reduction was observed with rimsulfuron. In both years, the combinations of 2,4-D + MCPA and alachlor +2,4-D + MCPA controlled 80 to 100% of *C. arvensis*, whereas rimsulfuron controlled only 17 to 70% of this weed. All treatments provided satisfactory control of *A. retroflexus*, with an efficiency of 60 to 100%. Control of Chinese lantern (*Physalis alkekengi* L.) in 2000 showed the lowest biomass reduction with some combinations of 2,4-D + MCPA and primisulfuron + prosulfuron. The combination of rimsulfuron + primisulfuron + prosulfuron controlled Johnson grass [*Sorghum halepense* (L.) Pers.] with 96 to 100% efficiency, and the effect of rimsulfuron was enhanced when combined with primisulfuron + prosulfuron. Results from both years indicated that all herbicide treatments increased corn grain yield compared to the weedy control, with the highest yield achieved by the atrazine + alachlor combination.

Collectively, these findings underscore the critical importance of selecting appropriate herbicide combinations and timing their application according to environmental conditions and accumulated thermal time to maximize weed control and optimize corn yield in Iran. Early herbicide applications and the use of tank mixtures markedly enhanced weed control efficiency, reduced weed pressure, and improved crop performance, thereby providing a framework for sustainable and effective herbicide management strategies.

One of the primary challenges in chickpea production in Iran is the widespread and diverse presence of weeds. Reports indicate that yield reductions in chickpeas can range from 24% to 63%, depending on the level of weed infestation (Muhammad et al. 2011). Effective weed control has the potential to increase chickpea yields by as much as 50% (Nosratti et al. 2020). In a study conducted on chickpea fields,

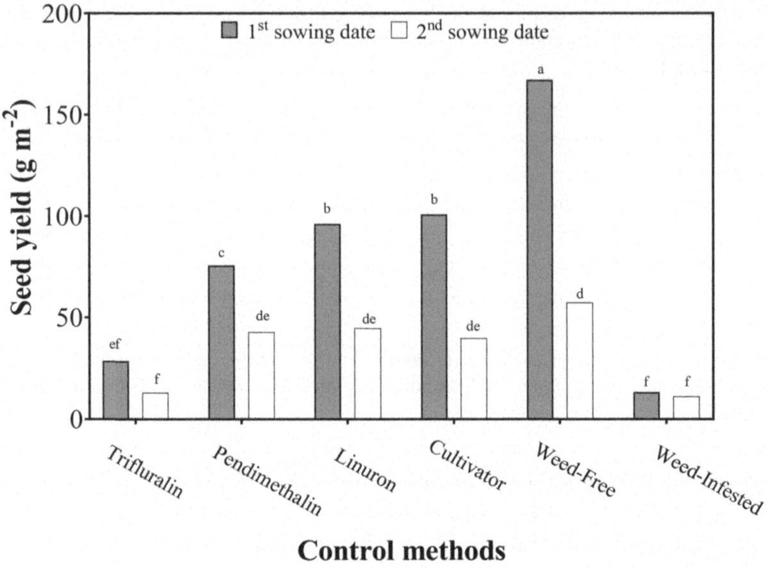

Fig. 3.4 Effects of sowing date (March 1 and March 15) and treatments on chickpea seed yield. Different letters above the bars indicate significant differences among sowing dates and treatments. (Modified from Nabati et al. 2023)

Nabati et al. (2023) reported that the most significant weed species for the March 1 planting date was common knotgrass (*Polygonum aviculare* L.), while for the March 15 planting date, the predominant species was fumitory (*Fumaria officinalis* L.). According to their findings, the seed yield with pendimethalin treatment was 82.5%, whereas the average seed yield with linuron application and inter-row cultivation was 86.4% higher than that of the weed-infested control (Fig. 3.4).

Currently, the most effective strategy for controlling weeds in chickpea fields in Iran involves the application of various herbicides, including those previously mentioned, as well as aclonifen, flumioxazin, and oxyfluorfen, which are also cited in some Iranian literature. Veisi et al. (2019) reported that weed damage to chickpea in Tabriz, Kermanshah, and West Azerbaijan was estimated at 49%, 57%, and 36%, respectively. A total of 64 weed species were identified in chickpea fields, with *C. arvensis* and *G. tricornutum* being the most prevalent. The only registered herbicides for use in chickpea fields in Iran are pyridate and linuron; however, research findings indicate that fomesafen and isoxaflutole are the most suitable herbicides for these fields. Other major herbicides studied include oxyfluorfen, imazethapyr, metribuzin, trifluralin, simazine, terbutryn, and pendimethalin. The combination of herbicide application with mechanical methods, including hand weeding and inter-row cultivation, has been recognized as the most effective approach to reduce weed populations.

The study conducted by Moradi et al. (2020) found that the application of clopyralid in conjunction with grass herbicides effectively controlled broadleaf weeds and wheat, which is regarded as a weed in rapeseed cultivation. The combination of clopyralid and haloxyfop-methyl demonstrated efficacy comparable to that of weed-free conditions for controlling wheat. Although no treatment achieved the effectiveness of weed-free conditions, the application of clopyralid with grass herbicides significantly reduced overall weed density. Hand weeding proved to be the most effective method for preserving rapeseed traits, with potential yield losses ranging from 26% to 81% when weeds were allowed to compete. The study suggests that the combination of clopyralid and haloxyfop-methyl can serve as an effective strategy for weed management in rapeseed fields, particularly those previously cultivated with wheat, as a yield reduction of 5% to 10% is considered acceptable with herbicide application.

As previously stated, weeds are among the major factors reducing rice yield worldwide, particularly in Iran, and can cause substantial economic losses if not effectively controlled. Among them, monochoria [*Monochoria vaginalis* (Burm. f.) C. Presl] is recognized as a noxious weed in transplanted rice fields, and its management is essential for sustainable rice production. Two field experiments were conducted to evaluate the efficacy of soil-applied and foliar-applied herbicides for controlling *M. vaginalis* (Hazrati et al. 2023). Results indicated that prepackaged herbicides such as triafamone + ethoxysulfuron and pyrazosulfuron-ethyl + pretilachlor (each reducing *M. vaginalis* biomass by 100%), as well as single applications of flucetosulfuron (100%) and pendimethalin (99%), achieved complete or near-complete control. Tank-mixed bensulfuron-methyl applied with pretilachlor, thiobencarb, or pendimethalin also provided 100% control. Herbicide applications significantly increased rice height, straw, and grain yield compared with untreated and hand-weeded controls, whereas full-season interference of *M. vaginalis* reduced rice grain yield by 32%. In the second study, foliar-applied herbicides including triafamone + ethoxysulfuron, flucetosulfuron, 2,4-D, dicamba +2,4-D, bispyribac-sodium, bentazon + MCPA, and pyribenzoxim provided ≥97% control of *M. vaginalis* and completely prevented its regrowth. These findings demonstrate that effective management of *M. vaginalis* in transplanted rice fields of Iran can be achieved using a wide range of soil-applied and foliar-applied herbicides with different modes of action.

3.5.4.3 Application of Adjuvants in Herbicides

Adjuvants enhance the efficacy of herbicides, particularly in conditions characterized by hard water and abiotic stress. Hard water's polyvalent cations can diminish the effectiveness of weak acid post-emergence herbicides. To counteract these effects, water conditioning adjuvants, particularly nitrogen compounds like ammonium sulfate, are commonly used, though their effectiveness can vary with different herbicides and weed species (Mirzaei et al. 2023). They improve solubility, adhesion, and absorption, effectively counteracting the binding of herbicides to minerals present in hard water. For instance, research has demonstrated that the use of adjuvants, such as ammonium sulfate and urea ammonium nitrate, mitigates the adverse effects of water hardness on the herbicide mixture of mesosulfuron, iodosulfuron, and mefenpyr-diethyl in controlling *S. arvensis*, a significant weed that impacts major winter crops such as wheat, chickpeas, and rapeseed (Mirzaei et al. 2022). Under abiotic stress conditions, adjuvants help maintain the effectiveness of herbicides, resulting in improved weed control and increased crop yields.

3.6 Cultural Management

Cultural practices such as delayed crop seeding, cover cropping, intercropping, black fallow, crop rotation, and the use of competitive crops can help mitigate the selection pressure for herbicide resistance in weeds. Alternating between different herbicides and implementing these cultural methods can further delay the development of resistance, depending on the genetic composition and survival traits of the weed populations (Nalewaja 1999). In this context, Koocheki et al. (2009) reported that various crop rotations, including those with differing life cycles, such as winter wheat-corn and winter wheat-sugar beet (*Beta vulgaris* L.), can provide additional benefits by reducing the weed seed bank.

A study investigated the effects of tillage systems and crop rotation on weed populations and crop yields in Iran from 2002 to 2005 (Sarani et al. 2014). The research focused on three tillage methods and three crop rotation strategies. The dominant weed species identified included Japanese brome (*Bromus japonicus* Thunb. ex Murr.), woolly distaff thistle (*Carthamus lanatus* L.), *P. aviculare*, darnel ryegrass (*Lolium temulentum* L.), and *A. ludoviciana*. The study concluded that integrating specific crop rotations with tillage methods could effectively manage problematic weed species. Safflower (*Carthamus tinctorius* L.) and rapeseed were particularly successful in reducing grass weed populations and also contributed to increased overall revenue due to their higher market prices.

Mulching is recognized as a cultural method of weed management. This technique involves applying a layer of organic or inorganic material to the soil surface to suppress weed growth, conserve moisture, and enhance soil health. Amini et al. (2024) introduced straw mulch as an effective method for reducing the emergence of *C. glandulosum*, a weed that has recently become problematic in wheat fields in northwestern Iran. A study conducted by Jalali et al. (2021) investigated the effects of intercropping potato (*Solanum tubersum* L.) and quinoa (*Chenopodium quinoa* Willd.), along with various weed management techniques, on potato yield and characteristics. The results indicated that intercropping significantly reduced weed density while enhancing plant height, leaf count, leaf area index, and tuber yield. Increasing the density of quinoa in the intercropping system further improved these traits. Manual weeding proved to be more effective than herbicide application in reducing weed density and increasing tuber yield. The optimal combination for achieving the highest potato yield was identified as intercropping potato and quinoa at a density of 5:20 plants/m^2 under manual weeding conditions in the Kahnooj region of Iran. Therefore, intercropping can be considered as an effective strategy for managing weeds in Iran's major crops.

An example of a successful experience with the use of cover crops to control weeds in Iran is a study that examined the effectiveness of various cover crops in managing weeds in cornfields and their impact on corn yield (Yeganehpoor et al. 2015). The results indicated that both the type of cover crop and the sowing date significantly influenced corn yield and its associated traits. Weed biomass was affected by the interaction between cover crop type and sowing date, with higher weed biomass leading to a reduction in grain yield. When clover (*Trifolium pratense* L.) was grown alongside corn, it resulted in the highest corn yield and optimal traits. Synchronous sowing of cover crops with corn improved grain length and yield compared to delayed sowing. Clover exhibited the strongest weed suppression due to its rapid growth and competitive nature.

3.7 Physical Management

Physical weed management in Iran frequently involves practices such as manual weeding and hoeing. These methods are designed to decrease reliance on herbicides while enhancing soil health. Manual weeding and hoeing are generally effective methods for managing weeds; however, high labor costs and a shortage of workers render these techniques economically unfeasible, especially in large fields. Consequently, the use of these methods in the expansive fields of key agricultural crops in Iran has significantly decreased. As illustrated in Fig. 3.4, the highest chickpea yield was achieved when weeds were manually removed throughout the growing season, resulting in a weed-free environment (Nabati et al. 2023).

3.8 Biological Management

Biological weed control involves utilizing insects, mites, or pathogens that specifically target weeds (Nega 2014). The primary objective is to reduce weed populations and their detrimental effects on crop production by leveraging natural enemies and competitors. Although biological control is primarily employed to manage pests, such as insects, there have been successful reports from Iran regarding the use of this method to control specific weeds and invasive plant species. For instance, in a study, the biological potential of rhizosphere *Fusarium* species (*Fusarium* spp.) for controlling significant weeds in wheat fields was investigated (Nooralvandi et al. 2020). The results indicated that an isolate of *Fusarium oxysporum* effectively controlled *S. arvensis* and *R. rugosum*, while an isolate of *Fusarium brachygibbosum* selectively managed *H. spontaneum*. Furthermore, an isolate of *Fusarium equiseti* demonstrated selective control over both *H. spontaneum* and wild oat.

A study investigated the potential of using tortoise beetles (*Cassida rubiginosa*) as a biological control method for Canada thistle [*Cirsium arvense* (L.) Scop.] in Iran (Asadi et al. 2013). Field experiments demonstrated a correlation between the density of beetle larvae and a reduction in *C. arvense* growth. Higher numbers of larvae significantly decreased *C. arvense* biomass and seed head production. In wheat fields, the introduction of 12 to 24 beetle larvae per thistle plant resulted in an increase in wheat ear weight of 46% to 82%. Overall, tortoise beetles show promise as an effective biological control agent against *C. arvense* in Iranian arable crops and grasslands.

3.9 Integrated Weed Management

Integrated weed management (IWM) in Iran utilizes a combination of strategies to sustainably control weed populations. This approach incorporates cultural practices, mechanical methods, and the careful application of herbicides, with the goal of reducing dependence on chemical solutions while enhancing crop productivity. By promoting biodiversity and improving soil health, IWM helps farmers tackle the challenges posed by weed resistance and environmental impacts. Investigating the effects of integrated weed control on yield, yield components, and qualitative traits of rapeseed revealed that conventional tillage, when combined with manual weeding, resulted in the highest seed yield, biological yield, and oil content in rapeseed crops (Atashbar et al. 2024). The greatest number of siliques per plant and seed weight were achieved through conventional tillage with a split application of 200 g ai of cycloxidim herbicide per ha. The highest harvest index was observed with no-tillage, which also utilized a split application of 200 g ai of cycloxidim herbicide per ha. The study on integrated methods for the chemical and cultural control of weeds in wheat in Iran concluded that the maximum grain yield and net income can be achieved by utilizing broad-spectrum herbicides that target both grass and broadleaf weeds, in conjunction with a narrow row spacing of 20 cm (Ashrafi et al. 2009). The study conducted by Ghorbani and Kulshreshtha (2013) assessed farmers' willingness to pay for IWM practices designed to mitigate the negative environmental impacts associated with herbicide use in crop production. The findings revealed that farmers were more inclined to invest in IWM compared to other weed management strategies. Factors that positively influenced the adoption of IWM included total annual income, the area of irrigated wheat, wheat yield loss attributed to weeds, the presence of perennial weeds, and awareness of herbicide resistance. Conversely, rain-fed wheat cultivation and a higher number of plots on the farm negatively impacted the adoption of IWM.

Integrated weed management in chickpea is of great importance as it reduces reliance on chemical herbicides, enhances production sustainability, improves yield, and alleviates the pressure of herbicide-resistant weeds. By combining cultural, mechanical, and chemical approaches, IWM provides more effective weed control. Ahmadi et al. (2019) reported that inter-row cultivation in rainfed chickpea significantly reduced the populations of *C. album* and *A. retroflexus*. While herbicides such as linuron and paraquat did not achieve acceptable control, pyridate, isoxaflutole, and metribuzin were found to satisfactorily suppress weed populations. None of the herbicide treatments effectively reduced *C. arvensis*, although the highest reduction was observed for *C. album* (100%) when sprayed with metribuzin. Total weed density reduction was greatest with pyridate (78%) and isoxaflutole (57%), while metribuzin reduced weed density by at least 71%. These results highlight the effectiveness of IWM strategies that integrate chemical and mechanical control methods in chickpea cultivation. Regarding grain yield of chickpea, hand weeding plus inter-row cultivation, Pyridate plus inter-row cultivation, and pyridate alone increased yield by 45%, 44%, and 41%, respectively, compared with the weedy check. Similarly, isoxaflutole plus inter-row cultivation and isoxaflutole alone improved yields by 37% and 34%, respectively. The highest phytotoxicity to chickpea (30%) was observed with linuron. Overall, the findings confirm that adopting IWM practices, particularly inter-row cultivation in rainfed chickpea in combination with selective herbicides, is a highly effective approach for sustainable weed management and yield improvement.

Considering the extensive area under corn cultivation in Iran, the strategy of IWM is of great importance, similar to other strategic crops. Findings by Feyzolahi et al. (2022) indicated that the highest grain yield was obtained in the hand weeding treatment (14.4 t ha^{-1}) and in the treatment combining 40 g ai ha^{-1} of nicosulfuron herbicide with straw mulch (13.2 t ha^{-1}). The highest biological yield was recorded in the hand weeding treatment (31.5 t ha^{-1}). Moreover, the highest harvest index was observed in the treatments of 60 g ai ha^{-1} of nicosulfuron (55.7%), 1012.5 g ae ha^{-1} of 2,4-D + MCPA (53.6%), and 1200 g ai ha^{-1} of atrazine (47.7%). In terms of weed population and dry weight, the greatest reduction was achieved by hand weeding and nicosulfuron combined with straw mulch, which consequently reduced crop–weed competition and enhanced productivity. Overall, the results demonstrated that the application of integrated weed management, combining chemical, cultural, and mechanical methods, plays a crucial role in increasing corn yield and reducing environmental pollution. At the same time, relying on any single method alone not only increases costs but can also reduce the long-term effectiveness of weed management, thereby lowering both economic and environmental efficiency, highlighting the importance of integrating multiple approaches.

3.10 Future Prospects for Weed Management in Iran

Weed management in Iran faces serious challenges due to the country's diverse climatic and ecological conditions, and since agriculture plays a central role in the national economy, improving weed control is crucial for enhancing crop productivity and sustainability. Modern technologies such as precision agriculture, the use of drones, and satellite imagery enable real-time monitoring and site-specific weed management, thereby reducing herbicide use, lowering costs, and minimizing environmental impacts. Among these innovations, artificial intelligence (AI) and machine learning are particularly valuable, as they allow predictive modeling of weed growth patterns and support smarter resource allocation and decision-making. In parallel, biological control methods and the development of resistant crop varieties are gaining importance in line with global trends toward sustainable agriculture. Future strategies should emphasize region-specific solutions tailored to Iran's environmental diversity, foster collaborations among universities, research institutions, and farmers, and carefully evaluate the ecological consequences of emerging technologies. Overall, the integration of AI with other advanced approaches can pave the way for more effective weed management in Iran, ultimately improving crop yields, food security, sustainability, and the economic resilience of the agricultural sector.

3.11 Conclusions

In conclusion, studying the biology and ecology of weeds and invasive plants is essential for developing effective management strategies that promote sustainable agricultural production in Iran, particularly concerning vital crops such as wheat, corn, rice, chickpeas, and rapeseed, which are crucial for food security. By understanding the life cycles, growth patterns, and ecological roles of weeds, farmers can devise targeted approaches that reduce reliance on herbicides and prevent the development of resistance. This knowledge facilitates the implementation of integrated weed management practices that enhance biodiversity, improve soil health, and sustain the productivity of these key crops. Ultimately, a comprehensive understanding of weed dynamics is critical for advancing sustainable agriculture in Iran, ensuring a stable food supply while preserving the ecological integrity of its diverse landscapes.

Acknowledgments Language editing support was provided using Wordvice AI. The authors bear sole responsibility for the final content of the chapter.

References

Ahmadi A, Veisi M, Aghaei Haji Abadi M, Mousavi SK (2019) Studying of integrated weed management in rainfed chickpea (*Cicer arietinum* L.). Iranian J Pulses Res 10(1):195–208

Alipour S, Amini Dehaghi M, Gholami S (2019) Allelopathic effect of different weeds extracts on germination and biochemical composition of three varieties of sesame (Sesamum indicum L.). J Seed Res 9(3):68–81

Alm DM, McGiffen ME, Hesketh JD, Hodges T (1991) Weed phenology. In: Predicting crop phenology, pp 191–218

Amini R, Alizadeh H, Yousefi A (2014) Interference between red kidneybean (*Phaseolus vulgaris* L.) cultivars and redroot pigweed (*Amaranthus retroflexus* L.). Eur J Agron 60:13–21

Amini R, Hasanfard A, Ahmadian N, Yuzband FZ (2024) Effect of environmental, seed burial depth, and straw mulch on germination and seedling emergence in *Cichorium glandulosum*. Weed Sci 72(2):164–171

Amiri M, Tarkesh M, Shafiezadeh M (2022) Modelling the biological invasion of *Prosopis juliflora* using geostatistical-based bioclimatic variables under climate change in arid zones of southwestern Iran. J Arid Land 14(2):203–224

Anafjeh G, Elahifard E, Derakhshan A, Zare A (2024) Yield gap analysis of rice in Shoaybiyeh, Khuzestan: yield-reducing factors have the greatest contribution to grain yield reduction. Plant Productions. Available Online from 28 September 2024

Anderson RL (2007) Managing weeds with a dualistic approach of prevention and control. A review. Agron Sustain Dev 27:13–18

Ansong M (2024) Unintentional human-vectored dispersal contributes to the spread of high risk weed species. Weed Res 64(2):149–157

Anwar MP, Islam AM, Yeasmin S, Rashid MH, Juraimi AS, Ahmed S, Shrestha A (2021) Weeds and their responses to management efforts in a changing climate. Agronomy 11(10):1921

Armengot L, Blanco-Moreno JM, Bàrberi P, Bocci G, Carlesi S, Aendekerk R, Sans FX (2016) Tillage as a driver of change in weed communities: a functional perspective. Agric Ecosyst Environ 222:276–285

Asadi G, Ghorbani R, Karimi J, Bagheri A, Mueller-Schaerer H (2013) Host impact and specificity of Tortoise Beetle (*Cassida rubiginosa*) on Canada thistle (*Cirsium arvense*) in Iran. Weed Technol 27(2):405–411

Ashrafi ZY, Sadeghi S, Mashhadi HR (2009) Study of integrate methods chemical and cultural control of weeds to wheat (*Triticum aestivum* L.). J Agric Sci 1(2):113

Atashbar N, Monsefi A, Fateh E (2024) Investigating integrated weed control on yield and yield components and qualitative traits of canola in rotation with corn. Plant Product 47(1):101–115

Bajwa AA, Mahajan G, Chauhan BS (2015) Nonconventional weed management strategies for modern agriculture. Weed Sci 63(4):723–747

Bajwa AA, Zulfiqar U, Sadia S, Bhowmik P, Chauhan BS (2019) A global perspective on the biology, impact and management of *Chenopodium album* and *Chenopodium murale*: two troublesome agricultural and environmental weeds. Environ Sci Pollut Res 26:5357–5371

Batlla D, Ghersa CM, Benech-Arnold RL (2020) Dormancy, a critical trait for weed success in crop production systems. Pest Manag Sci 76(4):1189–1194

Bayat M, Zargar M (2021) Postemergence herbicide combinations for effective Littleseed Canarygrass (*Phalaris minor*) control in winter wheat (*Triticum aestivum* L.). Acta Physiol Plant 43:1–5

Benvenuti S (2007) Weed seed movement and dispersal strategies in the agricultural environment. Weed Biol Manag 7(3):141–157

Bijanzadeh E, Ghadiri H (2006) Effect of separate and combined treatments of herbicides on weed control and corn (*Zea mays*) yield. Weed Technol 20(3):640–645

Buddenhagen CE, Hackell D, Henderson HV, Wynne-Jones B (2023) Factors impacting the detection of weed seed contaminants in seed lots. Pest Manag Sci 79(2):881–890

Buhler DD, Liebman M, Obrycki JJ (2000) Theoretical and practical challenges to an IPM approach to weed management. Weed Sci 48(3):274–280

Bullock JM, Pufal G (2020) Human-mediated dispersal as a driver of vegetation dynamics: a conceptual synthesis. J Veg Sci 31(6):943–953

Cechin J, Schmitz MF, Torchelsen JS, Durigon MR, Agostinetto D, Vargas L (2022) Winter cover crops reduce the soil seed bank and infestations of Italian ryegrass in no-tillage system. Crop Sci 62(1):479–488

Cherati FE, Bahrami H, Asakereh A (2011) Evaluation of traditional, mechanical and chemical weed control methods in rice fields. Aust J Crop Sci 5(8):1007–1013

Davies KW, Sheley RL (2007) A conceptual framework for preventing the spatial dispersal of invasive plants. Weed Sci 55(2):178–184

Ebrahimpour MH, Mokhtassi-Bidgoli A, Aghaalikhani M, Eyni-Nargeseh H (2022) Post-emergence herbicides efficacy as affected by factory and climatic conditions in wheat Triticum aestivum. J Crop Protect 11(1):121–131

Esmaeilzadeh Z, Eslami SV, Zand E (2012) Investigating the resistance of annual ryegrass (*Lolium rigidum*) biotypes collected from wheat fields of Fars province to pinoxaden herbicide. Iranian J Weed Sci 7(2):61–75

Evangelista D, Hotton S, Dumais J (2011) The mechanics of explosive dispersal and self-burial in the seeds of the filaree, *Erodium cicutarium* (Geraniaceae). J Exp Biol 214(4):521–529

Feyzolahi M, Monsefi A, Rahnama A, Farzaneh M (2022) Influence of chemical and integrated weed management on yield and yield components of grain maize and weed population in Northern Khuzestan. J Crops Improv 24(2):423–435

Frick B, Johnson E (2006) Livestock and other beneficial organisms for weed management. Research report 2006. Canada-Saskatchewan Agri-Food Innovation Fund

Frost RA, Launchbaugh KL (2003) Prescription grazing for rangeland weed management. Rangelands Arch 25(6):43–47

Garcia QS, Barreto LC, Bicalho EM (2020) Environmental factors driving seed dormancy and germination in tropical ecosystems: a perspective from campo rupestre species. Environ Exp Bot 178:104164

Ghorbani M, Kulshreshtha S (2013) An environmental and economic perspective on integrated weed management in Iran. Weed Technol 27(2):352–361

Hamidavi H, Eslami SV, Jami-Al-Ahmadi M (2021) Effect of environmental factors on rhizome bud germination and shoot emergence of invasive *Imperata cylindrica*. Weed Res 61(5):375–384

Hamidzadeh Moghadam S, Alebrahim MT, Mohebodini M, MacGregor DR (2023) Genetic variation of *Amaranthus retroflexus* L. and *Chenopodium album* L. (Amaranthaceae) suggests multiple independent introductions into Iran. Front Plant Sci 13:1024555

Hasanfard A, Izadi-Darbandi E (2020) Investigating the regrowth of wild barley (*Hordeum spontaneum* Koch.) and feral rye (*Secale cereale* L.) weeds compared to wheat after freezing stress. Cereal Res 10(1):73–85

Hasanfard A, Rastgoo M, Darbandi EI, Nezami A, Chauhan BS (2021) Regeneration capacity after exposure to freezing in wild oat (*Avena ludoviciana* Durieu.) and turnipweed (*Rapistrum rugosum* (L.) All.) in comparison with winter wheat. Environ Exp Bot 181:104271

Hasanfard A, Rastgoo M, Darbandi EI, Nezami A, Chauhan BS (2022) Freezing stress affects the efficacy of clodinafop-propargyl and 2, 4-D plus MCPA on wild oat (*Avena ludoviciana* Durieu) and turnipweed [*Rapistrum rugosum* (L.) All.] in wheat (*Triticum aestivum* L.). PLoS One 17(10):e0274945

Hasanfard A, Rastgoo M, Izadi-Darbandi E, Nezami A, Chauhan BS (2023) Freezing stress affects herbicide efficacy on sterile oat and turnipweed in wheat. Gesunde Pflanzen 75(5):1553–1560

Hazrati Z, Yaghoubi B, Hosseini P, Chauhan BS (2023) Herbicides for monochoria (*Monochoria vaginalis*) control in transplanted rice. Weed Technol 37(6):598–605

Honarmand SJ, Nosratti I, Nazari K, Heidari H (2016) Factors affecting the seed germination and seedling emergence of muskweed (*Myagrum perfoliatum*). Weed Biol Manag 16(4):186–193

Hosseini P, Karimi H, Babaei S, Mashhadi HR, Oveisi M (2014) Weed seed bank as affected by crop rotation and disturbance. Crop Prot 64:1–6

Humphries T, Florentine S (2022) Assessing seedbank longevity and seed persistence of the invasive tussock grass Nassella trichotoma using in-field burial and laboratory-controlled ageing. Plants 11(18):2377

Izadi-Darbandi E, Hasanfard A, Azimi M (2020) Evaluation the freezing tolerance of wild barley (*Hordeum spontaneum* Koch.) and feral rye (*Secale cereale* L.) compared to wheat (*Triticum aestivum* L.) at the two-leaf stage. Iran J Field Crop Sci 51(1):163–176

Jafari M, Akram W, Pang Y, Ahmad A, Ahmed S, Yasin NA, Hu X (2018) Genetic diversity and biogeography of *T. officinale* inferred from multi locus sequence typing approach. PLoS One 13(9):e0203275

Jalali M, Eslami SV, Mahmoodi S, Aein A (2021) Effect of potato (*Solanum tubersum* L.) and quinoa (*Chenopodium quinoa* Willd.) intercropping and weed management on yield and quantitive traits of potato. J Crop Ecophysiol 15(59):393–416

Jones PE, Consuegra S, Börger L, Jones J, Garcia de Leaniz C (2020) Impacts of artificial barriers on the connectivity and dispersal of vascular macrophytes in rivers: a critical review. Freshw Biol 65(6):1165–1180

Klupczyńska EA, Pawłowski TA (2021) Regulation of seed dormancy and germination mechanisms in a changing environment. Int J Mol Sci 22(3):1357

Koocheki A, Nassiri M, Alimoradi L, Ghorbani R (2009) Effect of cropping systems and crop rotations on weeds. Agron Sustain Dev 29:401–408

Kumar S, Bhowmick MK, Ray P (2021) Weeds as alternate and alternative hosts of crop pests. Indian J Weed Sci 53:14

Mirzaei M, Zand E, Rastgoo M, Hasanfard A (2022) Performance of 2, 4-D plus MCPA and Mesosulfuron plus Iodosulfuron plus Mefenpyr-diethyl as influenced by ammonium sulfate, urea ammonium nitrate, and carrier water hardness. Phytoparasitica 50(3):589–600

Mirzaei M, Zand E, Rastgoo M, Hasanfard A, Kudsk P (2023) Effects and mitigation of poor water quality on herbicide performance: a review. Weed Res 63(3):139–152

Moradi K, Mohammadvand E, Asghari J (2020) The efficacy of some herbicides and cultivation on weed control and rapeseed (*Brassica napus* L.) yield. J Plant Protect 34(2):243–258

Mousavi SK, Pezeshkpour P, Shahverdi M (2007) Weed population response to chickpea (*Cicer arietinum* L.) variety and planting date. J Crop Product Process 11(40):167–177. (In Persian)

Muhammad N, Sattar A, Ashiq M, Ahmad I (2011) Efficacy of pre and post emergence herbicides to control weeds in chickpea (*Cicer arietinum* L.). Pak J Weed Sci Res 17(1)

Nabati J, Hasanfard A, Rastgoo M (2023) Can different sowing dates affect weed control efficacy and chickpea yield? J Crop Protect 12(2):127–139

Nalewaja JD (1999) Cultural practices for weed resistance management. Weed Technol 13(3):643–646

Nasim G, Shabbir A (2012) Invasive weed species–a threat to sustainable agriculture. In: Crop production for agricultural improvement. Springer, pp 523–556

Nega A (2014) Review on concepts in biological control of plant pathogens. J Biol Agric Healthcare 4(27):33–54

Nezami A, Khazaie HR, Barjasteh AR, Zand E (2017) Effect of drought stress and maternal growth conditions on germination and seedling growth of wild oat (*Avena ludoviciana*). Seed Sci Technol 6(1)

Nooralvandi F, Alizadeh H, Saremi H, Salehi Jozani G (2020) Initial investigation on biological control of some important weed species of wheat fields using *Rhizosphere fusariums*. Iranian J Weed Sci 16(1):27–36

Norsworthy JK, Ward SM, Shaw DR, Llewellyn RS, Nichols RL, Webster TM, Barrett M (2012) Reducing the risks of herbicide resistance: best management practices and recommendations. Weed Sci 60(SP1):31–62

Nosratti I, Almaleki S, Chauhan BS (2019) Seed germination ecology of Soldier Thistle (*Picnomon acarna*): an invasive weed of rainfed crops in Iran. Weed Sci 67(2):261–266

Nosratti I, Sabeti P, Chaghamirzaee G, Heidari H (2020) Weed problems, challenges, and opportunities in Iran. Crop Prot 134:104371

Østrem L, Folkestad J, Solhaug KA, Brandsæter LO (2018) Frost tolerance, regeneration capacity after frost exposure and high photosystem II efficiency during winter and early spring support high winter survival in Juncus spp. Weed Res 58(1):25–34

Owen MD (2008) Weed species shifts in glyphosate-resistant crops. Pest Manag Sci 64(4):377–387

Pourmorad Kaleibar BP, Oveisi M, Alizadeh H, Mueller Schaerer H (2021) A thermal time model for optimising herbicide dose in maize. Weed Res 61(6):465–474

Rastgoo M, Hasanfard A (2022) Desertification in agricultural lands: approaches to mitigation, deserts and desertification. In: Zhu Y, Luo Q, Liu Y (eds) Deserts and desertification. IntechOpen, London

Rastgoo M, Izadi Darbandi E, Hasanfard A, Mijani S (2023a) Seed germination modeling of sterile oat biotypes susceptible and resistant to ACCase-inhibiting herbicides in response to temperature and drought stress. J Crop Protect 12(2):183–195

Rastgoo M, Nezami A, Hasanfard A, Nabati J, Ahmadi-Lahijani MJ (2023b) Freezing stress induces changes in the morphophysiological of chickpea and wild mustard seedlings. Legume Sci 5(2):e173

Ren Y, Huang C, Zhang J, Ma Y, Tian X (2021) Dispersal and germination of winged seeds of *Brandisia hancei*, a shrub in karst regions of China. Plant Divers 43(3):234–238

Rezvani M, Nadimi S, Zaefarian F, Chauhan BS (2021) Environmental factors affecting seed germination and seedling emergence of three *Phalaris* species. Crop Prot 148:105743

Salehian H, Jamshidi M, Karamzade H, Modaresi H (2013) An investigation of the phenology some weed species. Int J Agron Plant Product 4:1511–1520

Sarani M, Oveisi M, Mashhadi HR, Alizade H, Gonzalez-Andujar JL (2014) Interactions between the tillage system and crop rotation on the crop yield and weed populations under arid conditions. Weed Biol Manag 14(3):198–208

Sellers BA, Smeda RJ, Johnson WG, Kendig JA, Ellersieck MR (2003) Comparative growth of six Amaranthus species in Missouri. Weed Sci 51(3):329–333

Sheibany K, Baghestani Meybodi MA, Atri A (2009) Competitive effects of redroot pigweed (*Amaranthus retroflexus*) on the growth indices and yield of corn. Weed Biol Manag 9(2):152–159

Sohrabi S, Gherekhloo J, Kamkar B, Ghanbari A, Mohassel MHR (2016) The phenology and seed production of *Cucumis melo* as an invasive weed in northern Iran. Aust J Bot 64(3):227–234

Soltani E, Baskin CC, Baskin JM, Soltani A, Galeshi S, Ghaderi-far F, Zeinali E (2016) A quantitative analysis of seed dormancy and germination in the winter annual weed *Sinapis arvensis* (Brassicaceae). Botany 94(4):289–300

Sweeney AE, Renner KA, Laboski C, Davis A (2008) Effect of fertilizer nitrogen on weed emergence and growth. Weed Sci 56(5):714–721

Veisi M, Zand E, Moeini MM, Bassiri K (2019) Review of research on weed management of chickpea in Iran: challenges, strategies and perspectives. J Plant Protect Res 60(2):113–125

Vencill WK, Nichols RL, Webster TM, Soteres JK, Mallory-Smith C, Burgos NR, McClelland MR (2012) Herbicide resistance: toward an understanding of resistance development and the impact of herbicide-resistant crops. Weed Sci 60(SP1):2–30

Weston LA, Duke SO (2003) Weed and crop allelopathy. Crit Rev Plant Sci 22(3–4):367–389

Yeganehpoor F, Salmasi SZ, Abedi G, Samadiyan F, Beyginiya V (2015) Effects of cover crops and weed management on corn yield. J Saudi Soc Agric Sci 14(2):178–181

Zabihollahi V, Meighani F, Karaminejad M, Mirhadi M (2008) Study of chemical control of yellow foxtail (*Setaria glauca* (L.) Beauv.) and common dandelion (*Taraxacum syriacum* boiss.) in tall fescue (*Festuta arundinacea* Schreb.) turf. Iran J Weed Sci 4(1):1–13

Zand E, Baghestani MA, AghaAlikhani M, Soufizadeh S, Khayami MM, PourAzar R, Forouzesh S (2010) Chemical control of weeds in wheat (*Triticum aestivum* L.) in Iran. Crop Prot 29(11):1223–1231

Zhang X, Zhang Z, Stützel T (2014) Ontogeny of the ovule and seed wing in Catha edulis (Vahl) Endl.(Celastraceae). Flora Morphol Distrib Funct Ecol Plants 209(3–4):179–184

Chapter 4
Biology and Management of Weeds in Japan

Yoshiko Shimono, Saeko Matsuhashi, and Shunji Kurokawa

Abstract The wide range of agroecosystems in Japan, from flooded paddy rice to upland wheat and soybean fields, presents distinct weed management challenges. With 4.3 million hectares of cultivated land in 2023, over half is dedicated to rice and about a quarter to upland crops. This chapter reviews weed management in rice, wheat, and soybean, presenting quantitative insights into weed flora and control strategies. Japan harbors a remarkable weed diversity, including 129 species in paddy fields and 583 species in upland fields; nearly half of the weeds in upland fields are alien plants. Intensive herbicide use—such as one-shot sulfonylurea applications in rice—has led to herbicide resistance in 23 weed species as of 2024 (20 in flooded rice fields). In wheat fields, Poaceae species, such as *Alopecurus aequalis*, *Avena fatua*, and *Lolium multiflorum*, have become problematic owing to the lack of registered post-emergence graminicides. The spread of alien weeds, such as *Ipomoea* spp. and *Sicyos angulatus*, poses increasing challenges in soybean fields. Given these pressures, Integrated Weed Management (IWM) is paramount, incorporating chemical controls, cultural controls (e.g., water management in rice, crop rotation, and competitive cultivars), and mechanical controls (e.g., tillage and autonomous weeder robots). A precautionary approach to preventing the spread of alien weeds is crucial because many of the troublesome weeds in upland fields are alien species. Effective quarantine measures under the Plant Protection Act by the Ministry of Agriculture, Forestry and Fisheries in Japan must be established. Field monitoring technologies, such as drone image capture and AI identification technologies, promise to improve early detection and control of weeds.

Keywords Alien weeds · Crop rotation · Herbicide resistance · One-shot herbicide

Y. Shimono (✉) · S. Kurokawa
Graduate School of Agriculture, Kyoto University, Kyoto, Japan
e-mail: shimono.yoshiko.7s@kyoto-u.ac.jp

S. Matsuhashi
Institute for Plant Protection, National Agriculture and Food Research Organization (NARO), Tsukuba, Ibaraki, Japan

B. S. Chauhan, K. Jabran (eds.), *Biology and Management of Weeds in Asia and Africa*, Sustainability Sciences in Asia and Africa,
https://doi.org/10.1007/978-981-95-7314-1_4

4.1 Introduction: Overview of Japanese Agriculture and Agricultural Weeds

Japan's archipelago stretches approximately 3000 km from north to south, resulting in a wide range of climate conditions across the country (Fig. 4.1). Hokkaido, the northernmost prefecture in Japan, has a cool temperate climate with average annual temperatures between 5 and 10 °C. Key crops in Hokkaido include *Triticum aestivum* L. (wheat), *Solanum tuberosum* L. (potato), *Beta vulgaris* L. (sugar beet), *Zea mays* L. subsp. *mays* (corn), and *Glycine max* (L.) Merr. subsp. *max* (soybean), with dairy farming being prominent. In contrast, Okinawa, located in the subtropics, is the southernmost prefecture in Japan which experiences average annual temperatures of 23 to 25 °C and is known for crops suited to warmer climates, such as *Saccharum officinarum* L. (sugarcane), tropical fruits (e.g., *Ananas comosus* (L.) Merr. (pineapple), *Mangifera indica* L. (mango), *Hylocereus undatus* (Haw.) Britton et Rose (dragon fruit), *Carica papaya* L. (papaya)), as well as traditional crops like *Ipomoea batatas* (L.) Lam. (purple sweet potato) and *Momordica charantia* L. var. *pavel* Crantz (bitter gourds). Other regions of Japan have temperate or warm-temperate climates, where *Oryza sativa* L. (rice) cultivation predominates.

As of 2023, Japan's total cultivated land area is approximately 4.3 million ha. Of this, 54% area is dedicated to rice fields, 26% to upland field crops, 14% to pastureland, and 6% to orchards (Ministry of Agriculture, Forestry and Fisheries, MAFF 2024a). Wheat is the second most cultivated crop, followed by soybean. Rice remains the most important crop in Japan, though per capita consumption has steadily declined since its peak of 118.3 kg in 1962, falling to 50.9 kg in 2022 (MAFF 2024b). As rice production has outpaced demand, leading to excess

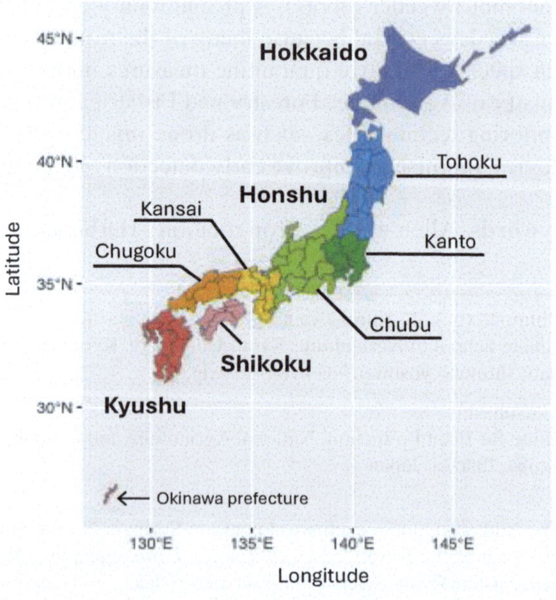

Fig. 4.1 Japan consists of four main islands (Hokkaido, Honshu, Shikoku, and Kyushu) and many small islands. It is divided into seven regions (Hokkaido, Tohoku, Kanto, Chubu, Kansai, Chugoku, Shikoku, and Kyushu). The boundaries in the map indicate the 47 prefectures

inventory, the government has encouraged using paddy fields for growing crops other than rice, such as wheat, soybeans, and feedstuffs, resulting in increased crop rotation.

Kasahara (1971) reported 191 species from 43 families as paddy field weeds and 302 species from 53 families as upland weeds based on surveys from 1942 to 1950 in Japan, most being archaeophytes introduced approximately 2400–2800 years ago (Mineta 2020). A more recent weed guide (Asai 2015) lists 129 species (28 families) of paddy field weeds and 583 species (54 families) of upland weeds, with many being neophytes introduced around the sixteenth century (referred to as alien plants). Approximately 15% of weeds in paddy fields are alien species, while approximately half of the weeds in upland crop fields are alien (Asai 2015). The expansion of international trade after the eighteenth century contributed to the spread of alien plants. Japan has the largest number of naturalized alien plant species, with more than 1700 species reported (Egawa and Koyama 2023). Approximately 30% of these were intentionally introduced as ornamental plants, aquatic plants, feed, or for revegetation, while 10% were unintentionally introduced as contaminants into specific commodities (Egawa and Koyama 2023), such as imported grains (Asai et al. 2007, Shimono and Konuma 2008; Ikeda et al. 2022). The introduction pathways for 60% of alien plants remain unknown, though most are believed to have arrived unintentionally (Egawa and Koyama 2023).

This chapter focuses on the cultivation systems and major weeds in rice, wheat, and soybeans—the primary crops cultivated in Japan (Table 4.1)—and introduces Integrated Weed Management (IWM) strategies for sustainable weed control in these crops.

Table 4.1 Major weeds in rice, soybean, and wheat cultivation

Crop	Family	Species	Life cycle[a]	Origin	Reproduction
Rice	Poaceae	*Echinochloa colona* (L.) Link	SA	Native	Seeds
	Poaceae	*Echinochloa crus-galli* (L.) P.Beauv. var. *crus-galli*	SA	Native	Seeds
	Poaceae	*Echinochloa crus-galli* (L.) P.Beauv. var. *formosensis* Ohwi	SA	Native	Seeds
	Poaceae	Echinochloa oryzicola (Vasing.) Vasing.	SA	Native	Seeds
	Poaceae	*Dinebra chinensis* (L.) P.M.Peterson et N.Snow	SA	Native	Seeds
	Poaceae	*Oryza sativa* L. (weedy rice)	SA	Native	Seeds
	Cyperaceae	*Eleocharis kuroguwai* Ohwi	P	Native	Seeds, Tubers
	Cyperaceae	*Schoenoplectiella juncoides* (Roxb.) Lye	P	Native	Seeds, Overwintering bud
	Pontederiaceae	*Monochoria vaginalis* (Burm.f.) C.Presl ex Kunth	SA	Native	Seeds
	Alismataceae	Sagittaria trifolia L.	P	Native	Seeds, Tubers
	Onagraceae	*Ludwigia decurrens* Walter	SA	Alien	Seeds
	Lythraceae	*Ammannia coccinea* Rottb.	SA	Alien	Seeds

(Continued)

Table 4.1 (continued)

Crop	Family	Species	Life cycle[a]	Origin	Reproduction
Wheat	Poaceae	*Alopecurus aequalis* Sobol.	WA	Native	Seeds
	Poaceae	*Avena fatua L.*	WA	Native	Seeds
	Poaceae	*Beckmannia syzigachne* (Steud.) Fernald	WA	Native	Seeds
	Poaceae	*Lolium multiflorum* Lam.	WA	Alien	Seeds
	Poaceae	*Poa annua* L.	WA	Native	Seeds
	Rubiaceae	*Galium spurium* L. var. *echinospermon* (Wallr.)	WA	Native	Seeds
	Fabaceae	*Vicia sativa* L. subsp. *nigra* (L.) Ehrh.	WA	Native	Seeds
	Brassicaceae	*Capsella bursa-pastoris* (L.) Medik	WA	Native	Seeds
	Brassicaceae	*Thlaspi arvense* L.	WA	Alien	Seeds
	Caryophyllaceae	*Stellaria media* (L.) Vill.	WA	Native	Seeds
	Polygonaceae	*Persicaria maculosa* Gray subsp. *hirticaulis* (Danser) S.Ekman et Knutsson var. *pubescens* (Makino) Yonek.	WA	Native	Seeds
	Poaceae	*Cynosurus echinatus* L.	WA	Alien	Seeds
	Asteraceae	*Cyanus segetum* Hill	WA	Alien	Seeds
Soybean	Poaceae	*Echinochloa crus-galli* (L.) P.Beauv. var. *crus-galli*	SA	Native	Seeds
	Amaranthaceae	*Amaranthus hybridus* L.	SA	Alien	Seeds
	Amaranthaceae	*Amaranthus retroflexus* L.	SA	Alien	Seeds
	Amaranthaceae	*Amaranthus blitum* L. subsp. *blitum*	SA	Alien	Seeds
	Amaranthaceae	*Amaranthus viridis* L.	SA	Alien	Seeds
	Amaranthaceae	*Amaranthus spinosus* L.	SA	Alien	Seeds
	Amaranthaceae	*Chenopodium album* L.	SA	Native	Seeds
	Asteraceae	*Xanthium occidentale* Bertol.	SA	Alien	Seeds
	Solanaceae	*Solanum nigrum* L.	SA	Native	Seeds
	Solanaceae	*Solanum americanum* auct. non Mill.	SA	Alien	Seeds
	Solanaceae	*Solanum nigrescens* M.Martens et Galeotti	SA	Alien	Seeds
	Solanaceae	*Physalis angulata* L. var. *angulata*	SA	Alien	Seeds
	Solanaceae	*Physalis angulata* L. var. *lanceifolia* (Nees) Waterf.	SA	Alien	Seeds
	Convolvulaceae	*Ipomoea coccinea* L.	SA	Alien	Seeds
	Convolvulaceae	*Ipomoea hederacea* Jacq.	SA	Alien	Seeds
	Convolvulaceae	*Ipomoea lacunosa* L.	SA	Alien	Seeds
	Convolvulaceae	*Ipomoea purpurea* (L.) Roth	SA	Alien	Seeds
	Convolvulaceae	*Ipomoea triloba* L.	SA	Alien	Seeds
	Cucurbitaceae	*Sicyos angulatus* L.	SA	Alien	Seeds

[a]*SA* summer annual, *WA* winter annual, *P* perennial

4.2 Paddy Rice

The dominant cultivation system in Japan is flooded paddy rice, which is well suited to the country's temperate, humid climate with abundant summer rainfall. In 2023, the planted areas for paddy rice totaled 1,344,000 ha for food and 289,000 ha for livestock feed (MAFF 2024c). Japan's annual rice production for food in 2023 reached 7,165,000 tons (MAFF 2024c), placing it 12th globally. While Japan's rice self-sufficiency rate is nearly 100%, per capita rice consumption has declined considerably, as mentioned above, reflecting the Westernization of dietary habits. Crop rotation between flooded paddy rice and upland crops, including wheat, soybean, and feedstuffs, is commonly practiced to address these issues.

4.2.1 Rice Type Grown in Japan

Among the five rice variety groups, two are indica (indica and aus), and three are japonica (temperate japonica, tropical japonica, and aromatic) (Garris et al. 2005; Huang et al. 2012), and temperate japonica is mainly cultivated in Japan. Traditionally, in Japan, the tropical japonica rice type was grown, and a few temperate japonica cultivars currently used for food partially contain genomes from tropical japonica (Sato 1992). Similarly, certain temperate japonica cultivars grown for livestock feed incorporate indica genomes (Maeda et al. 2019).

Japonica rice is characterized by short plant height and short, rounded grains. The amylose content ranges from less than 5% in glutinous rice to approximately 20% in non-glutinous rice, giving cooked japonica rice its moist and sticky texture. Over 800 Japanese rice brands are currently available, with Koshihikari being the most widely cultivated, covering 33.1% of the paddy rice area, followed by "Hitomebore" (8.3%) and "Hinohikari" (7.4%) (Rice Grain Organization 2024).

4.2.2 Cultivation System

In Japan, transplant cultivation is the dominant practice for rice production. Seedlings are grown in containers such as seedling pots for 15–30 days until they reach 10–15 cm in height. They are then transplanted into flooded fields from April to June, and harvesting is done from September to November. This method provides a significant competitive advantage over weeds that emerge after transplantation of rice seedlings.

Herbicide application plays a major role in weed control for conventional rice cultivation. Though the proportion of organic rice fields is increasing, it remains at 4.7% of the total cultivated area (MAFF 2024d). More than 60 herbicide active ingredients are registered to be used for weed control in rice (Table 4.2). Generally, "one-shot" herbicides, which combine acetolactate synthase (ALS)-inhibiting herbicides with grass herbicides (e.g., benthiocarb, cafenstrole, pretilachlor, and cyhalofop-butyl), are commonly applied simultaneously after rice transplanting (Watanabe 2011) (Fig. 4.2). If a field is heavily infested with weeds, early-stage, middle-stage, or late-stage herbicides are used (Fig. 4.2). Details on paddy field herbicides in Japan have been reported by Watanabe (2011) and Hamamura (2018).

While transplant cultivation is effective, it is labor-intensive due to the need to raise and transplant seedlings. To reduce costs and save labor, the adoption of direct-seeded rice has increased rapidly in Japan, though it accounted only for 2.7% of rice production by 2022 (MAFF 2024e). However, weed infestations pose a significant challenge in direct-seeded rice fields, as the lack of flooding fails to suppress weeds that emerge before or during seed sowing.

Weed management on paddy levees is critical. Traditionally, levees have been managed by mowing, with perennial plants such as *Imperata cylindrica* (L.) Raeusch. var. *koenigii* (Retz.) Pilg., *Rumex acetosa* L., and *Ixeris debilis* (Thunb.) A.Gray var. *salsuginosa* (Kitag.) Kitag. commonly dominating (Otsuka et al. 2006). Recently, herbicides like glyphosate and glufosinate have reduced labor costs but led to the dominance of annual weeds such as *Digitaria ciliaris* (Retz.) Koeler and *Cyperus microiria* Steud.

4.2.3 Weed Species in Paddy Rice, Weed Damage, and Control Measures

More than 100 weeds have been identified in rice fields of Japan. Among these, annual weeds such as *Echinochloa* species and *Monochoria vaginalis* (Burm.f.) C.Presl ex Kunth are the most troublesome. Perennial weeds, which reproduce vegetatively, are particularly challenging to manage with herbicides. The evolution of herbicide-resistant weed populations is a growing concern. Weedy rice has become a significant problem in Japan. While most weeds in rice fields are native or archaeophytes, several alien species, including *Ludwigia decurrens* Walter and *Ammannia coccinea* Rottb., have recently emerged as major issues (Table 4.1).

4.2.4 Herbicide-Resistant Weeds

The first herbicide, 2,4-dichlorophenoxyacetic acid (2,4-D), was introduced in Japan in 1950, followed by the commercialization of various herbicides. Bensulfuron-methyl, an ALS inhibitor, became commercially available in 1987. Due to their high crop selectivity, efficacy against a wide range of weeds at low doses, and low mammalian toxicity, ALS-inhibiting herbicides have become the

Table 4.2 Herbicides registered in rice in Japan as of October 2024

HRAC code	Mode of action	Chemical classification	Active ingredient	Application time	Effective weed spectrum
1	Inhibition of Acetyl CoA Carboxylase	Aryloxphenoxy-propionates	Cyhalofop butyl	Middle or Late stage	Grass
			Metamifop	Middle or Late stage	Grass
2	Inhibition of Acetolactate Synthase	Pyrimidinylbenzoates	Bispyribac-Na	Middle or Late stage	Broadleaf herb, Sedge, Grass
			Pyriftalid	One-shot	Grass
			Pyriminobac methyl	One-shot, Middle or Late stage	Grass
		Sulfonylurea	Azimsulfuron	One-shot	Broadleaf herb, Sedge, Monochoria, Sagittaria
			Bensulfuronmethyl	One-shot	Broadleaf herb, Sedge, Monochoria, Sagittaria
			Cyclosulfamuron	One-shot	Broadleaf herb, Sedge, Monochoria, Sagittaria
			Flucetosulfuron	Middle or Late stage	Grass, Sedge, Monochoria
			Halosulfuron methyl	One-shot	Broadleaf herb, Sedge, Monochoria, Sagittaria
			Imazosulfuron	One-shot	Broadleaf herb, Grass, Sedge, Monochoria, Sagittaria
			Metazosulfuron	One-shot	Broadleaf herb, Grass, Sedge, Monochoria, Sagittaria
			Propyrisulfuron	One-shot	Broadleaf herb, Grass, Sedge, Monochoria, Sagittaria
			Pyrazosulfuron ethyl	One-shot	Broadleaf herb, Sedge, Monochoria, Sagittaria
		Triazolopyrimidine	Penoxsulam	Middle or Late stage	Grass, Sedge, Sagittaria
		Sulfonanilides	Triafamone	One-shot	Grass, Sedge, Sagittaria
			Pyrimisulfan	One-shot	Broadleaf herb, Grass, Sedge, Monochoria, Sagittaria

(continued)

Table 4.2 (continued)

HRAC code	Mode of action	Chemical classification	Active ingredient	Application time	Effective weed spectrum
3	Inhibition of Microtubule Assembly	Phosphoroamidates	Butamifos	One-shot	Broadleaf herb, Grass
4	Auxin Mimics	Phenoxy-carboxylates	2,4–PA	Middle or Late stage	Broadleaf herb, Monochoria
			Clomeprop	One-shot	Broadleaf herb, Sedge, Monochoria
			MCPA, MCPB	Middle or Late stage	Broadleaf herb, Monochoria, Sagittaria
		Pyridine-carboxylates	Florpyrauxifen-benzyl	Early stage, One-shot	Broadleaf herb, Grass, Sedge, Monochoria, Sagittaria
5	Inhibition of Photosynthesis at PS II: D1 Serine 264 binders (and other non-histidine 215 binders)	Triazines	Dimethametryn	Early stage, One-shot	Broadleaf herb, Sedge, Monochoria
			Simetryn	Middle or Late stage	Grass, Broadleaf herb, Monochoria
6	Inhibition of Photosynthesis at PS II: D1Histidine 215 binders	Others	Bentazon	Middle or Late stage	Broadleaf herb, Sedge, Monochoria, Sagittaria
9	Inhibition of Enolpyruvyl Shikimate Phosphate Synthase	Glycine	Glyphosate	Before tillage, Before emergence of direct-seeded rice	
14	Inhibition of Protoporphyrinogen Oxidase	N-Phenyl-oxadiazolones	Oxadiazon	Early stage	Broadleaf herb, Grass, Sedge, Monochoria
		N-Phenyl-imides	Pentoxazone	Early stage, One-shot	Broadleaf herb, Grass, Sedge, Monochoria
		N-Phenyl-triazolinones	Carfentrazone ethyl	One-shot	Broadleaf herb, Monochoria
		Others	Pyraclonil	One-shot	Broadleaf herb, Grass, Sedge, Monochoria

15	Inhibition of Very Long-Chain Fatty Acid Synthesis	Azolyl-carboxamides	Ipfencarbazone	One-shot	Grass, Sedge, Monochoria
			Cafenstrole	One-shot	Grass, Sedge, Monochoria
			Fentrazamide	One-shot	Broadleaf herb, Grass, Sedge, Monochoria
		Benzofuranes	Benfuresate	One-shot	Sedge
		Isoxazolines	Fenoxasulfone	One-shot	Grass, Sedge, Monochoria
		Oxiranes	Indanofan	One-shot	Grass, Sedge, Monochoria
		Thiocarbamates	Benthiocarb	One-shot	Grass, Sedge
			Esprocarb	One-shot	Grass, Sedge
		Trifluoromethanesulfonanilide	Dimesulfazet	One-shot	Sedge
		α-Chloroacetamides	Butachlor, Pretilachlor	Early stage, One-shot	Broadleaf herb, Grass, Sedge
			Thenylchlor	Early stage, One-shot	Broadleaf herb, Grass, Sedge
27	Inhibition of Hydroxyphenyl Pyruvate Dioxygenase	Pyrazoles	Benzofenap	One-shot	Broadleaf herb, Monochoria, Sagittaria
			Pyrazoxyfen	One-shot	Broadleaf herb, Grass, Sedge, Monochoria, Sagittaria
			Pyrazolynate	Early stage, One-shot	Broadleaf herb, Grass, Sedge, Monochoria, Sagittaria
		Triketones	Benzobicyclon	One-shot	Broadleaf herb, Sedge, Monochoria
			Fenquinotrione	One-shot	Broadleaf herb, Sedge, Monochoria
			Lancotrione sodium	One-shot	Sedge, Monochoria
			Mesotrione	One-shot	Broadleaf herb, Sedge, Monochoria
			Tefuryltrione	One-shot	Broadleaf herb, Sedge, Monochoria, Sagittaria

(continued)

Table 4.2 (continued)

HRAC code	Mode of action	Chemical classification	Active ingredient	Application time	Effective weed spectrum
30	Inhibition of Fatty Acid Thioesterase	gem-Dimethylbenzamides	Bromobutide	One-shot	Sedge, Monochoria
			Cumyluron	One-shot	Sedge
			Oxaziclomefone	One-shot	Broadleaf herb, Grass, Sedge
33	Inhibition of Homogentisate Solanesyltransferase	Phenoxypyridazine	Cyclopyrimorate	One-shot	Broadleaf herb, Sedge, Monochoria, Sagittaria
	Unknown Mode of Action		ACN (Quinoclamine)	One-shot	Saggitaria
			Dymron	One-shot	Sedge
			Etobenzanid	One-shot	Grass
			Pyributicarb	One-shot	Grass, Sedge, Monochoria

Classification is based on HRAC Mode of Action Classification 2024

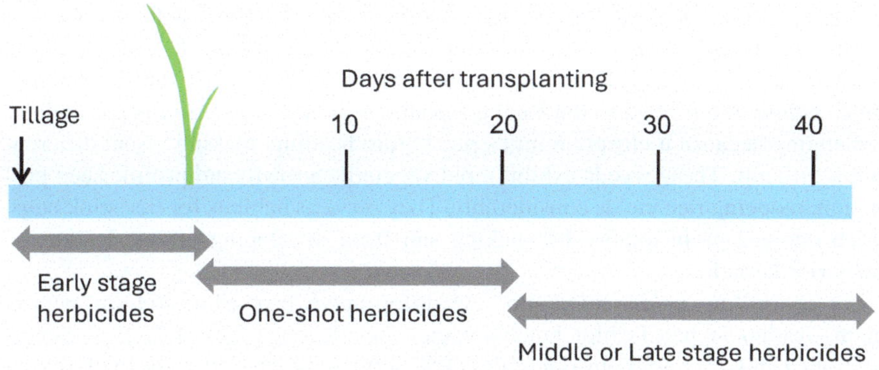

Fig. 4.2 Chemical weed control strategy in flooded paddy fields in Japan

most widely used in Japan (Watanabe 2011; Hamamura 2018). However, intensive herbicide use has led to the evolution of herbicide resistance in several weed species.

Resistance to ALS inhibitors was first recorded in *Monochoria korsakowii* Regel et Maack in Japan in 1996 and is now reported in 23 species (Uchino et al. 2024), 20 of which are weeds thriving in flooded rice fields. The most common mechanism of resistance is target-site resistance (TSR), caused by point mutations in the *ALS* gene that result in single amino acid substitutions (Sada and Uchino 2017; Tanigaki et al. 2021). Non-target-site-based resistance (NTSR) has been identified, as in *Sagittaria trifolia* L. (Iwakami et al. 2014).

Cyhalofop-butyl, an inhibitor of grass-specific chloroplastic acetyl-CoA carboxylase (ACCase), was introduced in Japan in 1996 (Hamamura 2018) and has been widely used to control *Echinochloa* species. Consequently, both TSR and NTSR have been observed in *Echinochloa crus-galli* (L.) P. Beauv. var. *crus-galli* (Iwakami et al. 2024), *E. crus-galli* var. *formosensis* Ohwi (Iwakami et al. 2015), and *E. oryzicola* (Vasing.) Vasing. (Uchino et al. 2021).

Glyphosate and glufosinate are commonly used to manage weeds in Japanese paddy levees. After nearly 20 years of glyphosate application, resistance evolved in *Lolium multiflorum* Lam. (Niinomi et al. 2013; Ichihara et al. 2016). Kurata et al. (2018) reported that glyphosate resistance in this species is linked to reduced herbicide translocation. Furthermore, *L. multiflorum* has evolved multiple resistances to both glyphosate and glufosinate (Ichihara et al. 2018).

4.2.5 *Troublesome Annual Weeds*

Echinochloa spp.

Echinochloa species are among the most problematic weeds in Japanese rice fields. Four species are widely distributed: the allohexaploids (*E. crus-galli* var.

crus-galli, *E. crus-galli* var. *formosensis*, and *E. colona* (L.) Link, with $2n = 6x = 54$ and the allotetraploid (*E. oryzicola*, with $2n = 4x = 36$). While *E. crus-galli* var. *crus-galli* and *E. oryzicola* are found across Japan, *E. crus-galli* var. *formosensis* and *E. colona* are limited to the southern Kanto region. *E. oryzicola* is particularly challenging because it closely mimics rice before heading, making visual differentiation difficult. These weeds exhibit rapid vegetative growth and prolific seed production, reducing rice yields considerably. They serve as habitats for rice stink bugs, which damage grain quality by sucking sap from developing grains, leading to pecky rice kernels.

For controlling *Echinochloa* spp., cyhalofop-butyl, bispyribac-sodium, penoxsulam, oxadiazon, pretilachlor, butachlor, etc., are effective (Table 4.2). However, as mentioned above, continuous use of the same herbicides can lead to the evolution of herbicide resistance in weeds, particularly ALS inhibitors (e.g., bispyribac-sodium) and ACCase inhibitors (e.g., cyhalofop-butyl). Rotating herbicides with different modes of action or integrating nonchemical methods is recommended to manage resistance. Maintaining water levels in flooded rice fields can substantially suppress seedling growth of *Echinochloa* spp.

Monochoria vaginalis (Burm.f.) C.Presl ex Kunth

Monochoria vaginalis is an allotetraploid ($2n = 4x = 52$) monocot belonging to the Pontederiaceae family. It is widely distributed in wetlands and rice paddies across tropical to temperate regions of Asia (Holm et al. 1977) and throughout Japan. Although its plant height is shorter than that of rice, the leaf nitrogen content is approximately twice that of rice, allowing it to compete more vigorously for nutrition. High densities of this weed can reduce rice yields. The seeds of *M. vaginalis* are initially dormant, and a few months of moist chilling conditions effectively break this dormancy.

To control *M. vaginalis*, several ALS-inhibiting herbicides, such as bensulfuron-methyl, pyrazosulfuron-ethyl, and penoxsulam, as well as herbicides with other modes of action, are effective (Table 4.2). However, intensive use of ALS inhibitors has led to the emergence of resistant biotypes of *M. vaginalis* throughout Japan (Tanigaki et al. 2021). Pyrimisulfan, penoxsulam, propyrisulfuron, metazosulfuron, and 4-HPPD inhibitors such as tefuryltrione, and mesotrione have been shown to control sulfonylurea-resistant weeds effectively (Watanabe 2011).

Dinebra chinensis (L.) P.M.Peterson et N.Snow

Dinebra chinensis, an annual weed of the Poaceae family, is distributed in rice fields, paddy levees, and upland crop fields rotated with rice throughout Japan, except in Hokkaido. It grows in paddy levees and spreads into rice fields by creeping. Recently, it has become a major problem in dry, direct-seeding rice fields in warm regions.

In the direct-seeding cultivation of paddy rice in dry fields, pre-emergence soil-applied herbicides such as benthiocarb and butachlor are effective. Cyhalofop-butyl has proven effective as a post-emergence foliar-applied herbicide.

4.2.6 Perennial Weeds

Sagittaria trifolia L.

Sagittaria trifolia is a diploid ($2n = 2x = 22$, Harada 1956) monocot from the Alismataceae family. It is widely distributed across tropical and temperate regions of Asia (Chen et al. 2007) and is prevalent throughout Japan. This monoecious species produces flowers sequentially, with female flowers opening before male flowers, though gender expression can vary depending on plant size (Huang et al. 2002). It reproduces sexually through selfed and outcrossed seeds and vegetatively via tubers. Germination of seeds requires hypoxic conditions, light exposure, and a day/night temperature regime of 30/20 °C (Ozaki et al. 2018). Controlling this species is challenging due to the uneven sprouting of tubers from varying soil depths and the larger size of its seedlings compared to its seeds.

Herbicides effective against *M. vaginalis* are often used to control *S. trifolia* (Table 4.2). However, biotypes resistant to ALS inhibitors were first identified in 1998 (Iwakami et al. 2014), and resistant populations have since been confirmed throughout Japan. Tubers are sensitive to dryness and cold, making crop rotation with upland crops or drying fields during winter highly effective in eliminating them.

Eleocharis kuroguwai Ohwi

Eleocharis kuroguwai, a monocot in the Cyperaceae family, is distributed throughout Japan, excluding Hokkaido. It has slender, erect stems and reproduces primarily through underground tubers that develop on rhizomes. Though seed production can occur, tuber-based reproduction is predominant, making control challenging due to the persistence and depth of the tubers. These tubers can survive in the soil for up to 4 years (Kusanagi 1984). Deep inversion plowing after harvest exposes tubers to unfavorable conditions, leading to desiccation and death. Herbicides such as pyrazosulfuron-ethyl, azimsulfuron, imazosulfuron, penoxsulam, bispyribac-sodium, and oxadiazon are effective in controlling *E. kuroguwai* (Table 4.2).

Schoenoplectus juncoides (Roxb.) Lye

Schoenoplectus juncoides is a perennial tetraploid ($2n = 4x = 74$, Iwasaki and Ueki 1979) monocot weed belonging to the Cyperaceae family. *S. juncoides* is widely distributed throughout Japan and is the most common paddy weed within the genus *Schoenoplectus*, which includes several other wetland species. This species reproduces vegetatively via shoots from overwintered plants or sexually via seeds. Despite its perennial nature, in rice paddies, it behaves like an annual, with seeds germinating annually. Seed dormancy is broken after exposure to winter conditions, and germination is more effective under anaerobic than aerobic conditions. With a 60–80 cm plant height, it competes with rice for nutrients and attracts rice stink bugs, further damaging crops.

Herbicides similar to those used for *E. kuroguwai* are effective against *S. juncoides* (Table 4.2). However, ALS inhibitor-resistant biotypes of *S. juncoides* have emerged across Japan (Sada and Uchino 2017). Several herbicides, such as bromobutide, clomeprop, and benzobicyclon, which have different modes of action from those of ALS inhibitors, are effective in controlling the resistant biotypes (Sada and Uchino 2017).

*4.2.7 Weedy Rice (*Oryza sativa *L.)*

Weedy rice, a conspecific weed of the genus *Oryza*, grows in and around rice fields. The characters that distinguish this weed from cultivated rice are its greater height, ease of seed shattering, and strong seed dormancy (Nadir et al. 2017). These traits, particularly grain shattering and seed dormancy, result in harvest losses and persistent seeds in paddy soils. The close resemblance of weedy rice to cultivated rice complicates its control through herbicides due to the limited options of selective herbicides.

Globally, weedy rice threatens rice production in direct-seeded rice systems (Nadir et al. 2017). In Japan, it was first identified in the 1970s in direct-seeded rice fields and recognized for its red grains. While initial control measures such as hand weeding, mechanical transplanting, and pre-emergent herbicides were largely effective, weedy rice resurged in direct-seeded and transplanted rice fields in the early 2000s (Imaizumi 2018). By 2015, weedy rice had been found in 16 of the 47 prefectures in Japan (Imaizumi 2018).

A recent whole-genome sequencing analysis of 55 Japanese weedy rice samples from various regions revealed that its populations have multiple origins. These include the BH strain, which evolved through the de-domestication of traditional temperate japonica rice that is no longer cultivated in Japan; the SH_TRJ strain, which evolved from tropical japonica landrace rice; and the SH_TEJ strain, which arose through hybridization between the BH strain and modern or landrace temperate japonica rice (Imaizumi et al. 2021). Seed dormancy levels vary among these strains, with SH_TEJ exhibiting the deepest dormancy, followed by BH and SH_TRJ (Imaizumi et al. 2022).

Weedy rice sheds seeds easily; therefore, hand weeding of weedy rice well before the seed set prevents the spread of weedy rice. Field operations in areas infested with weedy rice should only be conducted after completing work in rice fields free of weedy rice, to prevent the spread of weedy rice seeds that may be attached to the machinery and soil. If weedy rice seeds are added to the soil after shattering, avoiding tillage after harvest and leaving seeds exposed can help control them, as birds consume the seeds and cold winter conditions kill them. In direct-sowing systems, controlling weedy rice is particularly challenging. Cultivation through the transplanting method offers an effective alternative, utilizing growth differences between cultivated rice and weedy rice as part of an integrated weed control system.

4.2.8 *Toward Effective Weed Management*

4.2.8.1 Chemical Control

Herbicides remain a cornerstone of weed control in conventional agriculture, particularly in paddy rice systems. A wide range of herbicides and application techniques are available for use in these systems (Watanabe 2011; Hamamura 2018). To optimize herbicide effectiveness, careful preparation of the field is essential, including leveling the soil surface during plowing and ensuring uniform water distribution across the field. However, agricultural land ecosystems are now expected to play a role not only as a place for crop production, but also in preserving biodiversity (Katayama et al. 2023). Reliance solely on herbicides is not sustainable. Integrating chemical weed control with nonchemical practices, such as mechanical weeding or cultural methods, is crucial for minimizing herbicide use and mitigating the development of herbicide-resistant weed populations.

4.2.8.2 Cultural Control

Water Management Water management involves flooding fields during the critical period of weed-crop competition to suppress weed growth. Immediately after rice transplanting, the water level is maintained at 5 cm or more, then gradually increased to 10–15 cm as the rice grows. This practice suppresses the germination of weeds, which require oxygen to germinate, and inhibits their subsequent growth. Such deep-water management is highly effective in suppressing *Echinochloa* species, but less effective against *M. vaginalis*, which germinates under anaerobic conditions, and perennial weeds such as *S. trifolia* and *E. kuroguwai*, which regenerate from tubers.

Crop Rotation Crop rotation with upland crops, such as wheat and soybeans, helps reduce perennial weeds by reducing their tubers through exposing them to dry conditions, which are unsuitable for their survival.

Tillage Tillage, especially deep tillage or plowing during non-flooded winters, can bring weed tubers to the soil surface, increasing their susceptibility to desiccation.

4.2.8.3 Mechanical Control

In conventional Japanese rice farming, mechanical weed control during the cultivation period is uncommon due to reliance on herbicides and water management. However, there is potential for integrating innovative mechanical tools into modern systems, especially to address herbicide resistance and environmental concerns.

Recently, technological advancements have led to the development of the Aigamo robot, a fully automated, solar-powered field device designated for organic rice farming (Iseki 2024). "Aigamo" refers to a traditional integrated farming method used in Japan, particularly in paddy fields. It involves using a specific breed of domestic duck, often called Aigamo ducks, to assist in rice cultivation. Ducks are introduced to flooded rice paddies early in the season, where they consume weeds and disturb the soil surface, thereby preventing weed establishment. However, this method poses challenges, including the requirements such as labor and expertise to manage ducks effectively in rice paddies. Aigamo robots control weeds instead of ducks. The robot stirs up soil in flooded fields, creating turbidity that blocks sunlight and disrupts weed photosynthesis.

Weeds that escape early-season management or emerge late in the season may need to be removed manually to prevent them from maturing. Mature weeds can produce seeds, contributing to the soil seed bank and increasing infestation in subsequent seasons. They can interfere with mechanical harvesters, leading to uneven cutting, increased grain loss, or contamination of the harvested rice.

4.2.9 Breeding Rice to Enhance Competitiveness Against Weeds

Traditional rice breeding efforts have largely focused on traits such as increasing yield, improving grain quality (e.g., taste and texture), resistance to lodging, pest and disease resistance, and adaptability to various environments. Resistance to competition with weeds has generally not been a primary focus in rice breeding programs.

Recently, Inagaki et al. (2021) constructed a rice inbred line whose tillers spread during vegetative growth, forming a parabolic antenna-like structure. This architecture not only improved the light reception efficiency of individuals but also significantly suppressed weed growth under the canopy by reducing the amount of light incident on the ground, providing a significant competitive advantage against weeds. The development of rice varieties that have a competitive advantage against weeds could contribute to reducing weed control costs and align with sustainable agricultural practices by minimizing environmental impacts associated with chemical weed control.

4.3 Wheat

Wheat is an essential ingredient in bread, noodles, snacks, and soy sauce, with an average per capita consumption of 30 kg per annum in Japan (MAFF 2023b). However, the self-sufficiency rate for wheat remains low, at just 17% in 2023

(MAFF 2023a), with more than 80% of the supply imported from the United States, Canada, and Australia. Efforts to improve self-sufficiency have facilitated enhancing wheat quality and developing new varieties resistant to wet damage, diseases, and lodging. These initiatives have increased total wheat production, although the total planted area has remained relatively unchanged. As the number of management entities has drastically decreased (e.g., 46,000 in 2010; 31,000 in 2020; MAFF 2023b), the number of planted areas per management entity has increased (e.g., 4.4 ha in 2010; 6.7 ha in 2020; MAFF 2023b). Farmers increasingly introduce advanced tools for field management to efficiently handle large fields, including agricultural unmanned aerial vehicles (UAVs), which have improved productivity and streamlined operations.

4.3.1 Cultivation Systems

Japan's wheat self-sufficiency rate is low; however, wheat is cultivated in all Japanese prefectures. Hokkaido, a northern island, is the main production area (Fig. 4.1), accounting for approximately 60% of Japan's total production (MAFF 2024f). Most wheat grown in Japan is winter wheat (approximately 95%), with only a small amount of spring wheat produced in Hokkaido. The seeding season extends from September to early December, progressing from north to south, while the harvest season runs from May to early August, moving from south to north (Fig. 4.3).

Wheat cropping systems vary between Hokkaido and other regions (MAFF 2023b). In Hokkaido, wheat is considered a primary crop, and rotations such as four-crop/three-crop or three-crop/three-year systems involving potatoes, beans, sugar beets, and corn are common. In other areas, wheat is often cultivated as an interim crop with rice, using rotations such as two crops/year, three crops/two years, or four crops/two years with rice or soybeans.

Weed control strategies in wheat cultivation combine herbicide treatments, crop rotation, and inter-tillage. Herbicides are applied primarily at four stages: pre-plant burndown, pre-plant incorporation, pre-emergence, and post-emergence (Fig. 4.3). Pre-emergence soil-applied herbicides registered for wheat include diflufenican, trifluralin, prosulfocarb, linuron, metobromuron, benthiocarb, pendimethalin, prometryn, IPC (isopropyl-N-phenylcarbamate), esprocarb, flufenacet, and indanofan (Table 4.3). Post-emergence foliar-applied herbicides include ioxynil, pyraflufen-ethyl, bentazone, MCPA, and thifensulfuron-methyl.

As of 2024, no post-emergence herbicides are registered in Japan for controlling grass weeds in wheat. Therefore, IWM is critical to suppress outbreaks of grass species, such as *Avena fatua* and *Lolium multiflorum*.

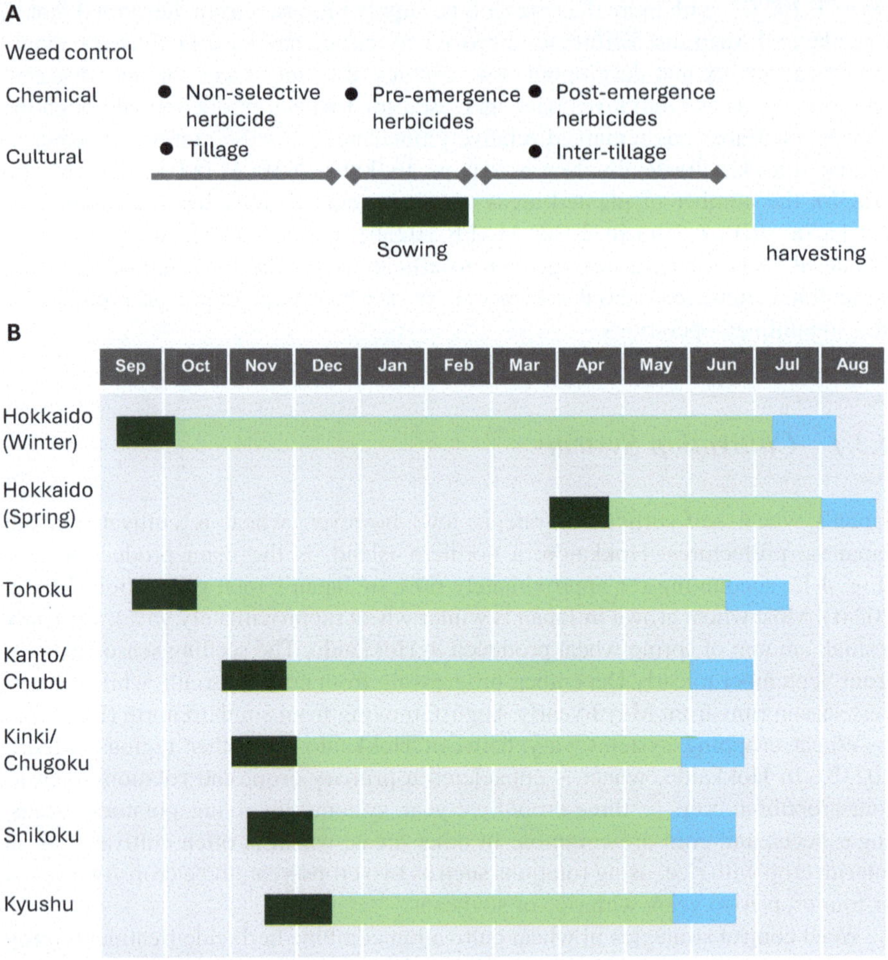

Fig. 4.3 Timing of weed control during wheat cultivation in Japan (**a**) and wheat cultivation calendar for each region (**b**)

4.3.2 Major Weeds, Weed Damage, and Control Measures

The main impact of weeds on wheat is yield loss due to competition. Most weeds are winter annual species, and weed vegetation in wheat fields varies from region to region. *Poa annua* L., *Capsella bursa-pastoris* (L.) Medik, *Stellaria media* (L.) Vill., *Chenopodium album* L., *Thlaspi arvense* L., and *Persicaria maculosa* Gray subsp. *hirticaulis* (Danser) S.Ekman et Knutsson var. pubescens (Makino) Yonek. are widely distributed, including in Hokkaido. *Alopecurus aequalis* Sobol., *Galium spurium* L. var. *echinospermon* (walls) Desp., *Avena fatua*, and *Lolium multiflorum* are more prevalent in Honshu and southern regions. *Vicia sativa* L. subsp. *nigra* (L.)

Table 4.3 Herbicides registered in wheat in Japan as of October 2024

HRAC code	Mode of action	Chemical classification	Active ingredient	Application time
2	Inhibition of Acetolactate Synthase	Sulfonylurea	Thifensulfuron-methyl	Post
3	Inhibition of Microtubule Assembly	Dinitroanilines	Trifluralin	Pre
			Pendimethalin	Pre
4	Auxin Mimics	Phenoxy-carboxylates	MCPA	Post
5	Inhibition of Photosynthesis at PS II: D1 Serine 264 binders (and other non-histidine 215 binders)	Ureas	DCMU (Diuron)	Pre
			Linuron	Pre
			Metobromuron	Pre
		Triazines	Prometryn	Pre
			IPC	Pre
6	Inhibition of Photosynthesis at PS II: D1 Histidine 215 binders	Nitriles	Ioxynil	Post
		Others	Bentazone	Post
9	Inhibition of Enolpyruvyl Shikimate Phosphate Synthase	Glycine	Glyphosate	Post
10	Inhibition of Glutamine Synthetase	Phosphinicacids	Glufosinate	Post
12	Inhibition of Phytoene Desaturase	Phenyl-ethers	Diflufenican	Pre
14	Inhibition of Protoporphyrinogen Oxidase	Others	Pyraflufen-ethyl	Post
15	Inhibition of Very Long-Chain Fatty Acid Synthesis	Thiocarbamates	Prosulfocarb	Pre
			Benthiocarb	Pre
			Esprocarb	Pre
		Isoxazolines	Pyroxasulfone	Pre
		α-Oxyacetamides	Flufenacet	Pre
		Oxiranes	Indanofan	Pre
22	PS I Electron Diversion	Pyridiniums	Diquat	Post
			Paraquat	Post

Classification is based on HRAC Mode of Action Classification 2024

Ehrh. is mainly found in Kanto and southern areas. *Beckmannia syzigachne* (Steud.) Fernald is primarily observed in the southwestern region. Additionally, alien species, such as *Cynosurus echinatus* L. in Hokkaido and *Cyanus segetum* Hill. (*Centaurea cyanus* L.) in Honshu, have shown recent increases.

Alopecurus aequalis Sobol.

It emerges in wheat fields in Honshu and southward, especially in Kyushu, which is Japan's second-largest wheat production area after Hokkaido. The germination period spans from October to April. Although two ecotypes (paddy and upland) exist in Japan, the paddy type causes problems in fields with paddy rice–wheat rotation. This leads to yield losses due to competition with wheat and complications during harvest caused by interference with the combine harvester. Resistance to the herbicides thifensulfuron-methyl and trifluralin, as well as multiple resistance to both herbicides, has been reported in paddy-type populations in southwest regions (Ohdan et al. 2014). IWM is widely adopted to address this issue by exploiting the shallow germination depth of most seeds (less than 2 cm) (Ono et al. 2014). To reduce the seedbank on the soil surface, seedlings emerging after rice harvest are removed using a non-selective herbicide before wheat sowing. Additionally, wheat is sown following either minimal or shallow tillage. This approach is commonly used to manage *A. aequalis* populations in wheat fields.

Vicia sativa L. subsp. *nigra* (L.) Ehrh.

The weed *Vicia sativa* is distributed in wheat fields in Kanto and southward, with germination occurring between September and March. Pre-emergence herbicides such as diflufenican and pendimethalin, as well as post-emergence herbicides like ioxynil, are effective in suppressing this species. However, chemical control becomes challenging for individuals germinating in or after March. The seed size and weight of this weed are similar to those of wheat, leading to severe problems with seed contamination in harvested wheat, as it is difficult to separate the two. The weed must be removed before harvesting to prevent contamination, but its vines complicate the weeding process.

Avena fatua L.

The weed *Avena fatua* has been recognized as a noxious weed in Japanese wheat fields since the 1990s (Asai & Yogo 2005), although it is present since ancient times. It germinates from September to April, and outbreaks cause severe yield losses and, in certain cases, complete harvest abandonment. Pre-emergence herbicides, including trifluralin, are relatively effective, but their efficacy is limited. With no registered post-emergence herbicides available as of 2024, crop rotation is a critical suppression strategy. The soil seedbank of this weed can be eradicated through continuous irrigation for approximately 20 days (Kida and Asai 2006), making paddy rice rotation a primary suppression technique in Japan. Other effective methods include vegetable cultivation or fallow with tillage during the booting to flowering stages to disrupt the weed's lifecycle (Matsuhashi et al. 2021), as its seed longevity is generally approximately 1 year in Japan.

This species is uncommon in wheat fields in Hokkaido, Japan's largest wheat production region, but serious outbreaks have been reported in spring wheat fields in the 2010s. Owing to the avoidance of repeated wheat cultivation, outbreaks were suppressed by the end of the 2010s.

Lolium multiflorum Lam.

The alien species *Lolium multiflorum* has caused yield losses in wheat fields by competing with crops and lowering work efficiency during harvest and field management. It was first reported as a significant weed in wheat fields in the central region of Honshu during the 1990s. When its density reaches 8–12 individuals/m^2, yield losses exceeding 20% can occur (Suzuki et al. 2010). Resistance to glyphosate and glufosinate has been confirmed around levees in rice fields in parts of central

Honshu, exacerbating the problem (Ichihara et al. 2018). Post-emergence graminicides are unavailable for this weed, but pre-emergence herbicides like flufenacet are relatively effective. The soil seedbank can be eradicated through continuous irrigation for approximately 60 days, making paddy rice rotation an effective control measure (Kida and Asai 2006).

Poa annua L.

The weed *Poa annua* is one of the major weeds in wheat fields in Hokkaido (Hokkaido 2024) and is distributed across Japan. Outbreaks are more frequently reported in northern areas than in southern ones. Although plant heights are lower than wheat, outbreaks cause yield losses (MAFF 2013). Pre-emergence herbicides are effective in controlling this weed.

Galium spurium L. var. *echinospermon* (Wallr.) Desp.

The weed *Galium spurium* is found in wheat fields from Honshu southward, germinating between November and March. Covered in hooked hairs, its stems can cling to wheat plants and harvesting equipment, disrupting crop growth, causing lodging, and reducing yields and harvesting efficiency. Prickled fruits with hooked hairs can contaminate harvested wheat. While pre-emergence herbicides are effective, their reliability is limited because a few seeds germinate from depths greater than 6 cm (Noda et al. 1964). Post-emergence herbicides for broadleaf weeds, such as pyraflufen-ethyl, are effective for control.

Beckmannia syzigachne (Steud.) Fernald

Beckmannia syzigachne tends to thrive in warm and wet soil conditions, making it a common weed in wheat fields with paddy rice–wheat rotations in western Japan. Improving field drainage is essential for suppressing outbreaks. As this weed grows taller than wheat during the flowering season, it can reduce wheat growth and severely impact yields. Its seed longevity is 1 or 2 years in Japan, and germination occurs from October to April, peaking approximately 50 days after sowing. Pre-emergence herbicides, such as trifluralin, prosulfocarb, and esprocarb, are effective, although their efficacy may vary due to deep-seated seeds and instances of resistance to trifluralin. Post-emergence herbicides, including thifensulfuron-methyl, are effective when applied by the two-leaf stage (Ohdan 2014).

4.3.3 Toward Effective Weed Management

4.3.3.1 Facilitating IWM Practices

Registered herbicides are limited for weed management in wheat in Japan; hence, IWM is essential for effective weed control and prevention of crop damage. Although it is basic and important to understand and utilize target weeds' characteristics, such as in the cases of *A. aequalis* or *A. fatua*, further application of cultural practices is expected to contribute to more efficient weed management.

Thus, crop rotation is an effective cultural practice. In Hokkaido, for example, rotation with crops that are different from wheat at the family level, such as potatoes and sugar beets, helps avoid outbreaks of noxious weeds (Hokkaido 2024). Additionally, for *A. fatua* in Honshu, a modeling study indicated the effects of crop rotation on the rapid suppression of outbreaks of a target noxious species (Matsuhashi et al. 2021). Thus, strategic crop rotation could be crucial for IWM.

In Japan, wheat production may increase by improving cultivation because the maximum yield per area is approximately twice the average yield (NARO 2020a, b). Improvements in wheat cultivation systems could potentially increase wheat's competitive ability against weeds. For instance, in Hokkaido and Aichi prefectures, which have the highest yield per unit area, changes in variety contributed to rapid yield improvement. This may imply that the varieties chosen in other areas of Japan have not yet been optimized. Wheat fields in Japan are not always suitable for cultivation because of the humid climate and wet soil conditions caused by rice rotation. Effective drainage during wheat cultivation is essential for improving wheat production in Japan (NARO 2020a, b), and achieving both drainage and rotation with rice is considered key to increasing wheat growth. Improvements in wheat cultivation systems have yet to show clear evidence of increasing wheat competition against weeds in Japan, but optimizing wheat cultivation to suit each locality and microenvironment, such as choosing suitable varieties and implementing effective drainage strategies, can create strong wheat fields resistant to weed invasion.

4.3.3.2 Strengthening Techniques for Prediction and Monitoring

Wheat fields in Japan face several changes, including farm enlargement, alien species invasion, and climate change, which are anticipated to alter weed species composition and introduce new weed problems. Strengthening prediction and monitoring techniques is necessary for early detection, risk assessment, and informed decision-making.

The recent increase in the handling areas per management entity is believed to heighten the risk of delayed identification of noxious weeds. Further progress in field monitoring technologies, such as UAVs and machine vision, has the potential to enable the efficient early detection of noxious weeds, even in larger fields. Additionally, nearly half of the weed species identified as problematic in Japan's wheat fields are alien. Repeated contamination of weed seeds in imported wheat (Asai et al. 2007; Shimono and Konuma 2008) shows the high risk of alien species invasion in these fields.

Proactive measures are essential to prevent new noxious weed invasions into wheat fields and reduce damage. These include field monitoring systems that provide early alerts for invasion, predictive methods such as simulations (Matsuhashi et al. 2021), and risk maps that help identify low-risk methods and support decision-making. Weed risk assessment frameworks are critical for prioritizing management efforts efficiently.

4.3.4 Soybean

According to the Ministry of Agriculture, Forestry and Fisheries, Japan's soybean demand in 2022 was as high as 3.89 million tons, with a self-sufficiency rate of only 6% (MAFF 2024g). Of this demand, 2.41 million tons are used as oilseeds, but domestic production accounts for only 0.01% of that amount (MAFF 2024g). Soybeans are a staple of traditional Japanese dietary culture (Intangible Cultural Heritage, UNESCO), and they are used in foods such as tofu, boiled beans, natto, miso, and soy sauce. Despite growing demand for domestic soybeans, Japan's average yield in 2022 was 1602 kg/ha, which is substantially lower than yields in the United States (3331 kg/ha) and Brazil (2952 kg/ha) (FAO 2024). Increasing soybean yield is critical. Factors contributing to low yields include water damage from cultivation in rice-rotated paddy fields and damage caused by pests, diseases, and invasive weeds such as *Ipomoea* species, *Sicyos angulatus*, and *Physalis angulata*, which have become increasingly problematic.

4.3.5 Cultivation Systems

Soybean cultivation systems vary by region. The total cultivated area of soybean in 2023 was 154,700 ha, of which 78.9% was grown as a rotation crop in paddy fields (MAFF 2023c). In Hokkaido, soybeans are grown in upland fields as a field rotation crop with wheat, sugar beets, and potatoes. In paddy fields across Honshu, Shikoku, and Kyushu, over 90% of soybeans are cultivated as part of a rotation with rice and wheat/barley. In areas along the west coast, where snowfall is heavy, soybeans are sown from mid-May to mid-June as a rotation crop without wheat or barley in winter. In regions with low snowfall, soybeans are sown from late June to late July after the wheat or barley harvest. Early growth rates and canopy closure timings can differ for varieties for different uses. Conventional weed control strategies combine chemical control with inter-tillage (Fig. 4.4). Recent labor-saving weed control strategies combine chemical control with the shielding effect of early closure by narrow-row spacing cultivation instead of inter-tillage (Fig. 4.4). In many cases, chemical control systems combine pre-emergence and post-emergence herbicides. Pre-emergence soil-applied herbicides registered for soybeans include linuron, trifluralin, alachlor, prometryn, dimethenamid, metolachlor, pendimethalin, flumioxazin, benthiocarb, pyroxasulfone, and DCMU (Table 4.4). Post-emergence foliar-applied herbicides registered for grass species' control include sethoxydim,

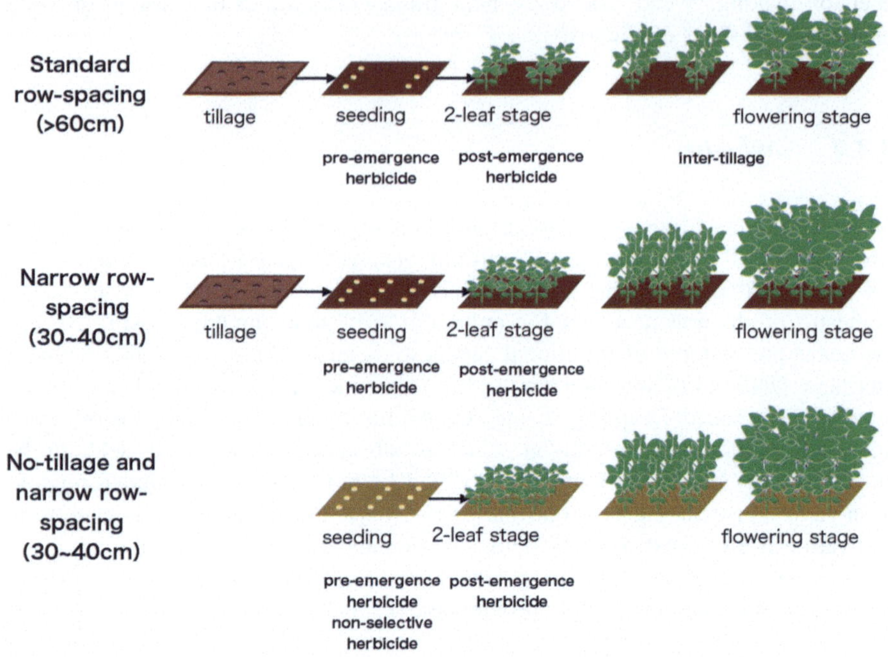

Fig. 4.4 Weed control strategies for three different soybean cultivation methods in Japan

quizalofop-ethyl, and clethodim. In soybean cultivation, bentazon is an important post-emergence herbicide for broadleaf weed control. A new post-emergence herbicide registered in 2018 for broadleaf weed control is fluthiacet-methyl (Shibuya 2019). Bentazon combined with fluthiacet-methyl may be a powerful tool for broadleaf weed control strategies. Direct spraying methods have been used to control late-emerging weeds. Even the non-selective herbicide glufosinate can be used as a direct post-emergence application method (Kubota et al. 2016). For inter-tillage, in addition to conventional rotary tillers, disk tillers with high work efficiency have been recently developed (Fig. 4.5). No-tillage cultivation systems that may affect the dynamics of weed emergence have been developed (Fig. 4.4).

4.3.6 *Weed Species in Soybean, Weed Damage, and Control Measures*

Weeds reduce the soybean yield and quality. Tall and climbing weeds significantly reduced yield. Berry-producing weeds can cause grain contamination at harvest, and weeds that may survive at harvest can cause grain contamination. Grain contamination is a problem in soybeans grown for boiled beans and natto. A few

Table 4.4 Herbicides registered in soybean in Japan as of October 2024

HRAC code	Mode of action	Chemical classification	Active ingredient	Application time
1	Inhibition of Acetyl CoA Carboxylase	Cyclohexanediones	Clethodim	Post
			Sethoxydim	Post
			Tepraloxydim	Post
		Aryloxphenoxy-propionates	Fluazifop-butyl	Post
			Quizalofop-ethyl	Post
2	Inhibition of Acetolactate Synthase	Imidazolinone	Imazamox	Post
3	Inhibition of Microtubule Assembly	Dinitroanilines	Trifluralin	Pre
			Pendimethalin	Pre
5	Inhibition of Photosynthesis at PS II: D1 Serine 264 binders (and other non-histidine 215 binders)	Triazines	Prometryne	Pre
		Ureas	Linuron	Pre
		Ureas	DCMU (Diuron)	Pre
6	Inhibition of Photosynthesis at PSII: D1 Histidine 215 binders	Others	Bentazon	Post
10	Inhibition of Glutamine Synthetase	Phosphinicacids	Glufosinate	Post
14	Inhibition of Protoporphyrinogen Oxidase	N-Phenyl-imides	Fluthiacet-methyl	Post
		N-Phenyl-imides	Flumioxazin	Pre
15	Inhibition of Very Long-Chain Fatty Acid Synthesis	Thiocarbamates	benthiocarb	Pre
		Isoxazolines	pyroxasulfone	Pre
		α-Chloroacetamides	Alachlor	Pre
		α-Chloroacetamides	Dimethenamid	Pre
		α-Chloroacetamides	Metolachlor	Pre

Classification is based on HRAC Mode of Action Classification 2024

large-seeded weeds are difficult to control using pre-emergence herbicides because they can emerge from deeper soils.

4.3.7 *Common Weeds*

Echinochloa crus-galli (L.) P. Beauv. var. *crus-galli*

It is an annual weed that grows to a height of 0.5–1.5 m. If it grows at a high density, it significantly reduces the yield. To control this weed, pre-emergence herbicides and foliar sprays of selective herbicides (quizalofop-ethyl, sethoxydim, fluazifop, clethodim, tepraloxydim, etc.) are effective.

Fig. 4.5 Disk tiller with high work efficiency. Photograph by Shunji Kurokawa

Amaranthus spp.

Several problematic *Amaranthus* species, including *A. hybridus* L., *A. retroflexus* L., *A. blitum* L. subsp. *blitum*, *A. viridis* L., and *A. spinosus* L., are annual broadleaf weeds. The most problematic species was *A. hybridus*, which grew to 2 m in height. At harvest, this can cause significant reductions in yield and grain contamination. This weed can be controlled with pre-emergence herbicides; however, late-emergent plants are difficult to control using bentazon. Fluthiacet-methyl was effective in controlling the late emergent plants.

Chenopodium album L.

It is an annual broadleaf weed that grows to 2 m in height. This can cause significant yield loss and grain contamination during harvest. This weed can be controlled with pre-emergence herbicides; however, late-emergent plants are difficult to control using bentazon. Excellent control was achieved using fluthiacetmethyl.

Xanthium occidentale Bertol.

It is an annual broadleaf weed that grows to 2 m height. Moist conditions are optimum for the growth of this weed. This can cause significant reductions in the yield. It produces large seeds that germinate deep into the soil and are difficult to control using pre-emergence herbicides. This weed can be controlled by foliar spraying with bentazon.

Solanum spp.

Solanum nigrum, *S. emulans*, and *S. nigrescens* are problematic weed species. These weeds produce berries, which can cause grain contamination during harvest. Their stems and leaves can remain green and moist until the first frost; therefore, stems and leaves containing high moisture content could cause grain contamination. Alachlor, dimethenamid, linuron, and flumioxazin are useful in controlling *Solanum* spp. as pre-emergence herbicides. *S. nigrum* could be controlled by foliar application of bentazone. However, *S. nigrescens* cannot be controlled using bentazon (Shibuya et al. 2006). Fluthiacet-methyl can be a post-emergence herbicide to control *S. nigrum* but is less effective against *S. nigrescens* (NARO 2019). Imazamox may be a useful herbicide for controlling *Solanum* species (BASF 2022).

4.3.8 Troublesome Weeds

Ipomoea spp.

Five annual *Ipomoea* species exist (*I. coccinea* L., *I. hederacea* Jacq., *I. lacunosa* L., *I. triloba* L., and *I. purpurea* (L.) Roth) that cause serious damage to soybean production in Japan. In control trials of *Ipomoea* species in soybean fields, yield reductions of 40–50% were observed on average (Endo et al. 2010). In the worst case, the yield can be reduced to zero. All five *Ipomoea* species are alien and difficult to control with pre-emergence herbicides because of their large seeds that emerge from deep soil. *I. coccinea* is the most widespread species and causes the greatest reduction in yield. Foliar sprays of bentazon can kill small seedlings of this weed, but the vines of late-emerging individuals wrap around and climb the soybean plants, eventually crushing them.

A previous study determined the period required for *I. coccinea* control (Kurokawa et al. 2015). This study showed that *I. coccinea* seedlings that emerged when the relative photosynthetic photon flux density (PPFD) was less than 49% were weakened or died. The soybean canopy height-to-row spacing ratio (CHRSR) at a relative PPFD of 49% was close to 1.0, which is a simple and practical field indicator for determining when to terminate *I. coccinea* control (Kurokawa et al. 2015). A control system implemented based on a CHRSR of 1.0 in an *I. coccinea*-infested field (*I. coccinea* density >250 m^2) completely controlled the weed (Fig. 4.6; Kurokawa et al. 2015; Kurokawa 2016). Although this index may be sensitive to the architecture of the cultivar, it can be used for more than 100 soybean core collections without the risk of weed control failure (Kurokawa et al. 2019). *I. hederacea* is the most widespread weed, causing significant yield losses. The efficacy of bentazon in controlling this weed was lower than that in controlling *I. coccinea* (Shibuya et al. 2006).

Chemical control of *I. hederacea* var. *integriuscula* using imazamox ammonium salt in narrow-row cultivation was examined (Asami et al. 2021). The results showed that a combination of imazamox and bentazon suppressed the growth of *I.*

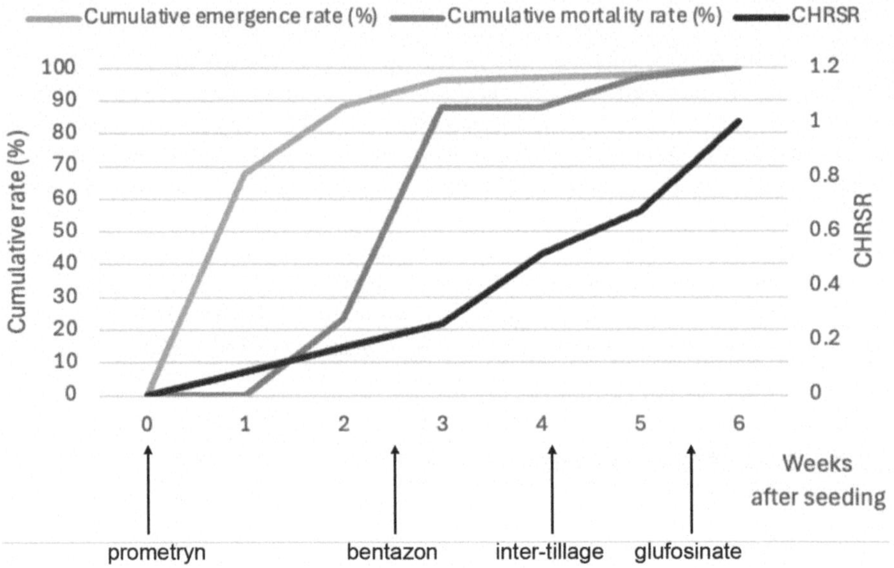

Fig. 4.6 Example of complete control of *Ipomoea coccinea* in an infected field based on the control system implemented under CHRSR (Canopy height-to-row spacing ratio) 1.0 (*I. coccinea* density >250 m^2)

hederacea by 87–97% (Asami et al. 2021). Areas infested by *I. lacunosa* and *I. triloba* are smaller than those infested by other *Ipomoea* species, but the yield loss they cause is as severe as that caused by other *Ipomoea* species. The efficacy of bentazon in controlling these effects was moderate. *I. purpurea* rarely occurs in soybean fields and has not caused any serious damage to soybean production in Japan so far. These five *Ipomoea* species cause significant yield reductions when they occur at high densities and are difficult to control. Therefore, preventive measures should be taken. Field margin management is important for preventing seed production and dispersal between fields when using agricultural machinery. Ihara and Kobayashi (2022) showed that weed management from early August to mid-September effectively reduced the population size of *I. hederacea* var. *integriuscula* at the field edge. The suppression of *I. hederacea* var. *integriuscula* and *Zoysia japonica* Steud. on ridges could be an effective strategy for suppressing weed invasion in soybean fields (Asami et al. 2024).

Sicyos angulatus L.

Sicyos angulatus causes the most damage to soybean fields. The spread of this weed species is closely related to water flow. A study of *S. angulatus* inhabiting the Abukuma River riparian zone in Japan identified five multilocus haplotypes (A–E) in the chloroplast DNA (Kobayashi et al. 2012). A comparison of haplotype composition among populations and a GIS simulation of water flow revealed that seed transfer from dairyland populations to riparian populations occurs upstream, midstream, and downstream (Kobayashi et al. 2012). This weed grows rapidly radially, and its vines, which are over 10 m long, enter soybeans and cause devastating damage. Additionally, controlling this weed with pre-emergence herbicides is challenging because its seeds are large and emerge from the deep soil. This weed can emerge from the soil at depths as high as 15 cm. An integrated weed control system was developed (Miyagi Prefecture 2015) that used flumioxazin as a pre-emergence herbicide, bentazon at the 2–3 leaf stage of soybeans, and inter-tillage, manual weeding, and direct spraying of DCMU at the 5-leaf stage. This system showed complete control of *S. angulatus* in soybeans (Miyagi Prefecture 2015). Foliar spraying of fluthiacet-methyl when *S. angulatus* is at the 2- to 3-leaf stage can cause the plant to die or severely suppress its growth (Shibuya 2019).

Physalis angulata L.

Physalis angulata var. *angulata* and *P. angulata* var. *lanceifolia* are serious weed problems in the western regions (Ikejiri et al. 2015; Kawanno et al. 2020; Ohmori et al. 2022; Togamori et al. 2024). Both species produce berries that can cause grain contamination during harvesting. Linuron may be useful in controlling them as a pre-emergence herbicide (NARO 2020a, b). *P. angulata* var. *lanceifolia* can be controlled by bentazon, whereas *P. angulata* var. *angulata* cannot be controlled by bentazon (Shibuya et al. 2006; NARO 2019, 2020a, b). Therefore, fluthiacet-methyl can control *P. angulata* as a post-emergence herbicide (NARO 2019, 2020a, b). Also, imazamox is a useful herbicide for controlling *P. angulata* var. *angulata* (BASF 2022).

4.3.9 *Toward Effective Weed Management*

4.3.9.1 Basic Weed Control System

Effective weed control relies on the proper implementation of fundamental crop cultivation practices. When the seedling establishment rate of soybeans is low, competitiveness with weeds is significantly reduced; therefore, it is necessary to take measures such as sowing techniques with as high a precision as possible and moisture damage prevention measures to prevent poor germination. In Japan, genetically modified soybeans that are resistant to non-selective herbicides are not cultivated; therefore, the basic herbicide system is a combination of pre- and post-emergence herbicides. Therefore, enhancing the effectiveness of pre-emerging herbicides is crucial to avoid the initial competition.

Many pre-emergence herbicides with species-specific suitability have been registered; therefore, selecting a pre-emergence herbicide based on the species is essential. However, 80% of soybean production is carried out in paddy fields, where increasing the soil crushing rate is challenging due to the clay soil. As a low soil crushing rate reduces the control effect of soil treatment with pre-emergence herbicides, increasing the soil crushing rate is an important factor. Because the seeds of

Ipomoea species and *Sicyos angulatus* are large and the effect of soil treatment with pre-emergence herbicides is low, initial competition causes serious weed damage. Some post-emergence herbicides are highly effective against grass species, but only a few options exist for broadleaf weed control. Since bentazon and fluthiacet-methyl are effective against different broadleaf weeds, selecting an appropriate herbicide according to the infested species is necessary. These herbicides may harm certain soybean varieties, warranting additional consideration while selecting herbicides.

4.3.9.2 Weed Management Based on the Precautionary Principle

The damage caused by weeds that cannot be controlled by the usual systemic application of pre- and post-emergence herbicides has increased in recent years. The damage caused by large, vine-like weeds such as *Ipomoea* species and *Sicyos angulatus* is serious. To prevent the damage caused by these weeds, soil seedbanks must be managed first to prevent explosive outbreaks based on the precautionary principle. The troublesome weeds mentioned above, *Ipomoea* species, *S. angulatus* and *Physalis angulatus* are all alien weeds. Avoiding the introduction of weed seeds into uncontaminated areas or fields is crucial. Further, preventing the entry of new invasive alien weeds into Japan requires effective quarantine measures. The 2023 revision of Japan's Plant Protection Act allows the designation of certain weed species as quarantine-harmful, enabling measures to prevent the introduction of weeds problematic for soybean cultivation abroad that have not yet reached Japan. Also, if weeds have already invaded a field, they must be weeded at the appropriate time to prevent the spread of seeds. Following the systematic treatment of pre- and post-emergence herbicides carried out in the early stages of soybean growth, a method of directly spraying glufosinate, linuron, and DCMU between soybean rows was registered.

Glyphosate is also registered as a herbicide for late defoliation treatment of soybeans. Furthermore, a method is registered in which glyphosate is applied directly to weeds growing above the soybean canopy using a special applicator. These techniques must be used to control weeds that persist after routine systematic treatment, preventing seed dispersal. Effective weed seedbank management requires preventing the spread of weeds between fields. Along with diligent management around the fields, prioritizing work in fields with the lowest weed density helps prevent seed transfer by machinery. Early detection of small infestations at early stages of invasion is crucial; however, monitoring is a challenge in the large-scale management of multiple fields. Technologies that can detect individuals with early invasion by using recent drone image capture and AI identification technologies will be important. *S. angulatus* spreads over a wide area through water systems. Therefore, thorough prevention of *S. angulatus* invasion based on precautionary principles, such as creating a risk map based on a model for spreading through water systems and establishing monitoring systems in high-risk areas, is crucial (Osawa et al. 2016).

4.3.9.3 Weed Risk Management Adapted to Complex Cultivation Systems

In Japan, soybean production has shifted to large-scale management in recent years; however, even large-scale farmers often manage more than 100 small, scattered fields of 0.3–0.5 ha each. The weed species that emerge and the seedbank size are different in each field, and the soil conditions, such as moisture, amount of gravel, and organic matter content, are different. Since precise management methods that match these conditions are crucial, a multi-field management system that can manage these differences centrally is necessary. Currently, various commercial information and communication technology (ICT) systems are available for managing multiple fields, but few incorporate weed risk-management systems.

Increasing diversity in cropping systems and crop varieties is altering population dynamics and the competitive interactions between crops and weeds, leading to varying levels of weed risk. In addition, certain areas are introducing labor-saving cultivation techniques to address labor shortages. In such places, narrow-row dense-planting cultivation, which omits labor-intensive intertillage, is being introduced. In such cases, producers may lack access to intertillage machines, making it impossible to perform intertillage even when necessary. Rake-type mechanical weeders that can be used in such fields have been developed; however, because they are used in the presence of crops, it is difficult to time them such that they can remove only weeds without affecting the crops. Introducing them into large-scale management systems that manage multiple fields during periods of unstable weather conditions may be challenging as well. As high-speed disc-type inter-tillage machines have been developed to replace traditional rotary-type machines, having a variety of weed control technologies that can be adapted to complex cultivation systems is essential.

4.4 Conclusions

Weed management in Japan's rice, wheat, and soybean systems must evolve to ensure long-term crop productivity and environmental sustainability. Crucially, no single control measure is sufficient—effective weed control requires combining multiple tactics. Farmers and policymakers should prioritize Integrated Weed Management strategies that reduce dependency on any one method. Agricultural land is also expected to play a role in conserving biodiversity, as many endangered species inhabit managed ecosystems. For example, 80 plant species found in paddy fields, including *Blyxa echinosperma* (C. B. Clarke) Hook.f., *Monochoria korsakowii* Regel et Maak, and *Marsilea quadrifolia* L., were listed as endangered or threatened species and were considered problematic weeds in the 1940s before the introduction of chemical control (Mineta 2020). Thus, food production systems that are in harmony with the natural environment while maintaining the essential function of crop production are required. Future prospects for weed management in Japan hinge on the continued development and implementation of IWM practices. Ongoing research and practices will be essential to address emerging weed challenges.

References

Asai M (2015) Weeds of Japan in colors. Zenkoku Noson Kyoiku Kyokai Co., Ltd. (in Japanese)

Asai M, Yogo Y (2005) Occurrence and background of wild oats and Italian ryegrass in winter cereals in the Kanto- Tokai region. J Weed Sci Technol 50:73–81. (in Japanese)

Asai M, Kurokawa S, Shimizu N, Enomoto T (2007) Exotic weed seeds detected from imported small cereal grains into Japan during 1990s. J Weed Sci Technol 52:1–10. (in Japanese)

Asami H, Tachibana M, Homma K (2021) Chemical control of *Ipomoea hederacea* var. *integriuscula* using imazamox ammonium salt in narrow-row soybean in southwestern Japan. J Weed Sci Technol 66:48–58

Asami H, Fushimi A, Homma K (2024) Suppression of entireleaf morning glory (*Ipomoea hederacea* Jacq. var. *integriuscula* A. Gray) growth by lawngrass (*Zoysia japonica* Steud.) on soybean field levees in southwestern Japan. Weed Biol Manag 24:62–67. https://doi.org/10.1111/wbm.12284

BASF (2022) Powergizer. https://crop-protection.basf.co.jp/sites/basf.co.jp/files/2022-03/202203070powergizer-tech-leaflet-tofuken.pdf. Accessed 15 Oct 2024

Chen JM, Gituru WR, Wang QF (2007) A comparison of the extent of genetic variation in the endangered *Sagittaria natans* and its widespread congener *S. trifolia*. Aquat Bot 87:1–6

Egawa C, Koyama A (2023) Temporal trends in the accumulation of alien vascular plant species through intentional and unintentional introductions in Japan. NeoBiota 83:179–196

Endo I, Hiraiwa K, Koide T, Koide N, Tani T, Hayashi M, Kuno C, Tanaka Y, Nomura Y, Inoue K, Sugiura K (2010) Effect of weeding of naturalized morningglories in soybean field by inter-ridge herbicide spraying. Research bulletin of the Aichi-ken Agricultural Research Center 42:51–56

FAO (2024) FAOSTAT: crops and livestock production. https://www.fao.org/faostat/en/#home. Accessed 15 Oct 2024. Licence: CC-BY-4.0

Garris AJ, Tai TH, Coburn J, Kresovich S, McCouch S (2005) Genetic structure and diversity in *Oryza sativa* L. Genetics 169:1631–1638

Hamamura K (2018) Development of herbicides for paddy rice in Japan. Weed Biol Manag 18:75–91

Harada I (1956) Chtological studies in Helobiae. I. Chromosome idiograms and a list of chromosome numbers in seven families. Cytologia 21:223–328

Hokkaido (2024) Agriculture Department, Technology Promotion Division, Production Promotion Bureau 2024. https://hokkaido-nosan.or.jp/manager/wp-content/uploads/r06_wheat-1.pdf. Accessed 15 Oct 2024. (in Japanese)

Holm LG, Plucknett DL, Pancho JV, Herberger JP (1977) The world's worst weeds: distribution and biology. East-West Center, University Press of Hawaii

Huang SQ, Sun SG, Takahashi Y, Guo YH (2002) Gender variation of sequential inflorescences in a monoecious plant *Sagittaria trifolia* (Alismataceae). Ann Bot 90:613–622

Huang X, Kurata N, Wei X, Han B (2012) A map of rice genome variation reveals the origin of cultivated rice. Nature 490:497–501

Ichihara M, Ishida Y, Koike K, Niinomi Y, Sawada H (2016) Distribution of glyphosate-resistant Italian ryegrass (*Lolium multiflorum* Lam.) in paddy field margins in Shizuoka prefecture. J Weed Sci Technol 61:17–20. (in Japanese)

Ichihara M, Miyata Y, Ishida Y, Koike K, Yamashita M, Sawada H (2018) Distribution of glufosinate-resistant Italian ryegrass (*Lolium multiflorum* Lam.) in paddy field margins in Chuen region of Shizuoka Prefecture. J Weed Sci Technol 63:109–112. (in Japanese)

Ihara N, Kobayashi H (2022) Populations of *Ipomoea hederacea* var. *integriuscula* in filed margins are maintained by seed production of individuals from a specific cohort. Agronomy 12:2392. https://doi.org/10.3390/agronomy12102392

Ikeda M, Nishi T, Asai M, Muranaka T, Konuma A, Tominaga T, Shimono Y (2022) The role of weed seed contamination in grain commodities as propagule pressure. Biol Invasions 24:1707–1723

Ikejiri A, Katayama M, Sugita M, Inoue K (2015) Remaining of weeds in soybean fields of Yamaguchi Prefecture in 2012–2013. J Weed Sci Technol 60:137–143

Imaizumi T (2018) Weedy rice represents an emerging threat to transplanted rice production systems in Japan. Weed Biol Manag 18:99–102

Imaizumi T, Ebana K, Kawahara Y, Muto C, Kobayashi H, Koarai A, Olsen KM (2021) Genomic divergence during feralization reveals both conserved and distinct mechanisms of parallel weediness evolution. Commun Biol 4:952

Imaizumi T, Kawahara Y, Auge G (2022) Hybrid-derived weedy rice maintains adaptive combinations of alleles associated with seed dormancy. Mol Ecol 31:6556–6569

Inagaki N, Asami H, Hirabayashi H, Uchino A, Imaizumi T, Ishimaru K (2021) A rice ancestral genetic resource conferring ideal plant shapes for vegetative growth and weed suppression. Front Plant Sci 12:748531

Iseki & Co., LTD (2024) Aigamo robot. https://products.iseki.co.jp/kanren/aigamo/. Accessed 22 Dec 2024 (in Japanese)

Iwakami S, Watanabe H, Miura T, Matsumoto H, Uchino A (2014) Occurrence of sulfonylurea resistance in *Sagittaria trifolia*, a basal monocot species, based on target-site and non-target-site resistance. Weed Biol Manag 14:43–49

Iwakami S, Hashimoto M, Matsushima K, Watanabe H, Hamamura K, Uchino A (2015) Multiple-herbicide resistance in *Echinochloa crus-galli* var. *formosensis*, an allohexaploid weed species, in dry-seeded rice. Pesticide Biochem Physiol 119:1–8

Iwakami S, Ishizawa H, Sugiura K, Kashiwagi K, Oga T, Niwayama S, Uchino A (2024) Syntenic analysis of ACCase loci and target-site-resistance mutations in cyhalofop-butyl resistant *Echinochloa crus-galli* var. *crus-galli* in Japan. Pest Manag Sci 80:627–636

Iwasaki K, Ueki K (1979) Chromosome numbers of three bulrushes (*Scirpus* Taxa). J Weed Sci Technol 24:240–242. (in Japanese)

Kasahara Y (1971) The species grouping of wild herbs, ruderals, naturalized plants, weeds and crops on the basis of their habitats. J Weed Sci Technol 12:23–27. (in Japanese)

Katayama N, Baba YG, Okubo S, Matsumoto H (2023) Taxon-specific responses to landscape-scale and long-term implementation of environmentally friendly rice farming. J Appl Ecol 60:1399–1408

Kawanno Y, Kakihara C, Chikanori T, Matsuo M, Nishiwaki A (2020) Occurrence of weeds in soybean fields of Oita Prefecture. J Weed Sci Technol 65:31–40

Kida Y, Asai M (2006) Effect of summer flooding on wild oat and Italian ryegrass seed viability. J Weed Sci Technol 51:87–90. (in Japanese)

Kobayashi H, Kurokawa S, Ikeda K (2012) Dairyland populations of bur cucumber (*Sicyos angulatus*) as a possible seed source for riverbank populations along the Abukuma River, Japan. Weed Biol Manag 12:147–155. https://doi.org/10.1111/j.1445-6664.2012.00447.x

Kubota J, Tachibana M, Ishioka G, Okuno R, Kurokawa S, Shibuya T, Takahashi K, Oki Y (2016) Effective and efficient weed control capitalizing on the residual effect of a soil-applied herbicide followed by glufosinate application as a post-emergence-directed spray in narrow-row soybean. J Weed Sci Technol 61:117–125

Kurata K, Niinomi Y, Shimono Y, Miyashita M, Tominaga T (2018) Non-target-site mechanism of glyphosate resistance in Italian ryegrass (*Lolium multiflorum*). Weed Biol Manag 18:127–135

Kurokawa S (2016) Field test case of morning glory control in soybean cultivation (in Japanese, translated by authors). Shokucho 50:279–283

Kurokawa S, Hajika M, Shibuya T (2015) Canopy height-to-row spacing ratio as a simple and practical onsite index to determine the time for terminating *Ipomoea coccinea* control in the Japanese soybean-growing systems. Weed Biol Manag 15:113–121. https://doi.org/10.1111/wbm.12074

Kurokawa S, Kaga A, Tsuda M, Sekine D, Shibuya T (2019) Evaluation of validity and limitations of the soybean canopy height-to-row spacing ratio as an onsite index to control weeds using diverse soybean accessions. Weed Biol Manag 19:103–110. https://doi.org/10.1111/wbm.12184

Kusanagi T (1984) Biology and control of perennial paddy weeds. J Pestic Sci 3:485–497. (in Japanese)
Maeda H, Murata K, Sakuma N, Tozawa Y (2019) A rice gene that confers broad-spectrum resistance to β-triketone herbicides. Science 365:393–396
MAFF (2013). https://www.maff.go.jp/j/seisan/ryutu/info/pdf/zassou-tyousa.pdf. Accessed 15 Oct 2024. (in Japanese)
MAFF (2023a) Food supply and demand table. https://www.maff.go.jp/j/zyukyu/fbs/attach/pdf/index-26.pdf. Accessed 15 Oct 2024. (in Japanese)
MAFF (2023b) Mugi no seisan wo meguru jijo (Circumstances surrounding wheat production). https://www.maff.go.jp/j/syouan/keikaku/soukatu/attach/pdf/mugi_kanren-180.pdf. Accessed 15 Oct 2024. (in Japanese)
MAFF (2023c) Statistical Survey on Crops. https://www.e-stat.go.jp/stat-search/file-download?statInfId=000040144990&fileKind=0. Accessed 15 Oct 2024. (in Japanese)
MAFF (2024a). https://www.maff.go.jp/j/tokei/kekka_gaiyou/sakumotu/menseki/r6/kouti/index.html. Accessed 12 Dec 2024. (in Japanese)
MAFF (2024b). https://www.maff.go.jp/j/heya/sodan/1808/01.html. Accessed 12 Dec 2024. (in Japanese)
MAFF (2024c) e-Stat. https://www.e-stat.go.jp/. Accessed 12 Dec 2024. (in Japanese)
MAFF (2024d) Agriculture and Forestry Census. https://www.e-stat.go.jp/stat-search/files?page=1&layout=datalist&toukei=00500209&tstat=000001032920&cycle=7&year=20200&month=0&tclass1=000001147146&tclass2=000001155386&tclass3=000001155387. Accessed 12 Dec 2024. (in Japanese)
MAFF (2024e). https://www.maff.go.jp/j/syouan/keikaku/soukatu/chokuha.html. Accessed 12 Dec 2024. (in Japanese)
MAFF (2024f) Statistical Survey on Crops. https://www.e-stat.go.jp/stat-search/file-download?statInfId=000040165941&fileKind=0. Accessed 15 Oct 2024. (in Japanese)
MAFF (2024g). https://www.maff.go.jp/j/seisan/ryutu/daizu/attach/pdf/index-21.pdf. Accessed 15 Oct 2024. (in Japanese)
Matsuhashi S, Asai M, Fukasawa K (2021) Estimations and projections of Avena fatua dynamics under multiple management scenarios in crop fields using simplified longitudinal monitoring. PLoS One 16:1–14
Mineta T (2020) Why have once paddy field weeds become endangered species? In: Ohtsuka Y, Mineta T (eds) Why do so many different creatures live in rice fields? Kyoto University Press, pp 214–231
Miyagi Prefecture (2015) Countermeasures against bur cucumber in soybean cultivation (in Japanese, translated by authors). https://www.pref.miyagi.jp/documents/20269/300672.pdf. Accessed 15 Oct 2024
Nadir S, Xiong HB, Zhu Q, Chen LJ (2017) Weedy rice in sustainable rice production. A review. Agron Sustain Dev 37:46
NARO (2019) Efficacy against weeds and early stage phytotoxicity to soybeans of fluthiacet-methyl, a new foliar herbicide for soybeans (in Japanese. translated by authors). https://www.naro.go.jp/publicity_report/publication/files/furutiruaseto190314ver2.pdf. Accessed 15 Oct 2024
NARO (2020a) Shindan ni motoduku komugi/ohmugi no saibaikaizengijutu donyu shien manual (Manual for supporting the introduction of diagnostic-based cultivation improvement techniques for wheat and barley). https://www.naro.go.jp/publicity_report/publication/files/sindan-saibaimanmanualmugi202107.pdf. Accessed 15 Oct 2024. (in Japanese)
NARO (2020b) Control of naturalized morning glories and cut-leaved ground cherry in soybean fields by systematic treatment with herbicides (in Japanese, translated by authors). https://www.naro.go.jp/publicity_report/publication/files/daizu_josou_boujo.pdf. Accessed 15 Oct 2024
Niinomi Y, Ikeda M, Yamashita M, Ishida Y, Asai M, Shimono Y, Tominaga T, Sawada H (2013) Glyphosate-resistant Italian ryegrass (*Lolium multiflorum*) on rice paddy levees in Japan. Weed Biol Manag 13:31–38

Noda K, Eguchi S, Ibaragi K (1964) Some ecological characteristics of cleaver (Galium aparine L.) and a consideration on its chemical control in winter drained paddy field. J Weed Sci Technol 1964(3):84–88. (in Japanese)

Ohdan H (2014) In: Morita H, Asai M (eds) Kazunokogusa (*Beckmannia syzigachne*), Zassou shindan boujo jiten (Encyclopaedia of weed diagnosis and control). Rural Culture Association, pp 165–166. ISBN 9784540131233. (in Japanese)

Ohdan H, Sumiyoshi T, Koarai A (2014) Occurrence of herbicide resistant biotypes of Alopecurus aequalis in northern Kyushu and their response to several herbicides. J Weed Sci Technol 59(3):157–166. (in Japanese)

Ohmori A, Tsuji K, Sugimoto M, Tominaga T (2022) Effects of light and temperature on the germination of alien *Physalis* spp. J Weed Sci Technol 67:137–142

Ono Y, Sato H, Odahara K, Hirata T, Ohdan H (2014) Integrated weed control of herbicide-resistant water foxtail *Alopecurus aequalis* in winter wheat using shallow tillage as a core technique. J Weed Sci Technol 59(4):195–203. (in Japanese)

Osawa T, Okawa S, Kurokawa S, Ando S (2016) Generating an agricultural risk map based on limited ecological information: a case study using *Sicyos angulatus*. Ambio 45:895–903. https://doi.org/10.1007/s13280-016-0782-9

Otsuka H, Nemoto M, Masuda S (2006) The floristic composition of paddy fields and their levees under different weed management techniques. J Weed Sci Technol 51:229–238. (in Japanese)

Ozaki Y, Shimono Y, Tominaga T (2018) Germination characteristics of Sagittaria trifolia. Weed Biol Manag 18:160–166

Rice Grain Organization (2024). https://www.komenet.jp/data/jishuchousa/hinsyu/. Accessed 12 Dec 2024 (in Japanese)

Sada Y, Uchino A (2017) Biology and mechanisms of sulfonylurea resistance in *Schoenoplectiella juncoides*, a noxious sedge in the rice paddy fields of Japan. Weed Biol Manag 17:125–135

Sato Y (1992) The routes of introduction of rice in Japan. J Brew Soc Japan 87:732–738. (in Japanese)

Shibuya T (2019) Efficacy of broadcast applicable herbicide, bentazon and fluthiacet-methyl for burcucumber, *Sicyos angulatus* L. in soybean. J Weed Sci Technol 64:155–160

Shibuya T, Asai M, Yogo Y (2006) Differential interspecific responses of broadleaved weeds to bentazon. J Weed Sci Technol 51:159–164

Shimono Y, Konuma A (2008) Effects of human-mediated processes on weed species composition in internationally traded grain commodities. Weed Res 48:10–18

Suzuki T, Adachi Y, Ichihara M, Yamashita M, Sawada H, Inagaki H, Ishida Y, Kida Y, Asai M (2010) A simple method to assess wheat yield loss by Italian ryegrass (*Lolium multiflorum* Lam.) in the fields. J Weed Sci Technol 55:174–182. (in Japanese)

Tanigaki S, Uchino A, Okawa S, Miura C, Hamamura K, Matsuo M, Yoshino N, Ueno N, Toyama Y, Fukumi N, Kijima E, Masuda T, Shimono Y, Tominaga T, Iwakami S (2021) Gene expression shapes the patterns of parallel evolution of herbicide resistance in the agricultural weed *Monochoria vaginalis*. New Phytol 232:928–940

Togamori Y, Tsuji K, Ohmori A, Sugimoto M, Yasukawa H, Tominaga T (2024) Effects on weeding period on the subsequent emergence and growth of alien *Physalis* in adzuki bean fields in Kyoto Prefecture. J Weed Sci Technol 69:19–24

Uchino A, Yoshida K, Kawai Y, Azuma S, Hattori M, Aoki M, Kashiwagi K, Oga T (2021) Cross-resistance and response to several alternative herbicides in the cyhalofop-butyl resistant biotypes of *Echinochloa* spp. occurred in the paddy fields of Japan. The 60th WSSJ Annual Meeting Abstracts, 58. (in Japanese)

Uchino A, Iwakami S, Tominaga T, Itoh K (2024) Herbicide resistant weeds in Japan. Online http://www.wssj.jp/~hr/2024.11.22

Watanabe K (2011) Development of lowland weed management and weed succession in Japan. Weed Biol Manag 11:175–189

Chapter 5
Biology and Management of Weeds in Malaysia

Tse-Seng Chuah, Norhafizah Md Zain, Rabiatuladawiyah Ruzmi, and Muhammad Saiful Ahmad-Hamdani

Abstract Weeds represent one of the most significant challenges to agricultural productivity in Malaysia. This chapter offers insights into the biology of key problematic weeds in major field crops such as oil palm, rice, and pineapple, together with the management of the problematic weeds through integrated approaches. In oil palm plantations, herbicide-resistant weeds such as *Eleusine indica* (goosegrass) and *Clidemia hirta* (soapbush) pose serious threats, demanding strategies like legume cover crops, mulching, and precision herbicide applications. Rice production suffers substantial yield losses (5–85%) due to invasive *Oryza* spp. (weedy rice) and *Echinochloa crus-galli* (barnyardgrass), necessitating interventions such as stale seedbed techniques integrated with pre-emergence herbicides, optimized water management, and herbicide-tolerant varieties like Clearfield® rice. Meanwhile, pineapple cultivation faces 25–35% yield reductions from aggressive weeds like *Ageratum conyzoides* (billy goat weed) and *Digitaria ciliaris* (southern crabgrass), with soil-specific solutions including plastic mulch in peat soils and targeted herbicides in mineral soils. The chapter explains the critical role of integrated weed management in sustaining agricultural productivity, emphasizing strategies tailored to crop life cycle from planting to harvest to minimize weed competition while preserving environmental and economic viability.

Keywords Oil palm · Pineapple · Rice · Biology · Management

T.-S. Chuah (✉)
Faculty of Plantation and Agrotechnology, Universiti Teknologi MARA, Perlis Branch, Arau, Malaysia
e-mail: chuahts@uitm.edu.my

N. M. Zain
Faculty of Agro-based Industry, University of Malaysia Kelantan, Jeli Campus, Jeli, Kelantan, Malaysia

R. Ruzmi · M. S. Ahmad-Hamdani
Department of Crop Science, Faculty of Agriculture, Universiti Putra Malaysia, Serdang, Selangor, Malaysia

B. S. Chauhan, K. Jabran (eds.), *Biology and Management of Weeds in Asia and Africa*, Sustainability Sciences in Asia and Africa,
https://doi.org/10.1007/978-981-95-7314-1_5

5.1 Introduction

Elaeis guineensis (Jacq.), commonly known as the oil palm, is the leading global oil crop, accounting for around 40% of the total vegetable oil traded worldwide (Murphy et al. 2021). Palm oil plays a key role in many diets and is widely used in non-food sectors, particularly in the production of cleaning and sanitizing products. With growing demand for palm oil in both food and industrial applications, oil palm is cultivated across Southeast Asia, Africa, and South America, with Indonesia and Malaysia dominating the global market (Ritchie 2021). During 2023 and 2024, Malaysia contributed more than 32% of the world's palm oil exports (FAOSTAT 2024a), and this share is expected to grow. As the second-largest palm oil producer after Indonesia, Malaysia spans 5.67 million ha of oil palm plantations currently (MPOB 2022). The Sarawak region leads in oil palm plantation area, covering 1.62 million ha (28.6% of the total), followed by Sabah with 1.51 million ha (26.6%) and Peninsular Malaysia with 2.54 million ha (44.8%) (MPOB 2022). In the year 2022, crude palm oil production totaled 18.45 million tonnes, with an average fresh fruit bunch yield of 15.49 tonnes per ha. Total palm oil exports, including palm-based products, reached 24.72 million tonnes (MPOB 2022). The crop is cultivated on both large plantations and smallholdings for various food and industrial uses. In 2021, Malaysia's total oil palm plantation area was 5.7 million ha, with 73.2% of this area managed by private or government estates, 15.1% by individual smallholders, and 11.7% by organized smallholders (Parveez et al. 2022).

Rice, a staple food in Malaysia, is the country's third most important crop after oil palm and rubber. In 2021, paddy fields covered over 647,859 ha, supporting more than 322,830 farmers (Zairy and Rozhan 2023). The consumption of rice in Malaysia reached about 2.9 million metric tonnes in 2022 (Siddharta 2024), but paddy production has steadily declined, from 2.74 million tonnes in 2005 to 2.43 million tonnes in 2021, highlighting challenges in the industry. Due to this situation, Malaysia had to import about 1.24 million tonnes of rice in 2022 to meet market demand (FAOSTAT 2024a). Approximately 64% (416,415 ha) of paddy is grown in granary areas such as Muda Agricultural Development Authority (MADA), Kemubu Agricultural Development Authority (KADA), and Integrated Agriculture Development Project (IADA), while the remaining portion is cultivated outside of these areas (DOA 2022a; Zairy and Rozhan 2023). Among these areas, MADA, which is situated in Kedah, is the largest granary area in Malaysia, with more than 201,347 ha, and produces more than 843,947 tonnes of rice. Rice cultivation began on a small scale in the early 1960s and has expanded to become Malaysia's most important food crop. Initially, manual transplanting was common, but since 1984, farmers have largely shifted to direct seeding. Today, over 90% of lowland rice is direct-seeded (DSR), while the remaining is transplanted. Direct seeding is gaining popularity due to the rising cost of labor and other inputs. However, weed infestations have intensified due to the wet-seeded rice system (Dilipkumar et al. 2021).

Pineapple production is increasing in Malaysia, contributing significantly to the economy through both domestic consumption and exports. Malaysia, known for high-quality pineapples, produced 537,231 metric tonnes, accounting for 1.8% of the world's total output (DOA 2023; FAOSTAT 2024a). The country ranks 24th globally in pineapple cultivation area (FAOSTAT 2024a) and has experienced a 10% annual growth in production since 2010 (MPIB 2019). Pineapples, following durian and bananas, are among Malaysia's top fruit crops, with a production value of USD 220,385, representing 7.6% of the total fruit sector output (DOA 2022b). Malaysia's pineapple production is concentrated in Johor (74%), Pahang (7.2%), and Sarawak (6.4%) (DOA 2023). The government has prioritized pineapples in the National Agrofood Policy 2021–2030 (NAP 2.0) due to their export potential (DOA, 2022b). The industry has shown consistent growth up to 2020, particularly with the cultivation of the MD2 variety, which promises potential earnings of up to USD 46,637 per ha. The sector supports 4779 pineapple growers, with monthly incomes ranging from USD 779 to USD 3, 542 (MPIB 2019).

Weeds have become a significant hindrance in achieving optimal crop outputs in Malaysia, threatening the productivity of oil palm, rice, and pineapple. The presence of weeds puts the crops in competition for space, nutrients, moisture, light, and carbon dioxide. Weeds also act as the hosts of several insect pests and plant disease pathogens. Hence, weed management is essential to ensure optimal plant growth and prevent losses in crop production. The weed management approach in these cropping systems is undeniably the use of herbicides. Figure 5.1 shows herbicide use trends in Southeast Asia from 2020 to 2022, with Indonesia, Malaysia, and Thailand leading in herbicide application. Indonesia recorded the highest herbicide use from 2020 to 2022, totaling 175,338 tonnes, followed by Malaysia (73,722 tonnes) and Thailand (23,409 tonnes). In 2022, Malaysia used 25,608 tonnes of pesticides, with herbicides accounting for 62.6%, followed by insecticides (16.9%), fungicides and bactericides (14.2%), and others (6.3%) (FAOSTAT 2024b).

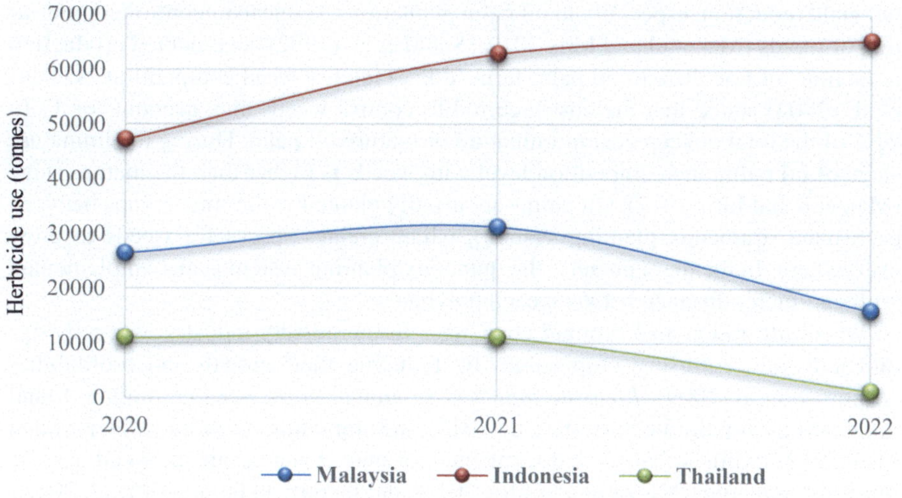

Fig. 5.1 Herbicide use in Indonesia, Malaysia, and Thailand from 2020 to 2022 (FAOSTAT 2024b)

The widespread use of herbicides has led to the evolution of herbicide resistance in several weed species due to intense selection pressure. Malaysia faces the highest prevalence of multiple and cross-resistance issues in Southeast Asia, with 17 weed species evolving resistance to different herbicide modes of action (Heap 2024). For example, *Eleusine indica* (L.) Gaertn. (goosegrass) has evolved resistance to multiple herbicides, including glyphosate, paraquat, glufosinate, and fluazifop (Jalaludin et al. 2014), while *Clidermia hirta* (L.) D. Don (soapbrush) has shown resistance against metsulfuron-methyl (Ramadzan et al. 2012) in oil palm plantations. In the rice fields, *Oryza sativa* L. (weedy rice) has evolved cross-resistance to imazapic and imazapyr (Dilipkumar et al. 2018), and *Echinochloa crus-galli* (L.) P. Beauv. (barnyardgrass) has become resistant to propanil, quinclorac, and cyhalofop-butyl (Rahman et al. 2010), with recent findings confirming bispyribac-sodium resistance (Gobi et al. 2025).

Weed management is a major prerequisite for improved crop productivity and production using different methods of crop establishment. Development of weed control strategies that are friendly to the environment and the economy requires an understanding of weed biology. The study of weed biology is connected to various aspects of plants, including their morphology, seed dormancy and germination, physiology of growth, competitive ability, and reproductive biology. Detailed descriptions of weed impact, weed biology, and management of the significant weed species in oil palm, rice, and pineapple are discussed in this chapter in order to achieve precise and efficient implementation of weed control measures.

5.2 Weed Impacts on Oil Palm

Oil yield reduction, especially in oil palm plantations, is due to uncontrolled infestation of weeds (Maizatul and Idris 2012). Sahid et al. (1992) documented production reductions of 6 to 20% in oil palm farms due to intense weed competition. Azahari et al. (2004) stated that the cost incurred to control weeds may account for 17 to 27% of the total upkeep cost in immature or mature oil palm. During the immature stage of oil palm, the competition level with weeds is higher than the mature stage (Maizatul and Idris 2012). Oil palms are usually planted with 9-meter gaps between each plant (quincunx planting system), which creates spaces for weeds to grow everywhere. In the main nursery, the quincunx planting system is also implemented with one-meter distance between each polybag.

Weed infestation also brought challenges in the nursery industry as weeds significantly affect nursery crop values by reducing their growth and profitability (Amoroso et al. 2009). *Eleusine indica* is an annual grass weed commonly found dominating oil palm nursery (Chung 2013) and immature oil palm (Maizatul and Idris 2012), while *C. hirta* is documented as one of the dominant weed species infesting immature (Maizatul and Idris 2012) and mature oil palm (Ali et al. 2021). *Clidemia hirta*, a woody broadleaf plant, is regarded as a noxious perennial weed species due to its undesirable characteristics. It significantly depresses growth, reduces yield, and hinders agronomic operations, including manuring and harvesting. A recent survey conducted by Chuah et al. (2024) documented that both *E. indica* and *C. hirta* have emerged as a problem in Malaysian oil palm plantations as a result of inadequate weed control.

5.2.1 Biology of Eleusine indica

Eleusine indica, being a C_4 plant, quickly establishes itself in areas with wet conditions, full sunlight, high temperature, and occasional periods of drought (Holm et al. 1979), but the plant is less competitive under shaded conditions (Ismail et al. 2003). This species characteristically forms clump-like growth patterns, originating from primary fibrous roots that branch into secondary roots (Breeden and Brosnan 2009) (Fig. 5.2). A single plant of *E. indica* could produce up to 50,000 seeds having a very hard seed coat. Based on phenotypic data analysis, Kerr et al. (2019) identified four *E. indica* ecotypes: dwarf, intermediate 1 (int_I), intermediate 2 (int_II), and wild. The authors found that on maintained surfaces, *E. indica* exhibited a dwarf growth habit, whereas in unmaintained areas, it displayed a larger

Fig. 5.2 *Eleusine indica* infestation in an oil palm nursery

growth habit. The dwarf ecotypes usually had the fewest seedheads compared to the wild ones, 5 and 17, respectively. In addition, int_I and int_II had the highest number of seedheads, which were 22 and 34, respectively. Subsequent analysis by Kerr et al. (2019) revealed significant height variation among ecotypes, with dwarf specimens measuring just 6 cm compared to 36 cm for wild-type plants.

Eleusine indica was observed to emerge from various soil depths, including 7 cm with 0 to 10% emergence, 5 cm with 1 to 18%, and 0–2 cm with 8 to 94% emergence (Ismail et al. 2002). This species showed significant resilience in the soil, with its seed viability declining from 79% to 39% after two years of dormancy when the burial depth was increased from 0 to 20 cm. Seeds in enforced dormancy contributed to maintaining the population during this burial period (Chuah et al. 2004). *Eleusine indica* can thrive in loamy and clay-loam soils and germinates across a broad range of alternating temperatures, such as 35/25, 35/20, 30/25, and 30/20 °C with a 12-h photoperiod (Ismail et al. 2002), and constant temperatures between 20 and 35 °C in the dark (Chuah et al. 2023). At 35 °C, water potential had no significant effect on the time required to reach 50% germination, suggesting that *E. indica* is tolerant to water stress at this temperature. The estimated base temperature and base water potential for *E. indica* were 10.6 °C and −1.28 MPa, respectively (Chuah et al. 2023).

Additionally, *E. indica* endured environmental stresses, remaining dormant during dry seasons but reemerging in rainy seasons when moisture levels reached −0.8 MPa (Ismail et al. 2002), withstanding up to 50 mM NaCl, and thriving in pH levels ranging from 5 to 10 (Chauhan and Johnson 2008a, b). Its smooth leaves are also coated with a substantial wax layer, which likely reduces water loss and provides further protection (Ngah et al. 2011). All these characteristics enable it to survive in harsh conditions and complicated control measures.

5.2.2 *Biology of* Clidermia hirta

Clidemia hirta, which originated in South America, spreads rapidly and has colonized undisturbed tropical forests in Southeast Asia. It is more prevalent outside of its native range than inside of it (Peters 2001). This adaptable understory shrub thrives in open and moist environments with humus-rich soils at altitudes of up to 1000 meters (Ismaini 2015). It typically grows to an average height of around 50 cm (Moravcová et al. 2015). Its reproductive and dispersal capabilities are substantial, with abundant fruit production and seeds dispersed mainly by birds and pigs, though other animals may also contribute (Peters 2001; Waterhouse 2002). The plant is frequently found in agricultural areas, ex-mining sites, open spaces, and along roadsides. Its invasive success is driven by its aggressive nature, high germination rates, and ability to tolerate shade (Fig. 5.3).

A mature plant can produce over 500 blue-black berries per year, each containing over 100 seeds. Its seeds can remain viable in the soil for more than four years (Clergeau and Mandon-Dalger 2001). At 25 °C, seeds of *C. hirta* germinate at rates

Fig. 5.3 *Clidemia hirta* infestation in mature oil palm plantations

exceeding 90% under alternating light and dark conditions, though none germinate in complete darkness or under plant shade. The species quickly reaches 50% germination, indicating its seeds are non-dormant while being desiccation tolerant. Across light levels from 20% to 100%, *C. hirta* maintains high seedling survival rates (Chandima et al. 2022).

Clidemia hirta is an invasive species that grows fast and accumulates biomass due to its efficient utilization of light (Idris et al. 2019). Light intensity does not significantly affect the growth and distribution of *C. hirta*, as it adapts well even in shaded conditions. In areas with dense canopy cover, the species can survive and thrive. In open environments, it tends to grow tall but does not form dense colonies. Conversely, in shaded environments, it grows more slowly but forms dense, high-population colonies. As a fast-growing, shade-tolerant pioneer shrub, *C. hirta* often occupies primary forests and steep slopes. As an invasive species, it flourishes in disturbed areas and forest gaps, forming dense thickets that outcompete native vegetation due to its large leaves (Le et al. 2018). Furthermore, *C. hirta* is a highly dangerous invasive alien species because it produces allelopathic compounds such as flavonoids, triterpenoids, and quinones, which inhibit the growth of native species. Research has demonstrated that extracts from its leaves and roots suppress the growth of *Ageratum conyzoides* L., *E. crus-galli*, and *Cyperus iria* L. (Fendiyanto et al. 2024).

5.3 Weed Management in Oil Palm

Weed management is a crucial agronomic practice in oil palm cultivation due to the crop's perennial nature and its lifespan of 25 to 30 years (Kuan et al. 1991), during which it encounters weed pressure at various growth stages. Effective weed control necessitates an integrated approach that incorporates chemical, cultural, physical, mechanical, biological, and preventive methods, spanning from the nursery phase to replanting and throughout immature (0 to 2.5 years post-transplanting in the field) to mature stages (>2.5 years post-transplanting in the field) (Fig. 5.4). Integrated weed management in oil palm reduces crop losses, hence improving the economic efficiency of the oil palm plantation (Ismail and Chan 2000). The integrated approach involves both preventive strategies and active control measures. Preventive strategies aim to reduce the introduction, colonization, spread, and succession of weeds. If preventive weed control is unsuccessful, being insufficient, or impractical, active control measures become necessary. These measures encompass manual weeding, ground cover, mechanical weeding, herbicide application, and organized livestock grazing.

In a centralized permanent nursery designated for preparing seedlings for field planting or commercial sale, land preparation is done following the dispatch of the previous batch of seedlings. In nurseries affected by herbicide-resistant weeds, such as *E. indica*, the potential for weed growth from the seedbank is substantial. The weed seedbank is typically found within the top 0 to 2 cm of soil (Chuah et al. 2004). To prevent the deeper burial of seeds of the herbicide-resistant weeds,

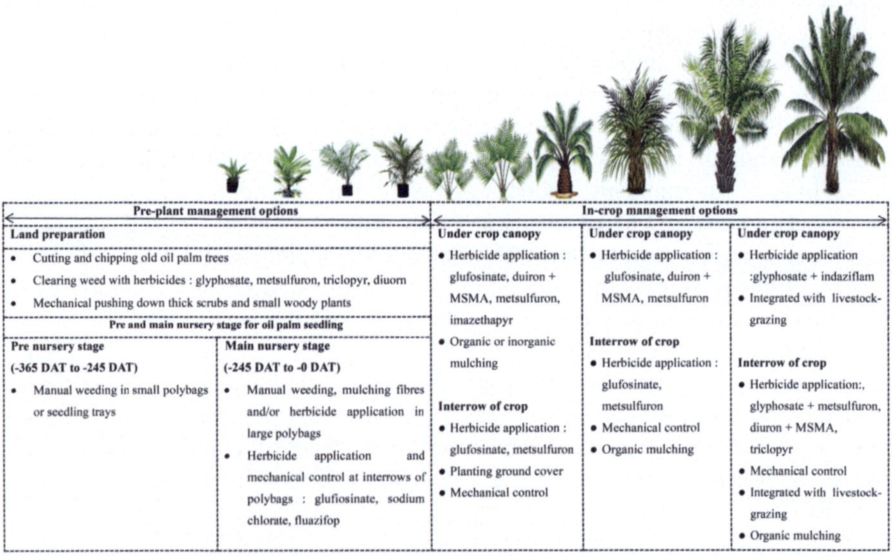

Fig. 5.4 Integrated weed management in oil palm. *DAT—day after transplanting to field, YAT—year after transplanting to field

plowing should be avoided, as this could lead to their later emergence and subsequent management challenges. It may be advantageous to consider scraping away the top layer of soil in the nursery to eliminate the weed seedbank. This operation not only aids in leveling the nursery ground but also facilitates the effective implementation of non-chemical measures, such as mulching and the placement of empty fertilizer bags. Equally important is the necessity for the nursery to commence and maintain clean, weed-free areas, necessitating the removal of weeds through all available means (Chung 2013).

Producing high-quality oil palm planting material relies heavily on nursery management, which involves strict adherence to established standards and procedures. Weeding is a vital maintenance task during both the pre-nursery and main nursery stages. Weeds, which compete for nutrients, water, and sunlight, must be kept out of polybags as they can also harbor pests and serve as hosts for diseases. Although minimal weeding is needed during the pre-nursery phase, when oil palm seedlings (OPS) are kept for three months in small polybags or seedling trays, hand weeding should be performed monthly if necessary. In addition, mulching can be done on all the exposed areas with a layer of broken shells or empty fertilizer bags. Herbicide application is not recommended at this stage, as any error could damage the OPS (Tan and Heriansyah 2011). During the primary nursery stage, OPS are transplanted and kept in large polybags for a period of nine months. Any emerging weeds can easily be removed by hand (Tan and Heriansyah 2011). Alongside manual weeding, mulching with fibers around the OPS is employed to inhibit weed growth in the polybags. Careful spraying of herbicides such as glufosinate and sodium chlorate is practiced to manage weeds in the polybags (Yahya et al. 2024). Weeds in the inter-row spaces are removed using a hand-operated hoe. Grass cutting using a mechanized rotary slasher is carried out. Ideally, this should be done before the weeds have flowered and produced seeds. The soil surface could be mulched with broken shells or cut weed residues to prevent soil compaction and suppress weed growth. Additionally, herbicides like glyphosate, fluazifop, glufosinate, metsulfuron, sodium chlorate, and sodium chlorate plus metsulfuron are spot-applied for weed control using a spray shield (Chung 2013). Heavy infestation of *E. indica* could occur due to its light-loving characteristics (Ismail et al. 2003). The weed could be controlled by sodium chlorate if it has evolved resistance to glyphosate, glufosinate, and fluazifop.

Prior to the planting stage, the complete removal of weeds is necessary for the planting of oil palm seedlings and sowing of leguminous cover crops (LCC) seeds. The combination of land preparation and spraying of herbicides is an effective means for the complete removal of weeds. During land preparation, an excavator is used for plowing or rotovating to clear existing vegetation, while old oil palm trees are cut, chipped, and left in situ. In some cases, involving rehabilitation of abandoned plantations, mechanical clearing through pushing down dense shrubs and small woody vegetation provides rapid land access, facilitating subsequent agricultural operations, including follow-up weed control measures (Chung 2013). Blanket weeding and spot weeding using pre-emergence herbicides such as diuron and indaziflam, as well as post-emergent herbicides like glyphosate plus metsulfuron, sodium chlorate plus metsulfuron, glyphosate, glufosinate, metsulfuron, and triclopyr, are also practiced (Chung 2013).

When the OPS reaches 10 months of age in polybags, these are transplanted into the field. During the initial planting phase, establishing ground cover, such as LCC, is a highly effective strategy for suppressing weed growth. In most cases, pure legume covers or at least more than 80% coverage are preferred. LCC have many agronomic advantages, including reduced soil erosion, addition of organic matter, improving soil aeration and structure, improving soil nutrient status, conserving moisture, and fixing atmospheric nitrogen (Sharma et al. 2018). Complete weed removal is essential to give the LCC seedlings a strong start and ensure they cover the interrow spaces within 6–9 months. LCC seeds are drill-sown at precise spacing to enable targeted herbicide application against weeds emerging between seedlings. Selective spot spraying of a contact herbicide is required each month to suppress isolated colonies of weeds within the LCC drills. Such herbicide application helps to ensure optimum growth of LCC and to achieve complete coverage of the interrow areas within nine months after planting (Chung 2013). Herbicides for selective spot spraying after planting of LCC include fluazifop, glufosinate, sodium chlorate, and glyphosate. Traditionally, a mixture of two or three species of creeping legumes is used, including *Neustanthus phaseoloides* (Roxb.) Benth., *Centrosema pubescens* Benth., *Calopogonium mucunoides* Desv., *Calopogonium caeruleum* (Benth.) C. Wright. The most common combination is *N. phaseoloides*, *C. mucunoides*, and *C. caeruleum* in a 3:3:1 ratio of the seed of each species, with 7 kg of seeds per ha. Many plantations now prefer the more vigorous and shade-tolerant *Mucuna bracteata* DC. Ex Kurz. *Mucuna pruriens* var. *utilis* (Wall. ex Wight) Baker ex is also planted for rapid coverage during the first ten months (Chung 2013).

In the absence of LCC, herbicides are frequently sprayed to control weeds during circle weeding under the oil palm canopy. These herbicides include metsulfuron, fluazifop, glufosinate, sodium chlorate with metsulfuron, and diuron mixed with monosodium methyl arsenate (MSMA). For example, a previous study demonstrated that diuron in combination with MSMA provided excellent control of *E. indica* that had evolved resistance against fluazifop, glufosinate, and glyphosate (Sim et al. 2018). Severe weed encroachment can sometimes lead to the scorching of lower fronds or the breaking of the spear and new fronds. Herbicides like imazethapyr, metsulfuron, and glufosinate are effective in controlling LCC. However, establishing LCC and using metsulfuron to control legume encroachment reduce the risk of crop injury compared to spraying other conventional herbicides. In addition to herbicide use, clean-weeded circles under oil palm can be mulched with plastic sheets, fiber mats, empty fruit bunches (EFB), or biodegradable mulch to suppress weeds. This also creates a protective barrier that minimizes herbicide scorching. Doubling the EFB mulching area around seedlings in vacant spots is a good practice. It prevents weeds from smothering young palms and significantly reduces the risk of herbicide drift during routine weeding. Manual "decreeping" is used to remove creeping weeds that encroach on immature plantings. Mechanical slashing, using tractor-mounted rotary slashers, helps to quickly remove thick weed growth in interrows, facilitating herbicide spraying while avoiding high costs and the risk of scorching young palms (Chung 2013).

Plants commonly accepted as natural ground covers include *Axonopus compressus* (Sw.) P. Beauv., *Cyrtococcum oxyphyllum* (Hochst. ex Steud.) Stapf, *Ottochloa nodosa* (Kunth) Dandy, *Paspalum conjugatum* P.J. Bergius, and *Nephrolepis biserrata* (Sw.) Schott. In some cases, erect broadleaf species like *Asystasia gangetica* L., *A. conyzoides*, and *Spermacoce latifolia* Aubl. are also used as natural cover crops. These natural covers help reduce soil erosion by dissipating the impact of raindrops and acting as barriers to surface runoff (Chung 2013). In areas where establishing leguminous cover crops is challenging due to flooding or other factors, conserving and maintaining natural ground covers is essential to prevent bare soil. During the immature phase, *P conjugatum* is favored, as it allows for circle weeding (weed control around the canopy of an oil palm plant) using fluazifop-butyl with minimal risk of crop injury. Generally, grasses serve as an alternate to legumes in the early years, with a lower impact on oil palm yield. *Nephrolepis biserrata* only thrives as a cover once significant shade has developed under the oil palm (Chung 2013).

As oil palm nears maturity (2.5 years post-transplanting) and its canopy develops and shading increases, *E. indica* and most legume species decrease in number, with the exceptions of *C. caeruleum* and *M. bracteata*. Consequently, natural ground cover should include selected indigenous plant species or a combination of *C. caeruleum* with indigenous flora. In mature oil palm plantations, natural ground cover typically consists of one or two dominant species, often grasses or ferns, distributed across the field or forming extensive patches of mixed soft grasses and broadleaf plants in the interrow areas. For young mature palms (2.5 to 5 years post-transplanting), herbicide application must be conducted with caution to avoid spray drift that could damage lower fronds. As palms reach full maturity (6 years and older post-transplanting), they exhibit greater tolerance to herbicides, enabling a less restrictive approach to chemical weed management.

Sodium chlorate is commonly employed to suppress natural ground covers. Circle and strip weeding are carried out using herbicides such as sodium chlorate in combination with metsulfuron, glyphosate with metsulfuron, glufosinate, and diuron mixed with MSMA. Glyphosate combined with indaziflam is also utilized in circle spraying to prolong the duration of weed control. Additionally, herbicides such as triclopyr, glyphosate, fluroxypyr, and 2,4-D, along with combinations of sodium chlorate and 2,4-D, and various glyphosate and metsulfuron mixtures, are frequently applied for selective spot spraying. Triclopyr, particularly when mixed with diesel, and glyphosate in combination with metsulfuron are especially effective against epiphytes and broadleaf woody weeds such as *C. hirta* (Chung 2013). For the metsulfuron-resistant biotype of *C. hirta*, it has been demonstrated that triclopyr was effective in controlling the weed (Ramadzan et al. 2012). *Clidemia hirta* should be controlled at the immature stage since its control becomes more difficult during later growth stages when the *C. hirta* stems become woody. Interrow natural ground covers (soft grasses and soft ferns) on flat terrains are usually kept in check by mechanical slashing using a tractor-mounted rotary slasher. This is useful during prolonged wet weather with more rainy days. Interrow areas are mulched with empty fruit bunches (EFB), which is a beneficial practice for nutrient recycling and weed suppression. Additionally, careful spreading of cut fronds during harvesting helps reduce weed growth in interrow spaces, preventing the formation of thick weed mats while also recycling nutrients (Berger 2003). In addition, it has been documented that crude extracts and powders of oil palm fronds have the potential to be an alternative to commercial herbicides for the inhibition of *E. indica* (Chuah and Lim 2015).

Another approach to weed management in mature oil palm is organized livestock grazing. Livestock such as sheep, goats, and cattle are allowed to graze in mature plantations, consuming palatable weeds such as *A. gangetica* and *C. hirta* (Nobilly et al. 2022). Grazing is typically only practiced in mature plantings; it is unsuitable for immature or young mature palms (less than five years old) as the livestock can severely damage the lower fronds (Tohiran et al. 2017). In addition, targeted grazing would reduce the use of herbicides from 75 sprayings in the typical plantation cycle down to 15 sprayings (Tohiran et al. 2017).

5.4 Impacts of Weeds on Rice

Weed surveys have been performed in different rice-growing areas throughout the country (Azmi and Mashhor 1995), and it was estimated that rice yield losses caused by weeds reach 5–85% in Malaysia (Dilipkumar et al. 2012). Weed infestation in rice depends on the planting method, season, location, predominant weed flora, weed density, management practices, and infestation duration. Malaysia has moderately warm to high temperatures and high humidity, which are conducive to year-round luxuriant weed growth, such as *E. crus-galli* complex, *Echinochloa colona* (L.) Link, and *Echinochloa oryzicola* (Vasinger) Vasinger (Itoh 1991). It was estimated that the rice yield loss caused *by E. crus-galli* was 41% in Malaysia (Azmi 1994). Besides, *Oryza* spp., a type of weed that resembles cultivated rice, has been rated as one of the four most noxious weeds infesting rice fields worldwide (Delouche et al. 2007). In Malaysia, *Oryza* spp. is locally known as *padi angin* and well known as red rice in many temperate countries due to the red pigmentation of the pericarp (Sudianto et al. 2016; Mardiana et al. 2019). In 1988, *Oryza* spp. was first found at Tanjung Karang, Selangor, and then spread to the northern part of the Muda area, Kedah, Malaysia, in 1990 (Abdullah et al. 1994). By the year 2000, *Oryza* spp. had spread to almost all of Peninsular Malaysia, with at least a 10% infestation in every rice field (Mispan et al. 2019). Moody (1994) found that the direct seeding method (DSR) was reported to be the main cause of the emergence of *Oryza* spp. in the majority of rice agroecosystems in Southeast Asia rice. Azmi (1994) reported that Malaysian rice production has decreased by 54% due to weed problems in DSR. Furthermore, Azmi and Karim (2008) reported the rice yield losses of 12–15% in the paddy area with moderate *Oryza* spp. infestations (15–20 panicles/m^2), 15–22% yield losses under high infestations (21–30 panicles/m^2), and

Oryza spp. might result in complete yield loss under a heavily infested area (>50 panicles/m^2). To date, *Oryza* spp. has become a major problem wherever direct seeding is practiced across rice fields in Malaysia.

5.4.1 Biology of Echinochloa crus-galli

Echinochloa crus-galli was reported in Malaysia as early as 1925 (NWGIAS 2014), and it was introduced unintentionally through contaminated seeds (Moody 1989). Department of Agriculture, Malaysia, has listed *E. crus-galli* as one of the invasive alien species (IAS), which seriously disturbed the rice fields in Peninsular Malaysia, and its domination can threaten all sectors such as agriculture, forestry, fishery, marine, and animals (Hakim et al. 2013). *Echinochloa crus-galli* is a warm-season, annual grass species that is characterized by wide and more upright green leaves than most weedy annual grasses (Fig. 5.5). The leaf evolved to a purple-red-brown near the base and a pronounced white midrib. For the young plant, *E. crus-galli* may have a more prostrate appearance and become more upright toward maturity. Mature plants grow up to 5 feet (1.5 m) tall, have many tillers, and grow in clumps. *E. crus-galli* has a fibrous root system and can grow new roots where tillers touch the ground.

The propagation of this weed is achieved through its seeds, which are compressed dorsoventrally. When growing under suitable environmental conditions (i.e., moist soils and optimal soil temperatures of 20–30 °C for germination), it

Fig. 5.5 *Echinochloa crus-galli* infestation in rice fields

spreads rapidly and produces large amounts of seeds. A single plant of *E. crus-galli* may produce as many as 40,000 seeds (FAO 2024). The seeds are noticeably dormant for 3–4 months and do not germinate in water deeper >12 cm (CABI 2024). Several subspecies have been described within *E. crus-galli*. Although the morphological features of these varieties are similar, they may exhibit differences in response to management practices. The main differences between the two major *E. crus-galli* types found in Malaysia are: *E. crus-galli* var. *formosensis* possesses short awns, open panicles, shiny spikelets, while *E. crus-galli* var. *crus-galli* has a long awn (sometimes the awn is absent) and closed panicles (Itoh et al. 1992).

5.4.2 Biology of Oryza *spp. L. (Weedy Rice)*

Oryza spp. possesses a number of common features with other weeds and distinctive characteristics such as phylogenetic and morphological resemblance to cultivated rice (*Oryza sativa* L.) (Fig. 5.6). This situation occurs because of the existence of wild-to-wild gene flow in the evolution of rice seeds through natural hybridization between weedy rice and cultivated rice (Song et al. 2003). Dilipkumar et al. (2018) stated that the genotypic and morphological variations and plasticity of adaptation allow the weed to successfully reproduce and persist in rice field ecosystems in Malaysia. Chauhan (2013) reported that the *Oryza* spp. has weedy characteristics, such as a short life span, taller plant, weak culms, small seeds, easy shattering, and a red pericarp. In another study by Delouche et al. (2007), it was documented that *Oryza* spp. generally grows higher (90–150 cm) than cultivated rice (60–70 cm) with better use of the available nitrogen (N). Compared to the cultivated rice, *Oryza* spp. produces more tillers, panicles, and biomass, shatters early, has better resistance to unfavorable dry conditions, and possesses longer and varied seed dormancy in the soil. Furthermore, weedy rice is more susceptible to lodging

Fig. 5.6 *Oryza* spp. infestation in Pasir Mas rice area in Malaysia's East Coast Region

and exhibits more rapid seedling growth when compared to the cultivated rice to which it infests. In Malaysia, different *Oryza* spp. biotypes were reported with their distinctive characteristics such as (1) taller, shorter, and similar height as the cultivated rice; (2) erect, horizontal, and descending flag leaves; (3) open and closed panicles; (4) easy shattering and non-shattering seeds; and (5) awned and awnless seeds (Dilipkumar et al. 2020; Sudianto et al. 2016).

Ferrero (2003) stated that *Oryza* spp. requires a minimum temperature of 10 °C which is similar to cultivated rice, and can spread rapidly and produce large amounts of seeds. Abd Hamid et al. (2007) found that *Oryza* spp. produces abundant seeds, i.e., as high as 200–220 spikelets per panicle for awn type, 190–210 spikelets for open panicle type, and 210–230 spikelets for compact panicle type. All these variants complete their life cycles within 85–95 days after emergence. In addition, Qistina et al. (2022) reported that seeds of *Oryza* spp. in Malaysia displayed a high degree of seed dormancy and can retain their viability >200 days once imbibed, thus enhancing the survivability of *Oryza* spp. in the tropical rice agroecosystems. Moreover, Ferrero (2003) documented that *Oryza* spp. emergence was between 60% and 80% when submerged below 2–3 cm of water.

5.5 Weed Management in Rice

Malaysia adopted various integrated weed management strategies, mainly about land preparation and pre-harvest period controls to manage *E. crus-galli* and *Oryza* spp. This includes weed control by cultural, mechanical, and chemical methods (Fig. 5.7). One week after harvest, the process of decomposition and returning nutrients to the soil can be done by spreading chopped straw evenly across the field and to suppress or prevent the pests that may act as a host for diseases (Azmi and Karim 2008). The rice residues (e.g., stubble and straw) were also burnt to destroy weedy rice seed and to promote new emergence from the soil surface or seedbank (Mispan et al. 2019). Since both cultivated and *Oryza* spp. belong to the same species and have similar anatomical and physiological traits, and hence, the control efforts must be done before rice planting. The stale seedbed technique is one of the cultural methods to control *Oryza* spp. before rice sowing (Chauhan 2013). In the stale seedbed technique, weeds are forced to germinate and then killed by nonselective herbicides. During this period, glufosinate or glyphosate were applied pre-plant at 7 to 14 days before tillage (Dilipkumar et al. 2022). Seedbed preparation consisted of deep tillage under dry soil conditions, followed by shallow tillage under flooded soil conditions to destroy weedy rice seedlings. Chauhan (2013) reported that the tilling operations could bury the weed seeds below the depth of their emergence, and it was recommended that subsequent tilling of land should be shallow in the next few seasons to avoid bringing back weedy rice seeds to the surface of the soil. Additionally, pre-emergence herbicide pretilachlor is applied immediately after wet tillage. In this situation, the water depth is maintained at 5.0 to 7.5 cm for 7 days before draining. Alternatively, pretilachlor can be applied at 0 to 3 days after sowing (DAS)

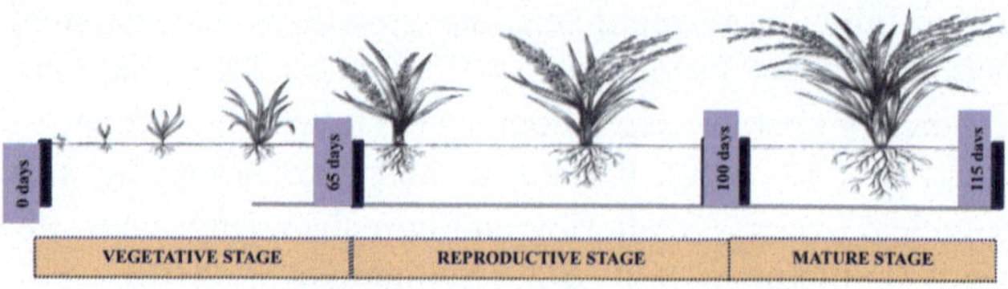

Pre-plant management options		In-crop management options		
After harvesting previous crop ***Cultural control***		**Sowing day**		**Harvest**
1-3 d after harvest • Cut stubble; the straw and stubble is spread evenly.	**3-7 d after harvest** • Burn the straw to destroy weedy rice seed and to promote new emergence from the seedbank.	***Cultural control*** **0 DAS** • Use good quality certified seed, competitive rice cultivars, and broadcast the pre-germinated seeds with higher seed rate.	***Chemical control*** **10-14 DAS** • Apply post-emergence herbicide (eg: bentazone/ propanil/ cyhalofop penoxsulam).	***Harvest*** **100-115 DAS** • Panicle roguing Slashing the weedy rice panicles before harvest
Pre-planting		***Chemical control***	**21 – 30 DAS**	**• Harvesting**
Chemical control **7-14 d before tillage** • Apply pre-planting herbicides: glufosinate ammonium/ glyphosate. ***Mechanical control*** **14 d before sowing** • Dry tillage (deep) **7 d before sowing** • Wet tillage (shallow)	***Chemical control*** **7 d before sowing** • Apply pre-emergence herbicide, pretilachlor (*Cultural control:* the water depth was maintained at 5 to 7.5cm for 7d) ***Mechanical control*** **1-2d before sowing** • Land levelling: - removed emerged weedy rice. - *cultural control*: water level at 3cm distribution is used as a standard reference for land levelling.	**0-3 DAS** • Apply pre-emergence herbicide (eg: pretilachlor/oxadiazon) **0-5 DAS** • Clearfield® Paddy Production System - apply OnDuty (imazapic+imazapyr) (imidazolinone) based on the planting manual of Clearfield® rice technology (CRT) to control weedy rice. **5-7 DAS** • Apply early post-emergence herbicide (eg: quinclorac/ bispyribac/ thiobencarb) to control weeds. ***Cultural control*** **7-14 DAS** • Flooding	• Apply late post-emergence herbicide (eg: MCPA + bentazone/ MCPA/2,4-D/ fenoxaprop). ***Mechanical control (weedy rice)*** **>20 DAS** - Early manual or mechanical weeding to avoid competition. - Prevent weedy rice seed production by roughing and chopping weedy rice panicles before seed setting.	- Clean harvester before use to avoid the spread of weed seeds.

Fig. 5.7 Integrated weed management in rice. *DAS: day after sowing

under saturated conditions. Application of pretilachlor could deplete the soil seedbank of *Oryza* spp. Previous studies have confirmed the success of pre-emergence herbicides, such as pretilachlor (Olajumoke et al. 2016) and oxadiazon (Eleftherohorinos and Dhima 2002), in suppressing *Oryza* spp. before its emergence. Besides, land leveling needs to be carried out prior to the sowing process (1–2 days before sowing) to ensure the uniform distribution of irrigation water in the field, which helps in managing weeds and destroying the emerged *Oryza* spp. plants and encourage the germination of weed seeds.

During the sowing period (0 DAS), certified rice seeds must be used to prevent the introduction and spread of *Oryza* spp. Rao et al. (2007) stated that the use of clean and certified seeds could avoid continuous infestation of *Oryza* spp., whereas agricultural equipment or tools also cause the spread of weedy rice within and between agricultural fields (Saha et al. 2014). Cleaning equipment while moving to the next field for a tillage operation can significantly reduce the movement of weed seed and other material from one field to another. Vaughan et al. (1995) have reported that combine-harvesting is considered a major cause for the spread of *Oryza* spp. in several Southeast Asian countries.

Chemical methods during the sowing period effectively control the majority of weed species in the early crop establishment stage, except for *Oryza* spp. Weed control starts as early as 0–3 DAS for pre-emergence and 5–7 DAS for early post-emergence application of herbicides. Instead of pretilachlor, pendimethalin is another type of pre-emergence herbicide to control *E. crus-galli* in rice fields. Meanwhile, herbicides such as quinclorac, bispyribac-sodium, and thiobencarb are among the early post-emergence herbicides to control *E. crus-galli.*

Chemical control of *Oryza* spp. can only be performed through Clearfield® Rice Technology (CRT). This rice production system has been developed through a collaboration between the Malaysian Agricultural Research and Development Institute (MARDI) and BASF Malaysia Sdn. Bhd. It consists of imidazolinone (IMI)-tolerant rice varieties (MR 220CL1 and MR 220CL2) and a premixed IMI herbicide (imazapic plus imazapyr, OnDuty™) (Dilipkumar et al. 2021). Imazapic + imazapyr are usually applied at 0–5 DAS for weedy rice control. Recently, this system has become unsustainable in Malaysia, where farmers ignored crop rotation guidelines and planted Clearfield® rice continuously for over nine seasons (Dilipkumar et al. 2021). As a result, many *Oryza* spp. populations have now survived IMI herbicide applications, with some confirmed to be resistant to imazapic and imazapyr (Dilipkumar et al. 2018). The herbicide-resistant *Oryza* spp. issue could be tackled successfully by integrating the stale seedbed technique with pre-emergence herbicides prior to rice sowing to eliminate the emergence of weedy rice seedlings. This integrated weed management approach could help to maintain the rice weed populations below the economic threshold level (Dilipkumar et al. 2022).

After the application of pre-emergence and early post-emergence herbicides at the vegetative growth stage ranging from 0 to 7 DAS, water management in a rice field is also crucial in order to prevent weed infestation at 7 to 14 DAS. According to Chauhan (2013), the timing, duration, and depth of floods are crucial for cultural weed management in rice, as they affect the growth of weeds and *Oryza* spp. It was also reported that weed infestation of rice fields could be reduced by flooding at the early stages (Chauhan and Johnson 2011). Azmi and Karim (2008) suggested a flooding depth of 5 to 10 cm to reduce the emergence of *Oryza* spp. in Malaysian rice fields. Proper water management was found to help in *Oryza* spp. management since flooding suppresses seedling emergence and biomass of weedy rice (Olajumoke et al. 2016). During the flooding period, the weed management practices in rice should be continued with post-emergence herbicides at 10 to 14 DAS. Herbicides such as bentazone, propanil, cyhalofop, and penoxsulam are categorized as post-emergence herbicides to control *E. crus-galli* in rice fields. For late post-emergence (21 to 30 DAS), several herbicides are usually employed including MCPA + bentazone, MCPA, 2,4-D, and fenoxaprop. Generally, the effectiveness of herbicides is low on weeds if applied at their mature stage. The herbicide degradation rate or metabolism could be faster in mature plants; thus, herbicide rates may be increased to achieve the same level of weed control (Singh and Singh 2004). Furthermore, during the rainy season, farmers may not be able to apply post-emergent herbicides in a timely manner. These problems could be solved through the application of late post-emergent herbicides.

Conversely, during the application of post- and late post-emergence herbicides for control of *E. crus-galli* and other weeds, there was no selective herbicide available to control weedy rice during the post-emergence stages. Due to this issue, methods for reducing weedy rice infestations have been concentrated on roughing and chopping the panicles before seed setting at >20 DAS. Baki and Shakirin (2010) found that most farmers in Malaysia used manual weeding in their direct-seeded fields between the early 1990s and the middle of the 2000s, by roguing the *Oryza* spp. Farmers may easily identify *Oryza* spp. by slashing the plants before harvesting, as they are typically taller than cultivated rice types. However, this practice no longer seems to be effective because of high labor costs (Mispan et al. 2015) and the emergence of new *Oryza* spp. biotypes, which are morphologically similar to commercial cultivated varieties (e.g., MR220 and MR219), thus making *Oryza* spp. almost unrecognizable (Mispan et al. 2015).

5.6 Impacts of Weeds on Pineapple

Weed infestation poses a significant threat to pineapple farming in Malaysia, causing a significant decline in yield, which is 25 to 35%. Weeds compete with pineapple plants for essential resources such as nutrients, light (canopy), water, and carbon dioxide (Fakayode et al. 2012). Moreover, Abouziena et al. (2016) highlighted that weed infestation can lead to damage and loss of up to 80% of the pineapple crop. Pineapple plants, characterized by their slow growth and shallow root systems, are particularly vulnerable to competition from weeds. If left uncontrolled, weeds significantly disrupt the growth of the pineapple plants, ultimately affecting both the quality and quantity of the fruit produced (Zubir and Ghani 2023). The challenge of managing weeds is exacerbated in scenarios of high planting density, inadequate mulch coverage, inefficient irrigation practices, excessive rainfall, and the presence of weed seeds in the field and surrounding areas (Noorsuliyana et al. 2023). The absence of effective weed control has been shown to significantly impact the size of pineapples. Eshetu et al. (2007) demonstrated that areas that were not weeded exhibited markedly smaller plant sizes in comparison to those subjected to more frequent weeding. Additionally, Pegoraro et al. (2014) noted that the slow growth of both the fruit and root systems in pineapples is largely attributable to intense competition with weeds, leading to delays in fruit development and a consequent reduction in yield and quality.

Ageratum conyzoides and *Digitaria* spp. are prevalent weed species identified in pineapple cultivation systems, not only in Malaysia but also in various regions worldwide (Py 1959; Pandey and Johnson 2006; Maia et al. 2018; Girija and Menon 2019; Jabit et al. 2021; Zubir and Ghani 2023). Effective integrated management of these weeds through physical control (manual and mechanical control), cultural, biological, and chemical control is crucial for improving yields and ensuring the sustainability and profitability of pineapple farming in Malaysia. *Ageratum conyzoides*, commonly known as the billy goat weed, is a ubiquitous and aggressive weed species prevalent in tropical and subtropical regions (Yadav et al. 2019). Its ability to thrive in diverse environments makes it a significant agricultural pest, notably in pineapple farms in Malaysia. Meanwhile, *Digitaria ciliaris* is a very common, yet problematic weed found in tropical, subtropical, and temperate regions worldwide, rapidly colonizing disturbed areas such as agricultural fields, roadsides, and wastelands due to its prolific seed production and vigorous growth, and it can cause significant yield losses in crops even at low densities (Chauhan and Johnson 2008a, b).

5.6.1 *Biology of* Ageratum conyzoides

Ageratum conyzoides is native to tropical America, with its distribution ranging from Mexico through the Caribbean and into South America (Ming 1999; Kohli et al. 2006). It has since spread globally, largely due to its adaptability and the movement of agricultural goods. Its introduction to regions like Southeast Asia is a consequence of both its deliberate and inadvertent dispersal through trade and agricultural practices (Kaur et al. 2023). *Ageratum conyzoides* is a fast-growing, herbaceous plant that can reach a height of up to 1 meter (Okunade 2002; as cited by Wang et al. 2023). It has a sprawling growth habit, characterized by its broad, ovate leaves and small, densely clustered flowers (Thorat et al. 2018). The plant demonstrates remarkable adaptability to various climates, soil textures, moisture conditions, and altitudes (Kaur et al. 2023). Its optimal growth occurs at temperatures between 20 and 25 °C, although it can also thrive in conditions ranging from 15 to 30 °C (Arora 1999). This adaptability accounts for its widespread presence in both higher altitudes (temperate climates) and plain areas (tropical climates) (Kosaka et al. 2010). Additionally, its extensive root system aids in nutrient and water absorption, which supports its vigorous growth and competitiveness (Thorat et al. 2018; Batish et al. 2017). The competitive ability of *A. conyzoides* is marked by its aggressive colonization and rapid growth (Kohli et al. 2006; Kaur et al. 2012). It often outcompetes other plant species for light, water, and nutrients, especially in disturbed environments like agricultural fields (Banyikwa and Rulangaranga 1985). In pineapple farms of Malaysia, this weed can form a dense canopy that shades the crop and hinders its growth. Its competitive nature is further enhanced by its ability to rapidly re-establish following disturbance and grow in a variety of soil conditions (Kaur et al. 2023).

The fruits of *A. conyzoides* are achene and disperse easily, while the seeds require light to germinate (photoblastic) and often disappear within 12 months (Thorat et al. 2018). The seeds exhibit a relatively short period of dormancy, often germinating within a few days to weeks under favorable conditions (Kaur et al. 2023). *Ageratum conyzoides* germinated across a broad temperature range, achieving its highest germination rate (over 90%) at 30/20 °C (Desai et al. 2024). However, as salinity, osmotic potential, and burial depth increased, germination and emergence of *A. conyzoides* decreased, with complete inhibition occurring at 300 mM salinity, −0.8 MPa osmotic potential, and a burial depth of 1 cm (Desai et al. 2024). The high

Fig. 5.8 *Ageratum conyzoides* infestation on pineapple farms

germination rate possessed by the seeds contributes to their rapid spread and establishment in new areas (Ekeleme et al. 2005; Kaur et al. 2023). *Ageratum conyzoides* possesses several reproductive strategies to ensure its persistence and spread. It prolifically reproduces primarily through seeds (Kaur et al. 2012). The plant's flowers are small and inconspicuous, but they are highly effective at producing copious amounts of seeds (Wang et al. 2016). Each plant can produce thousands of seeds (averaging 94,772 seeds per plant), with a seed-shedding period of 5 to 8 months and up to 50% of the seeds germinating (Holm et al. 1977), leading to the abundant emergence of plants (approximately 1000 plants per m^2) (Ekeleme et al. 2005). This abundance of seed production, coupled with the seeds' ability to remain viable in the soil for extended periods, ensures that *A. conyzoides* can quickly recolonize and dominate new areas. Additionally, the plant's ability to propagate vegetatively, although less common, also contributes to its persistence (Kohli et al. 2006). In summary, *A. conyzoides* is a formidable weed due to its rapid growth, competitive ability, and prolific reproductive strategies. Its impact on pineapple farms in Malaysia underscores the challenges of managing such invasive species, highlighting the need for effective control measures to protect agricultural productivity (Fig. 5.8).

5.6.2 Biology of Digitaria ciliaris

Another important weed species that causes considerable challenges to pineapple farms in Malaysia is *D. ciliaris*, commonly known as the southern crabgrass; this invasive species is a major concern due to its aggressive growth and competitive nature (Souza et al. 2012). *Digitaria ciliaris* is a fast-growing, perennial grass that thrives in tropical and subtropical climates (Souza et al. 2012). Its adaptability to various environmental conditions and rapid growth make it a troublesome weed in agricultural settings, including pineapple plantations in Malaysia (Souza et al. 2012; Jabit et al. 2021). *Digitaria ciliaris* is native to tropical and subtropical regions of the Old World, including parts of Africa and Asia. Its introduction to Southeast Asia is a result of both intentional and unintentional human movements, often as a contaminant in crop seeds or soil (Chauhan and Johnson 2008a, b).

The seeds of *D. ciliaris* exhibit minimal dormancy, allowing them to readily germinate when conditions are suitable (McIvor and Howden 2000). Germination typically occurs rapidly within a few days to a week after seed shattering (Oreja et al. 2020). Optimal conditions for germination include various alternating temperatures (25/15, 30/20, and 35/25 °C day/night), with germination rates decreasing at the lowest temperature (25/15 °C) (Chauhan and Johnson 2008a, b). While light naturally enhanced germination, a small percentage of seeds still germinated in darkness (Chauhan and Johnson 2008a, b). The seeds are capable of germinating across a wide pH range from 5 to 10 (Chauhan and Johnson 2008a, b). Seedling emergence can reach up to 98%, with the highest rates observed for seeds placed on the soil surface. In contrast, burying seeds at a depth of 8 cm completely prevented the emergence of *D. ciliaris* (Chauhan and Johnson 2008a, b).

Digitaria ciliaris is characterized by its vigorous growth and extensive vegetative spread (Chauhan and Johnson 2008a, b; Jones et al. 2021). The plant forms a dense mat of grass with sprawling stems and broad, strap-like leaves. It can reach heights of 30 to 60 cm, and its growth is optimized in warm, sunny conditions with ample water (Jones et al. 2021). The plant's root system is extensive and fibrous, allowing it to effectively absorb nutrients and moisture from the soil. This physiological adaptation enables *D. ciliaris* to compete effectively with other plants, including crops (Souza et al. 2012). In pineapple farms, this grass can quickly overtake large areas, shading out crops and reducing their access to sunlight. Its ability to form a thick, dense cover also helps it outcompete other plant species for resources such as water and nutrients. The competitive nature of *D. ciliaris* makes it particularly problematic in agricultural fields, where it can significantly impact crop yields and overall farm productivity (Itoh and Froud-Williams 2014).

The reproductive strategies of *D. ciliaris* are key to its success as a weed. It reproduces primarily through seeds. The lightweight seeds are easily dispersed by wind, water, and human activities, contributing to their wide distribution. Additionally, *D. ciliaris* can propagate vegetatively through its extensive network of stolons and rhizomes, which can produce new plants and establish colonies even when the above-ground portions are removed or damaged. This ability to reproduce both sexually and asexually enhances its resilience and persistence in disturbed environments (Chauhan and Johnson 2008a, b). In summary, *D. ciliaris* represents a significant challenge in pineapple cultivation due to its aggressive growth, competitive abilities, and effective reproductive strategies. Managing this weed requires targeted control measures to mitigate its impact on agricultural productivity and ensure the successful cultivation of pineapple crops (Fig. 5.9).

Fig. 5.9 *Digitaria ciliaris* infestation on pineapple farms

5.7 Weed Management in Pineapple

Weed management in pineapple farms is critical for maintaining optimal crop growth and yield, as weeds vigorously compete with pineapple plants for nutrients, water, and light. In Malaysia, weed management in pineapple becomes challenging because of high planting density, diverse soil types in each cultivation area, inadequate mulch coverage, poor irrigation practices, excessive rainfall, and a substantial weed seedbank in pineapple fields and their vicinity (Noorsuliyana et al. 2023). Thus, effective control of weeds requires a comprehensive strategy that integrates various methods, including chemical, cultural, physical, biological, and prevention approaches.

Pineapple cultivation in Malaysia is typically done on three types of soil: peat soil (58%), mineral soil (40%), and bris soil (2%) (MPIB 2023). Each soil type exerts a distinct influence on weed composition, establishment, and thus management, particularly in the two predominant soil types, namely, peat and mineral soil. Peat soil, characterized by its high organic matter content exceeding 65% and excellent moisture retention, often supports a diverse array of weed species that thrive in wet conditions; consequently, weed invasion is typically denser in peat soil compared to mineral soil, which has an organic matter content of nearly 35% and lower water-holding capacity (Borowy et al. 2018). In contrast, mineral soil, which is comprised of inorganic materials and tends to have lower organic matter, presents a

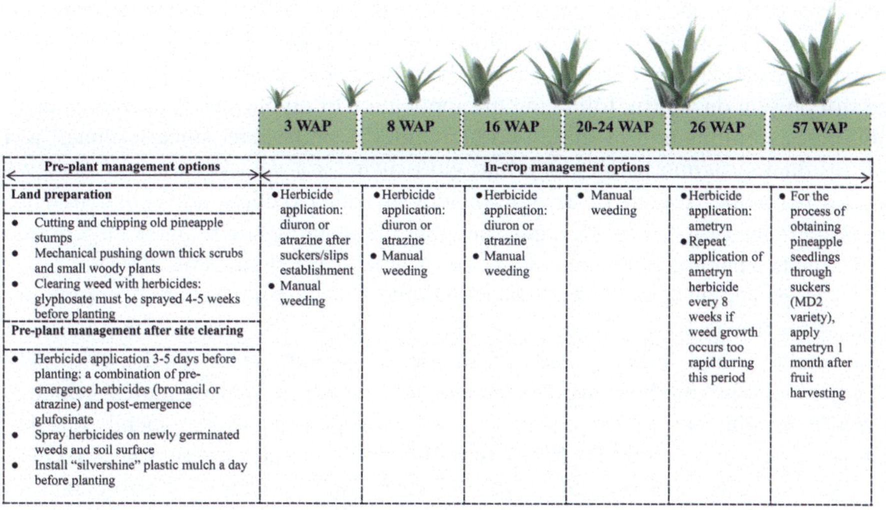

Fig. 5.10 Integrated weed management in pineapple farm * WAP—week after planting to field

different challenge. The lower water retention capacity of mineral soil may limit weed growth, but certain species (i.e., invasive plant species) adaptable to drier conditions can still flourish (Zimdahl 2018). The influence of soil type on weed infestation in pineapple farms is significant, making it essential to understand these differences for the development of sustainable weed management practices tailored to each specific soil condition.

Various herbicides are routinely employed as essential elements of pineapple weed management programs, as weeds serve as hosts for significant populations of mealybugs, ants, and aphids, which are known vectors for the pineapple wilt virus (Botella and Fairbairn 2005). Weed management in the pineapple farm begins as early as the land preparation phase (Fig. 5.10). To control the emerged weeds and ensure a weed-free planting site, glyphosate is applied as pre-plant during the land preparation (4–5 weeks before planting) (MPIB 2017; Kogan and Alister 2010). Glyphosate is effective in controlling a broad spectrum of weed species, including the perennial *Imperata cylindrica* Raeusch., which is common in pineapple cultivation areas (Prusty et al. 1992; Huang et al. 2012). The entire planting site must be sprayed thoroughly to eliminate existing weeds and old pineapple stumps (for areas that have been planted with pineapple) and prevent their competition with the pineapple seedlings. In addition to chemical control, manual weeding methods can be employed to manage weeds effectively during land preparation. This includes slashing, physical/hand removal, tillage, and mulching to suppress weed growth (MPIB 2013), although these approaches do not completely eliminate weeds and can be labor-intensive and less cost-effective.

Repeated herbicide applications must be conducted prior to planting to ensure that the planting area is thoroughly free from weeds. Three to five days before planting, application of a pre-emergence herbicide, such as PSII-inhibiting bromacil or atrazine, and/or post-emergence glutamine synthetase-inhibiting glufosinate, is essential (MPIB 2017). These herbicides burn down the remaining weed plants that are still present on a farm, following the application of glyphosate (Suwanarak et al. 1998). The combination of these two herbicides is preferable, since bromacil is a soil residual systemic herbicide, while glufosinate is a well-known non-selective contact herbicide. Together, these herbicides provide excellent and robust defense against a range of weed species, ensuring that the planting site remains largely free from weeds when the pineapple plants are established (Eshetu et al. 2007).

At the pineapple farms in Malaysia, a double-row planting system is commonly used to maximize land utilization, providing shade for the plants to protect them from sunburn, and reduce weed infestations (Laishram et al. 2012). A widely adopted mulching method involves the use of “silvershine” plastic mulch, which is carefully installed after fertilizers are applied to the planting beds, typically one day before planting begins (MPIB 2013). Initially, plastic mulch was employed to cover the planting beds, as it not only helped conserve soil moisture but also enhanced root development by directing moisture toward the root zone and controlling weed growth (Bunroj et al. 2011). To accommodate the use of plastic mulching for weed suppression, fertilization in large-scale pineapple farms in Malaysia is often carried out through foliar feeding (Zubir et al. 2023).

Weed control continues at the post-planting stage to manage weeds that might escape the first two herbicide applications or newly emerge from the seedbank. At 3 weeks succeeding pineapple suckers/slips establishment, it is highly recommended to again apply the PSII-inhibiting herbicide diuron or atrazine (MPIB 2013). These herbicides are effective in controlling a variety of broadleaf weeds and grasses and, when in contact with the soil, will act as pre-emergence herbicides as well (Shimabukuro and Swanson 1969; Jones et al. 2003). Therefore, it is recommended to spray them on weeds with 4 to 6 newly developed leaves and on the soil surface before the weeds flower and produce seeds (Malaysia Pineapple Industry Board 2013). Preventing flower development is essential to reduce the weed seedbank, for achieving long-term weed control and minimizing associated costs. It is also important to ensure that the soil is moist during application to enhance herbicide absorption and to avoid applying herbicides during rainy days to prevent herbicide runoff. The herbicide application should then be repeated at 8- and 16-weeks after planting (WAP) to maintain effective weed control throughout the growth period (Malaysia Pineapple Industry Board 2013; Murray and Hoffman 1998). To further ensure minimal weed disturbance prior to the application of flowering-induction hormones, applying ametryn herbicide around 26–35 WAP as per the recommended label rates is crucial (Malaysia Pineapple Industry Board 2013). Ametryn is effective against many weed species and helps control late-emerging weeds (Maia et al. 2018). For the process of obtaining pineapple seedlings through suckers (MD2 variety), the application of ametryn can be performed once again following the fruit harvesting (around 57 WAP/1 month after harvest) to ensure the healthy growth of the mother plant and to maintain the field free from any weeds (Malaysia Pineapple Industry Board 2013).

Effective weed management in peat soil typically requires more frequent monitoring and the application of targeted herbicides, as the moisture levels can facilitate rapid weed proliferation. A study from Zubir and Ghani (2023) also demonstrated that applications of post-planting selective herbicides (bromacil, ametryn, diuron, and atrazine) with a suitable frequency of application (8 or 10-week intervals starting from 12 WAP) provided better weed control, especially on peat soil of Malaysia's pineapple plantation, without affecting pineapple plant growth. Additionally, the weed control efficacy of the mesotrione and atrazine (e.g., Calaris®25SC) premix has also been evaluated, although further investigation on the crop phytotoxicity and appropriate spray windows is still needed (Ahmad-Hamdani, pers. comm.). It is a selective herbicide that targets specific weed species while minimizing damage to pineapple plants. The application should follow the manufacturer's recommendations regarding dosage and timing to ensure its effectiveness. Incorporating Calaris®25SC into the weed management strategy can enhance overall control and help in maintaining a healthy pineapple crop. Manual weeding must also be conducted periodically and consistently. However, the implementation of physical weed removal is limited to the early stage of pineapple growth (6 months or less) and is only applicable to small, low-density pineapple farms (MPIB 2017). Manual weeding is typically performed at 8, 16, 20, 25, and 36 WAP. If weed growth occurs too rapidly during this period, implementation of herbicide spray is necessary (MPIB 2017).

When addressing specific weed species, both *A. conyzoides* and *D. ciliaris* require tailored approaches for effective management. *Ageratum conyzoides* is a broadleaf weed that can spread rapidly and outcompete pineapple plants for resources. To manage this species, both pre-emergence and post-emergence herbicides are effective (Kaur et al. 2023). Herbicides such as diuron and atrazine can control *A. conyzoides* when applied according to the recommended schedule (Kaur et al. 2023). Additionally, manual weeding and physical removal of this weed can help reduce its seed bank and prevent future infestations. Mulching can also be beneficial in suppressing the germination and growth of *A. conyzoides*. Meanwhile, management of *D. ciliaris* can be more challenging due to its rapid growth and ability to spread through stolons and seeds. Glyphosate, applied during land preparation, is effective against this species. However, post-emergence herbicides such as ametryn that target grasses are crucial for managing established populations post-planting. Regular monitoring and manual removal of any escaped *D. ciliaris* plants can prevent them from becoming a persistent issue.

Integrated Weed Management (IWM) combines various weed control strategies to provide a holistic approach to weed management. In pineapple farming, IWM involves integrating the use of herbicides, cultural practices (like mulching and crop rotation), and physical methods (manual weeding and tilling). By employing a combination of these methods, pineapple farmers can achieve more effective and sustainable weed management. In conclusion, effective weed management in pineapple farms requires a multi-faceted approach that begins with thorough land preparation and extends through various stages of crop growth. The use of pre-planting and pre-emergence herbicides, along with timely post-emergence applications, is crucial.

Specific strategies for managing *A. conyzoides* and *D. ciliaris* must be exercised to address their unique growth characteristics. Integrating chemical, cultural, and physical methods within an IWM framework provides a comprehensive solution to managing weeds in pineapple cultivation.

5.8 Conclusions

Malaysia's agricultural sector faces significant challenges from weed infestations across its three major crops—oil palm, rice, and pineapple—each requiring tailored management approaches to ensure productivity and sustainability. As the country's leading agricultural export, oil palm plantations are particularly vulnerable to herbicide-resistant weeds like *E. indica* and *C. hirta*, necessitating integrated solutions including legume cover crops, strategic mulching, and precision herbicide applications. The rice sector, crucial for national food security, contends with substantial yield losses (5–85%) from invasive *Oryza* spp. and *E. crus-galli*, prompting farmers to adopt stale seedbed techniques integrated with pre-emergence herbicides, improved water management practices, and the Clearfield® Rice Technology system, though emerging herbicide resistance demands continuous innovation in control methods. Pineapple cultivation, especially of the high-value MD2 variety, suffers 25–35% yield reductions due to aggressive weeds like *A. conyzoides* and *D. ciliaris*, with effective management requiring soil-specific approaches, including plastic mulching in peat soils and targeted herbicide regimes in mineral soils.

Across all crops, the growing problem of herbicide resistance, coupled with economic pressures from rising labor costs and dependence on chemical controls, underscores the urgent need for integrated weed management strategies. Looking ahead, Malaysia's agricultural sustainability will depend on adopting advanced technologies like drone-assisted spraying and artificial intelligence-based weed detection while reducing reliance on synthetic herbicides through agroecological practices. By integrating these innovative approaches with traditional knowledge and ecological stewardship, Malaysia can address its weed management challenges while maintaining the productivity and profitability of its key crops in an environmentally sustainable manner, ensuring long-term food security and economic stability for its agricultural sector.

References

Abd Hamid ZA, Mashhor M, Azmi M (2007) Life cycle and morphological characteristics of weedy rice (*Oryza sativa* complex, locally called Padi Angin), a noxious weed of rice fields in Malaysia. Jurnal Biosains 18(1):55–79

Abdullah MZ et al (1994) Padi Angin: its probable origin and diversity revealed through DNA polymorphism. 1st National Congress on Genetics, November 1994, Kuala Lumpur, Malaysia, pp 164–167

Abouziena HF, Haggag WM, El-saeid HM et al (2016) Safe methods for weed control in fruit crops: challenges, and opportunities. Der Pharm Lett 8(5):325–339

Ali NBM et al (2021) Weeds diversity in oil palm plantation at Segamat, Johor. In: IOP Conference Series: 3rd Asia Pacific Regional Conference on Food Security Kelantan, Malaysia, March 2021. vol 756, IOP Publishing, London

Amalia Qistina Z, Nur Ardiyana R, Mispan MS (2022) Seed dormancy status of tropical weedy rice population in Malaysia. Sains Malays 51(6):1697–1705

Amoroso G, Fini A, Piatti R et al (2009) Mulching as an alternative to chemical weed control in nursery containerized crops. Adv Hort Sci 23(4):276–281

Arora V (1999) Comparative assessment of eco-physiological functions between *Ageratum conyzoides* L. and *Parthenium hysterophorus* L. Ph.D. Thesis. Panjab University

Azahari M, Samingin I, Seman IA (2004) Weed management. In: Ghani EA, Zakaria ZZ, Wahid MB (eds) Oil palm cultivation in Malaysia. MPOB, Selangor, pp 123–145

Azmi M (1994) Biology and control of *Echinochloa crus-galli* (L.) Beauv. in direct seeded rice. Dissertation. University Sains Malaysia

Azmi M, Karim SMR (2008) Weedy rice—biology, ecology and management. Malaysian Agricultural Research and Development Institute, Kuala Lumpur, p 56

Azmi M, Mashhor M (1995) Weed succession from transplanting to direct seeding method in Kemubu rice area. Malaysia J Biosci 6(2):143–153

Baki BB, Shakirin MM (2010) Spatio-temporal distribution pattern of new biotypes of weedy rice (*Oryza sativa* L.) in Selangor North-West Project, Malaysia. Korean J Weed Sci 30:68–83

Banyikwa FF, Rulangaranga ZK (1985) Growth analysis of groundnuts (*Arachis hypogea*) in competition with *Ageratum conyzoides*. Turrialba 35:3

Batish DR, Shalinder Kaur SK, Singh HP et al (2017) Effect of soil texture on phytotoxicity of an invasive weed, *Ageratum conyzoides* L. Int J of Trop Agr 35(1):141–146

Berger K (2003) Palm kernel oil. In: Wiley encyclopedia of food sciences and nutrition, 2nd edn. Wiley, New York, pp 4325–4331. https://doi.org/10.1016/B0-12-227055-X/01379-1

Borowy A, Kaplan M, Krawiec M (2018) Impact of effective microorganisms on weed infestation and yield of peppermint cultivated on muck-peat soil. Acta Agrobot 71(4)

Botella JR, Fairbairn DJ (2005) Present and future potential of pineapple biotechnology. In: Martinez AR (ed) Proceedings of the IVth International Pineapple Symposium, Brisbane, January 2005. International Society for Horticultural Science, vol 666. Acta Horticulturae, Belgium, pp 23–28

Breeden GK, Brosnan JT (2009) Goosegrass (*Eleusine indica*). Pest Manag Sci 65(10):1084–1088

Bunroj A, Saridnirun P, Shinawong S (2011) Changes in soil microorganisms, flowering, and yield of 'Smooth Cayenne' pineapple as affected by black plastic mulch in organic production. Thai J Agr Sci 44(4):219–224

CABI (2024) *Echinochloa crus-galli* (barnyard grass). https://www.cabidigitallibrary.org/doi/10.1079/cabicompendium.20367. Accessed 27 Sept 2024

Chandima BR, Samarasinghe P, Jayasuriya KMGG et al (2022) Seed biology and early seedling developmental traits of *Clidemia hirta,* an invasive species of Sri Lankan rainforests compared to two native species sharing the same habitat. Taiwania 67(1):73–82. https://doi.org/10.6165/tai.2022.67.73

Chauhan BS (2013) Strategies to manage weedy rice in Asia. Crop Prot 48:51–56

Chauhan BS, Johnson DE (2008a) Germination ecology of goosegrass (*Eleusine indica*): an important grass weed of rainfed rice. Weed Sci 56(5):699–706. https://doi.org/10.1614/WS-08-048.1

Chauhan BS, Johnson DE (2008b) Germination ecology of southern crabgrass (*Digitaria ciliaris*) and India crabgrass (*Digitaria longiflora*): two important weeds of rice in tropics. Weed Sci 56(5):722–728

Chauhan BS, Johnson DE (2011) Row spacing and weed control timing affect yield of aerobic rice. Field Crop Res 121:226–231

Chuah TS, Lim WK (2015) Assessment of phytotoxic potential of oil palm leaflet, rachis and frond extracts and powders on goosegrass (*Eleusine indica* (L.) Gaertn.) germination, emergence, and seedling growth. Malays Appl Biol 40:75–84

Chuah TS, Salmijah S, Ismail BS (2004) Changes in seed bank size and dormancy characteristics of the glyphosate-resistant biotype of goosegrass (*Eleusine indica* (L.) Gaertn.). Weed Biol Manag 4(2):114–121. https://doi.org/10.1111/j.1445-6664.2004.00127.x

Chuah TS, Hiew CN, Nursyafiqah IN et al (2023) Seed germination characteristics as affected by interaction of moisture stress and temperature in sethoxydim-resistant biotype of goosegrass (*Eleusine indica* (L.) Gaertn.) from Malaysia. Sains Malays 52(7):1985–1997

Chuah TS, Yusop N, Pauzi SA et al (2024) Survey of herbicide-resistant weed management in oil palm estates from peninsular Malaysia and Indonesia. Adv Weed Sci 42:e020240030

Chung GF (2013) Integrated management of weeds in oil palm plantations. In: Proceedings of 5th MPOB-IOPRI international seminar, November 2013. MPOB, Selangor

Clergeau P, Mandon-Dalger I (2001) Fast colonization of an introduced bird: the case of Pycnonotus jocosus on the Mascarene Islands. Biotropica 33:542–546

Delouche JC, Burgos NR, Gealy DR et al (2007) Weedy Rice: origin, biology, ecology and control. Food and Agriculture Organization of the United Nations, Rome

Desai DH, Desai HS, Chauhan BS (2024) Germination attributes of metsulfuron-resistant and metsulfuron-susceptible tropical ageratum (*Ageratum conyzoides*) populations under various environmental conditions. Weed Sci 72:1–8

Dilipkumar M, Adzemi MA, Chuah TS (2012) Effects of soil types on phytotoxic activity of pretilachlor in combination with sunflower leaf extracts on barnyardgrass (*Echinochloa crus-galli*). Weed Sci 60:126–132

Dilipkumar M, Burgos NR, Chuah TS et al (2018) Cross-resistance to imazapic and imazapyr in a weedy rice (*Oryza sativa*) biotype found in Malaysia. Planta Daninha 36

Dilipkumar M, Chuah TS, Goh SS et al (2020) Weed management issues, challenges and opportunities in Malaysia. Crop Prot 134:10434

Dilipkumar M, Ahmad-Hamdani MS, Rahim H et al (2021) Survey on weedy rice (*Oryza* spp.) management practice and adoption of Clearfield® rice technology in Peninsular Malaysia. Weed Sci 69(5):558–564

Dilipkumar M, Mohamad-Ghazali M, Shari E et al (2022) Integrated use of the stale seedbed technique with preemergence herbicides to control weedy rice in wet seeded rice. Weed Technol 36(3):373–378

DOA (Department of Agriculture Malaysia) (2022a) Rice Check: Padi. https://www.doa.gov.my/doa/resources/aktiviti_sumber/sumber_awam/penerbitan/pakej_teknologi/padi/rice_check_padi_2022.pdf. Accessed 25 Sept 2024

DOA (Department of Agriculture Malaysia) (2022b) Statistik Tanaman Buah-buahan 2022 (statistics of fruit crops 2022). Putrajaya, Malaysia

DOA (Department of Agriculture Malaysia) (2023) Booklet Statistik Tanaman (sub-sektor Tanaman Makanan) 2023 (crop statistics booklet (food crop sub-sector) 2023). Putrajaya, Malaysia

Ekeleme F, Forcella F, Archer DW et al (2005) Seedling emergence model for tropic ageratum (*Ageratum conyzoides*). Weed Sci 53(1):55–61

Eleftherohorinos IG, Dhima KV (2002) Red rice (*Oryza sativa*) control in rice (*O. sativa*) with preemergence and postemergence herbicides. Weed Technol 16:537–540

Eshetu T, Tefera W, Kebede T (2007) Effect of weed management on pineapple growth and yield. Ethiop J Weed Manag 1(1):29–40

Fakayode SB, Rahji MAY, Adeniyi ST (2012) Economic analysis of risks in fruit and vegetable farming in Osun state, Nigeria. Bangladesh J Agric Res 37(3):473–491

FAO (2024) *Echinochloa crus-galli* (L.) Beauv. https://www.fao.org/agriculture/crops/thematic-sitemap/theme/biodiversity/weeds/listweeds/ech-cru/en/. Accessed 27 Sept 2024

FAOSTAT (2024b) Pesticide use. https://www.fao.org/faostat/en/#data/RP. Accessed 5 Oct 2024

FAOSTAT (2024a) Food and agricultural commodities production by country. http://faostat3.fao.org/browse/rankings/commodities_by_country/. Accessed 29 Sept 2024

Fendiyanto MH, Kurniyanto IR, Pratami MP et al (2024) Allelopathic activity of leaf and root extracts of *Clidemia hirta* on morpho-physiological characters of rice (*Oryza sativa*) and some weeds in rice fields. Int J Agric Biol 32:87–93

Ferrero A (2003) Weedy rice, biological features and control. In: Labrada R (ed) Weed Management for developing countries, FAO plant production and protection paper 120 add. 1. UN-FAO, Rome

Girija T, Menon M (2019) Diversity of weed flora in pineapple plantations of Kerala. J Crop Weed 15(1):218–221

Gobi K, Ahmad-Hamdani MS, Zakaria N et al (2025) Tank mixture of propanil and quinclorac for inhibiting a bispyribac-sodium-resistant barnyardgrass (*Echinochloa crus-galli*) biotype in Malaysia. Adv Weed Sci 43:e020250107

Hakim MA, Juraimi AS, Ismail MR et al (2013) A survey on weed diversity in coastal rice fields of Sebarang Perak in Peninsular Malaysia. J Anim Plant Sci 23(2):534–542

Heap I (2024) The international survey of herbicide resistant weeds. http://www.weedscience.com/ Accessed: 29 Sept 2024

Holm LG, Plucknett DL, Pancho JV et al (1977) The world's worst weeds; distribution and biology. Honolulu, Hawaii

Holm LG, Pancho JV, Herberger JP et al (1979) A geographical atlas of world weeds. Wiley, New York

Huang J, Silva EN, Shen Z et al (2012) Effects of glyphosate on photosynthesis, chlorophyll fluorescence and physicochemical properties of cogongrass (*Imperata cylindrical* L.). Plant Omics 5(2):177–183

Idris A, Linatoc AC, Mohd Fadzellyab AB (2019) Effect of light intensity on the photosynthesis and stomatal density of selected plant species of Gunung Ledang, Johor. Malays Appl Biol 48(3):133–140

Ismail BS, Chan KW (2000) Integrated ground cover management in plantations. In: Yusuf B, Jalaini BS, Chan KW (eds) Advances in oil palm research. Chapter 16. MPOB, Kuala Lumpur, pp 632–652

Ismail BS, Chuah TS, Salmijah S et al (2002) Germination and seedling emergence of the glyphosate-resistant and -susceptible biotype of goosegrass (*Eleusine indica* (L.) Gaertn.). Weed Biol Manage 2:177–185

Ismail BS, Chuah TS, Salmijah S et al (2003) Effects of shade and watering frequency on growth and development of glyphosate-resistant and susceptible biotypes of goosegrass (*Eleusine indica* (L.) Gaertn.). Plant Prot Q 18(1):130–134

Ismaini L (2015) Allelopathic effect of invasive plant (*Clidemia hirta*) on seed germination of native plant (*Impatiens platypetala*). In: Prosiding Seminar Nasional Masyarakat Biodiversitas Indonesia, July 2015, vol 1. Smujo International, Surakarta, pp 832–837

Itoh K (1991) Integrated weed management of direct-seeded wet rice fields in Southeast Asia and Pacific regions with special reference to Malaysia, October 1991. 13th Asian-Pacific Weed Science Society Conference, Jakarta, pp 77–94

Itoh M, Froud-Williams RJ (2014) Roles of emergence time and interspecific competition on the dominance and coexistence of *Setaria faberi* and *Digitaria ciliaris* in an orchard weed community in Japan. Weed Biol Manag 14(1):31–42

Itoh K, Ho N, Man A (1992) *Echinochloa* weeds in peninsular Malaysia. Res Highl Trop Agric Res:20–22

Jabit NN, Aani SNA, Noor NM (2021) Survey of weed floral composition under pineapple plantation on peat and mineral soil. Int J Agr Forestry Plantation 11

Jalaludin A, Yu Q, Powles SB (2014) Multiple resistance across glufosinate, glyphosate, paraquat, and ACCase-inhibiting herbicides in *Eleusine indica* population. Weed Res 55(1):82–89. https://doi.org/10.1111/wre.12118

Jones RJ, Muller J, Haynes D, Schreiber U (2003) Effects of herbicides diuron and atrazine on corals of the great barrier reef, Australia. Marine Ecol Prog Ser 251:153–167

Jones EA, Contreras DJ, Everman WJ (2021) *Digitaria ciliaris, Digitaria ischaemum*, and *Digitaria sanguinalis*. In: Chauhan BS (ed) biology and management of problematic crop weed species. Elsevier Inc, pp 173–195. https://doi.org/10.1016/B978-0-12-822917-0.00014-8

Kaur S, Batish DR, Kohli RK, Singh HP (2012) *Ageratum conyzoides*: an alien invasive weed in India. In Bhatt JR, Singh JS, Singh SP, Tripathi RS, Kohli RK (eds) Invasive alien plants: an ecological appraisal for the Indian subcontinent, Wallingford, UK, p 57–76

Kaur A, Kaur S, Singh HP et al (2023) Ecology, biology, environmental impacts, and management of an agro-environmental weed *Ageratum conyzoides*. Plants 12(12):2329

Kerr RA, Zhebentyayeva T, Saski C et al (2019) Comprehensive phenotypic characterization and genetic distinction of distinct goosegrass (*Eleusine indica* L. Gaertn) ecotypes. J Plant Sci Phytopathol 3(2):95–100. https://doi.org/10.29328/journal.jpsp.1001038

Kogan M, Alister C (2010) Glyphosate use in forest plantations. Chilean J Agr Res 70(4):652–666

Kohli RK, Batish DR, Singh HP et al (2006) Status, invasiveness and environmental threats of three tropical American invasive weeds (*Parthenium hysterophorus* L., *Ageratum conyzoides* L., *Lantana camara* L.) in India. Biol Invasions 8:1501–1510

Kosaka Y, Saikia B, Mingki T et al (2010) Roadside distribution patterns of invasive alien plants along an altitudinal gradient in Arunachal Himalaya, India. Mount Res Dev 30:252–258

Kuan CY et al (1991) Crop loss by weeds in Malaysia. In: Lee SA, Kon KF (eds) Proceedings of the third tropical weed science conference, Kuala Lumpur, December 1990. Malaysian Plant Protection Society, Malaysia

Laishram M, Meitei WI, Singh NG (2012) Effect of double and single row system of planting on growth and yield of pineapple [*Ananas comosus* (L.) Merr] cv. Kew. Asian J Hortic 7(2):259–262

Le C, Fukumori K, Hosaka T et al (2018) The distribution of an invasive species, *Clidemia hirta* along roads and trails in Endau Rompin National Park, Malaysia. Trop Conserv Sci 11:1–9. https://doi.org/10.1177/1940082917752818

Maia VM, Aspiazú I, Pegoraro RF (2018) Sustainable weed control in pineapple. In: Weed control. CRC Press, pp 470–484

Maizatul SM, Idris AS (2012) Occurrence of common weeds in mature plantings of oil palm plantation in Malaysia. Planter 88(1037):537–547. https://doi.org/10.56333/tp.2012.006

Mardiana JK, Bajrai FSM, Muhamad-Safwan I et al (2019) Effect of Onduty herbicide on weed populations and rice yields in selected rice fields, Melaka, Malaysia. Sains Malays 48(11):2355–2365

McIvor JG, Howden SM (2000) Dormancy and germination characteristics of herbaceous species in the seasonally dry tropics of northern Australia. Aust Ecol 25(3):213–222

Ming LC (1999) *Ageratum conyzoides*: a tropical source of medicinal and agricultural products. In: Janick J (ed) Perspectives on new crops and new uses. ASHS Press, Alexandria, pp 469–473

Mispan MS, Jalaluddin A, Majrashi AA et al (2015) Weed science in Malaysia: an analysis. In: Rao VS et al (eds) Weed science in the Asian Pacific Region. Asian-Pacific Weed Science Society

Mispan MS, Bzoor M, Mahmod I et al (2019) Managing weedy rice (*Oryza sativa* L.) in Malaysia: challenges and ways forward. J Res Weed Sci 2(2):149–167

Moody K (1989) Weeds reported in rice in South and Southeast Asia. International Rice Research Institute, Manila, Philippines

Moody K (1994) Weed management in rice. In: Labrada R, Caseley JC, Parker C (eds) Weed management for developing countries, FAO plant production and protection paper 120. UN-FAO, Rome, pp 249–263

Moravcová L, Pyšek P, Jarošík V et al (2015) Getting the right traits: reproductive and dispersal characteristics predict the invasiveness of herbaceous plant species. PLoS One 10(4). https://doi.org/10.1371/journal.pone.0123634

MPIB (Malaysia Pineapple Industry Board) (2013) Malaysia pineapple industry board. Manual Tanaman Nanas Tanah Mineral (Mineral Soil Pineapple Plant Manual)

MPIB (Malaysia Pineapple Industry Board) (2017) Manual Pengurusan Penanaman nanas di Tanah Gambut (pineapple cultivation management manual on peat land). Official Portal of

Malaysia Pineapple Industry Board https://www.mpib.gov.my/pengurusan-tanaman-nanas/. Accessed 5 Oct 2024

MPIB (Malaysia Pineapple Industry Board) (2019) Perangkaan Industri Nanas Malaysia 2020 (Malaysia Pineapple Industry Statistics 2020). Johor Bahru, Malaysia

MPIB (Malaysia Pineapple Industry Board) (2023) Data Nanas 2022 (Pineapple Data 2022). https://www.mpib.gov.my/wp-content/uploads/2024/01/Data-Nanas-2022.pdf. Accessed 5 Oct 2024

MPOB (Malaysia Palm Oil Board) (2022) Overview of the Malaysian oil palm industry. https://bepi.mpob.gov.my/images/overview/Overview2022. Accessed 10 Dec 2023

Murphy DJ, Goggin K, Paterson RRM (2021) Oil palm in the 2020s and beyond: challenges and solutions. CABI Agric Biosci 2(39):1–22. https://doi.org/10.1186/s43170-021-00058-3

Murray DA, Hoffman JE (1998) To ascertain the ideal combinations of atrazine and/or diuron with thiazopyr (visor) in order that acceptable broad spectrum pre-emergent weed control in pineapples is achieved. In: III International Pineapple Symposium 529, pp 289–292

Ngah N, Omar D, Juraimi AB et al (2011) Leaf surface characteristics of selected Malaysian weed species of oil palm. J Agrobiotechnol 2(1):53–65

Nobilly F, Zulkifli R, Maxwell et al (2022) Application of targeted goat grazing in oil palm plantations: Assessment of weed preference, spatial use of grazing area and live weight change. J Oil Palm Res, 34(2):289–299. https://doi.org/10.21894/jopr.2021.0047Noorsuliyana

Noorsuliyana MN, Noor NM, Rahman SMA, Aani SNA (2023) Weed species composition of different peat areas under pineapple plantation in Muar, Johor. In: IOP Conference Series: Earth and Environmental Science, June 2023, vol 1182. IOP Publishing, p 012077

NWGIAS (National Working Group on Invasive Alien Species) (2014) National action plan for prevention, eradication, contaminant and control of invasive alien species (IAS). Crop Biosecurity Division, Department of Agriculture, Malaysia

Okunade AL (2002) *Ageratum conyzoides* L. (asteraceae). Fitoterapia 73(1):1–16. https://doi.org/10.1016/S0367-326X(01)00364-1

Olajumoke B, Juraimi AS, Uddin MK et al (2016) Competitive ability of cultivated rice against weedy rice biotypes: a review. Chil J Agric Res 76:242–251

Oreja FH, Batlla D, de la Fuente EB (2020) *Digitaria sanguinalis* seed dormancy release and seedling emergence are affected by crop canopy and stubble. Weed Res 60(2):111–120

Pandey RR, Johnson MW (2006) Weeds adjacent to Hawaiian pineapple plantings harboring pink pineapple mealybugs. Environ Entomol 35(1):68–74

Parveez GKA, Kamil NN, Zawawi NZ et al (2022) Oil palm economic performance in Malaysia and R&D progress in 2021. J Oil Palm Res 34:185–218. https://doi.org/10.21894/jopr.2023.0028

Pegoraro RF, Souza BAMD, Maia VM, Amaral UD, Pereira MCT (2014) Growth and production of irrigated Vitória pineapple grown in semi-arid conditions. Rev Bras Frutic 36:693–703

Peters HA (2001) *Clidemia hirta* invasion at the Pasoh Forest reserve: an unexpected plant invasion in an undisturbed tropical forest. Biotropica 33:60–68. https://doi.org/10.1111/j.1744-7429.2001.tb00157.x

Prusty JC, Behera B, Lenka PC (1992) Glyphosate – a new herbicide for pineapple orchard. Agr Sci Dig (Karnal) 12(2):113–116

Py C (1959) Weed control in pineapple plantations. Results of trials in Guinea. Fruits 14(8):329–340

Rahman MM, Sahid IB, Juraimi AS (2010) Study on resistant biotypes of *Echinochloa crus-galli* in Malaysia. Aust J Crop Sci 4(2):107–115

Ramadzan AMN, Ismail B, Chuah TS (2012) A preliminary report on the potential resistance of a soapbush (*Clidemia hirta* (L.) D. Don) biotype to metsulfuron-methyl in an oil palm plantation in Jerantut. Malaysia Plant Prot Q 27:64–69

Rao AN, Johnson DE, Sivaprasad B et al (2007) Weed management in direct-seeded rice. Adv Agron 93:153–255

Ritchie H (2021) Palm oil. Our World in Data https://ourworldindataorg/palm-oil Accessed 10 Sept 2023

Saha S, Patra BC, Munda S et al (2014) Weedy rice: problems and its management. Indian J Weed Sci 46(1):1422

Sahid I, Hamzah A, Aria PM (1992) Effect of paraquat and alachlor on soil microorganisms in peat soil. Pertanika 15:121–125

Sharma P, Singh A, Kahlon C et al (2018) The role of cover crops towards sustainable soil health and agriculture- a review paper. Am J Plant Sci 9(10):1935–1951. https://doi.org/10.4236/ajps.2018.99140

Shimabukuro RH, Swanson HR (1969) Atrazine metabolism, selectivity, and mode of action. J Agr Food Chem 17(2):199–205

Siddharta S (2024) Production of rice in Malaysia 2013–2021, Statista. https://www.statista.com/statistics/794700/rice-production-volume-malaysia/. Accessed 27 Sept 2024

Sim KC, Anthony Tan SH, Wong KJ et al (2018) Efficacy of MSMA-based premix herbicides on control of goosegrass (Eleusine indica) that evolved resistance across glyphosate, glufosinate, and fluazifop. Plant Prot Q 33(1):1–10

Singh S, Singh M (2004) Effect of growth stage on trifloxysulfuron and glyphosate efficacy in twelve weed species of citrus groves. Weed Technol 18(4):1031–1036

Song ZP, Xu X, Wang B et al (2003) Genetic diversity in the Northernmost *Oryza rufipogon* populations estimated by SSR markers. Theor Appl Genet 107:1492–1499

Souza RC, Dias AC, Figueiredo MRA (2012) Growth of the crabgrass species *Digitaria ciliaris* and *Digitaria nuda*. Planta Daninha 30:317–325

Sudianto E, Neik TX, Tam SM et al (2016) Morphology of Malaysian weedy rice (*Oryza sativa*): diversity, origin and implications for weed management. Weed Sci 64:501–512

Suwanarak K, Kongsaengdao S, Vasunun S (1998) Efficiency of pre-planting herbicides on weed control and growth of no tillage pineapple (*Ananas comosus* L.). III International Pineapple Symposium 529:293–302

Tan CC, Heriansyah (2011) Nursery practices for production of superior oil palm planting materials. In: Goh KJ, Chiu SB, Paramananthan P (eds), Planter, Kuala Lumpur 81(948): 159–171. https://doi.org/10.56333/tp.2005.003

Thorat VH, Ghorpade SS, Patole T (2018) *Ageratum conyzoides* Linn.: a review. Int J Pharmacog 5:213–221

Tohiran KA, Nobilly F, Zulkifli R et al (2017) Targeted cattle grazing as an alternative to herbicides for controlling weeds in bird-friendly oil palm plantations. Agron Sustain Dev 37:1–11

Vaughan DA, Watanabe H, Abdullah Zain M (1995) Evolution and genetic diversity of Malaysian weedy rice. 15th Asian-Pacific Weed Science Society Conference, July 1995, Tsukuba, Japan, pp 461–466

Wang J, Chen W, Ma R et al (2016) Role of short-term cold stratification on seed dormancy break and germination of alien species in southeastern China. Plant Ecol 217:383–392. https://doi.org/10.1007/s11258-016-0580-2

Wang Z, Wang H, Wang H et al (2023) Dual tolerance of ageratum (*Ageratum conyzoides* L.) to combined pollution of acid and cadmium: field survey and pot experiment. J Environ Manage 326:116677

Waterhouse BM (2002) Rainforest weeds: detecting and managing new incursions. In: Grice AC, Setter MJ (eds) Weeds of rainforests and associated ecosystems. CRC-TREM, Australia, pp 34–38

Yadav N, Ganie SA, Singh B et al (2019) Phytochemical constituents and ethnopharmacological properties of *Ageratum conyzoides* L. Phytother Res 33(9):2163–2178

Yahya A, Hassan H, Adam NM et al (2024) Mechanisation status of oil palm nursery and field cultivation in West Malaysia. J Oil Palm Res, MPOB, Selangor. https://doi.org/10.21894/jopr.2024.0014

Zairy A, Rozhan AD (2023) Sociological issues and challenges of rice production in Malaysia. https://ap.fftc.org.tw/article/3473. Accessed 27 Sept 2024

Zimdahl RL (2018) Fundamentals of weed science, 3rd edn, California, London

Zubir MN, Ghani NSA (2023) Management of weeds on pineapple farm using selective herbicide on peat soil. Int J Agr Env Res 9(6):1021–1029

Zubir MN et al (2023) Pineapple growth performance, yield and fruit quality influenced by foliar fertilization on clay soil. In: Proceedings of 12th Kuala Lumpur international agriculture, Forestry & Plantation Conference 2023 (KLIAFP12)

Chapter 6
Biology and Management of Weeds in Morocco

Asmae Baggar, Badr Hajjaj, and Mina Devkota

Abstract Morocco has a diverse and highly adaptable weed flora that has resulted from a long history of its crop production, diverse agro-ecological zones, and climatic diversity (ranging from Mediterranean conditions in the north to arid and semi-arid regions in the south). Weeds represent a critical challenge in Morocco's crop production systems, particularly in rainfed drylands. They compete for limited water and nutrients, reduce yield, and act as hosts for pests and diseases, causing yield losses of up to 30–60% when left unmanaged. More than 50% of the identified weeds originate from the Mediterranean region. The majority of the weed species occurring in cereal fields are dicots. Herbicides are widely used and are effective for controlling weeds; however, their overuse has resulted in the emergence of herbicide-resistant species like *Lolium rigidum* (annual ryegrass) and *Papaver rhoeas* (common poppy). These resistant species are now affecting large areas, posing a growing threat to crop production. Additionally, the environmental impacts of herbicides and the withdrawal of certain products from the market have further complicated weed management. This chapter has highlighted the importance of understanding weed biology and ecology as foundation for designing effective weed, management strategies; it emphasizes need to analyse weed dynamics under different cropping systems to implement integrated management approaches (such as crop rotation, optimized sowing techniques, and competitive crop varieties—with judicious herbicide application), and examine how those integrated strategies functions under conventional tillage and conservation agriculture (CA) practices. With increasing diverse weeds, including herbicide-resistant, perennial weeds under no-tillage systems, there is a need for special consideration, especially during the initial few years of adoption of CA. The study highlighted the inclusion of forage

A. Baggar · M. Devkota (✉)
International Center for Agricultural Research in the Dry Areas (ICARDA), Rabat, Morocco
e-mail: m.devkota@cgiar.org

B. Hajjaj
Regional Agricultural Research Center of Tangier, National Institute of Agricultural Research (INRA), Tangier, Morocco

B. S. Chauhan, K. Jabran (eds.), *Biology and Management of Weeds in Asia and Africa*, Sustainability Sciences in Asia and Africa,
https://doi.org/10.1007/978-981-95-7314-1_6

mixture in rotation and timely application of herbicides can minimize weed populations in CA-based systems. Further study on deep understanding of weed biology and their adaptive traits is essential for designing sustainable and effective management strategies to ensure the resilience and productivity of Morocco's cereal-based systems.

Keywords Weed biology · Conservation agriculture · Herbicide resistance · Wheat production · Yield losses · Rainfed drylands · Morocco

6.1 Introduction

The cereal sector in Morocco plays a crucial role in the country's agriculture and economy. It contributes more than 20% to the agricultural gross domestic product (GDP) and occupies over 5 million hectares (M ha) (approximately 75% of arable land) and provides livelihood to 40% of the agricultural workforce. Soft wheat is cultivated at an area of around 2 M ha, while durum wheat occupies approximately 0.9 M ha (Ferrahi and Bekkaoui 2020). Despite this large-scale production, Morocco's dependence on cereal imports remains high, exposing the country to food security risks, particularly due to environmental factors. Drought is the major abiotic stress affecting rainfed agriculture in Morocco, leading to reduced yields and agricultural instability. The increasing frequency of dry spells in the Mediterranean climate, along with fluctuations in precipitation and temperature, exacerbates this issue. In the rainfed areas, the average productivity of wheat is about 2.0 t/ha, while in areas with supplemental irrigation, the average yield is about 4 t/ha. However, overall yield potential is limited by irregular rainfall and temperature fluctuations. In addition to abiotic stresses, biotic factors such as insect pests, diseases, and weed infestation also significantly contribute to yield losses each year.

Weeds represent major challenges for agriculture by reducing crop yields, spreading diseases, and competing for resources, making their management a long-standing issue in agricultural practices. In drylands, weeds compete for limited water and nutrients, thereby significantly reducing yields. As in tropical regions, weeds in drylands are underestimated as crop pests (Hajjaj et al. 2018). Studies in Moroccan drylands indicate yield losses of 30–60% in wheat fields infested with unmanaged weeds (Shekhawat et al. 2020; Table 6.1). Table 6.1 presents grain yield losses due to weeds in different regions of Morocco. The results showed that yield losses can be up to 70%. The competition is particularly severe during the critical growth stages of wheat, such as tillering and grain filling. Furthermore, weeds can act as alternate hosts for pests and diseases, and also weeds degrade the quality of harvested seeds and hinder harvesting operations.

In addition to their genetic variability, weeds can easily develop adaptive and survival traits that enable them to respond rapidly to changes in the ecosystems they inhabit and to outcompete neighboring plants. Hence, studies on the biology and ecology of weeds form the foundation of weed science, as they are crucial for designing effective management strategies. Chauhan et al. (2017) emphasized that

Table 6.1 Yield loss due to weed competition on different cereal crops observed in different regions of Morocco.

Region	Crop	Grain yield loss due to weeds (%)
Chaouia	Durum wheat	40
	Bread wheat	13–36
	Barley	28–49
Gharb	Triticale	16
	Durum wheat	10–57
	Durum wheat	27
	Bread wheat	34
Sais	Durum wheat	67
	Bread wheat	41
Tadla	Bread wheat	42
	Bread wheat	31

Source: Adopted from Bouhache (2020)

even in the era of chemical weed control, knowledge of weed biology is indispensable and can serve as the basis of practical weed management. They noted that different approaches linking weed biology studies to practical management strategies are necessary. In general, all management methods aim to modify the balance of plant populations in an ecosystem in a direction that counteracts ecological evolutionary trends (Zimdahl, 2007; Varanasi and Jugulam 2017). For example, understanding the germination and persistence of weed seeds in the soil is essential to grasping the factors governing their dynamics in an ecosystem. Furthermore, identifying the factors influencing reproduction, multiplication, and dispersal of weeds is crucial for developing strategies to reduce or prevent the introduction or spread of weeds, especially invasive plants, to other regions or ecosystems.

6.2 History of Weed Evolution in Morocco

Humanity became aware of weeds since it began cultivating plants around 10,000 B.C. (Hay 1974), and almost simultaneously, these unwanted plants were recognized as a problem. This initiated the conflict between humans and weeds. Morocco has a diverse and highly adaptable weed flora that has resulted from a long history of its crop production (from the Neolithic period to modern practices), diverse agroecological zones, and climatic diversity (ranging from Mediterranean conditions in the north to arid and semi-arid regions in the south). This historical interaction, including the introduction of new crops and farming techniques, has led to the proliferation of persistent and evolving weed species, now posing significant challenges to contemporary weed management strategies in the country.

Pre-agricultural Era Before the introduction of agriculture, natural vegetation dominated the Moroccan landscape. Early weeds likely co-evolved with native plant species in grasslands and forests. Early human settlers may have inadvertently introduced some rudimentary weed species through foraging and pastoral activities.

Domestication and Early Agriculture (Neolithic Period) Agriculture in Morocco dates back to over 5000 years and started with the domestication of crops such as barley, wheat, and legumes. Early farming practices involved seed collection and sowing, inadvertently selecting for fast-growing and highly competitive weed species such as *Avena fatua* L. (wild oat) and *Lolium temulentum* L. (darnel ryegrass).

Influence of Agricultural Expansion (Middle Ages) The Arab expansion and the Islamic Golden Age introduced advanced irrigation systems and crop rotations in Morocco. This agricultural intensification brought with it new crops (e.g., citrus, olives) and associated weeds, some of which became naturalized. Key invasive weeds during this period included *Chenopodium album* L. (fat hen) and *Amaranthus retroflexus* L. (redroot pigweed), which thrived in irrigated systems.

Colonial Period and Globalization of Agriculture During the French Protectorate (1912–1956), modern agricultural practices were introduced, including large-scale monocropping and mechanization. These practices created new ecological niches for weeds, particularly those adapted to disturbed soils and chemical fertilizers. Similarly, trade and colonial agricultural experiments led to the introduction of exotic weed species, such as *Conyza canadensis* (L.) Cronquist (horseweed) and *Ambrosia artemisiifolia* L. (ragweed). Many of these exotic species outcompeted native flora due to their rapid reproduction and adaptability to changing agricultural environments.

Post-independence Era (1956–Present) Expansion of irrigated agriculture, introduction of high-yield crop varieties, and extensive use of herbicides have shaped the weed dynamics in Morocco's post-independence era. Herbicide resistance has emerged as a significant challenge, with species such as *Phalaris minor* Retz. (canary grass) and *Lolium rigidum* Gaud. (rigid ryegrass) evolving resistance to commonly available herbicides. Similarly, rapidly changing climatic conditions such as rising temperatures and erratic rainfall patterns have altered weed distribution and phenology. Highly adaptive species like *Parthenium hysterophorus* L. (parthenium weed) and *Cynodon dactylon* (L.) Pers. (Bermuda grass) are expanding their ranges into new regions.

Current Challenges and Ongoing Evolution Continuous selection pressure from herbicide use has led to the proliferation of super weeds resistant to multiple herbicide modes of action. Examples include *Amaranthus* spp. and *Convolvulus arvensis* (field bindweed). Similarly, the increasing global trade and climate variability have introduced new invasive weeds such as *Euphorbia esula* L. (leafy spurge) and *Prosopis juliflora* (Sw.) DC. (mesquite), which pose a threat to native ecosystems and agriculture. With Morocco's increasing focus on dryland agriculture, weeds like *Salsola kali* L. (Russian thistle) and *Setaria* spp. (foxtail grasses) are becoming dominant in marginal areas.

6.2.1 History of Weed Management in Morocco

The history of weed management in Morocco and internationally is characterized by three distinct eras. The prehistoric era (1945) where weeds were controlled mainly manually and then gradually with small tools, animals, and other towed tools. The second era, between 1940 and 1980, corresponds to the great development of weed science following the synthesis and use of herbicides. Finally, the current era, starting from 1980 to today, corresponds to the era where several methods are used according to the principle of integrated weed management following the appearance of the phenomenon of weed resistance to herbicides (examples are the resistance of *Lolium rigidum* and *Papaver rhoeas* L. (common poppy) in different agricultural regions of Morocco). Moreover, in the history of Morocco, the weeds have not been totally harmful and sometimes present virtues for the Moroccan population. Some of these weeds served as a staple food for the local population during the severe famines of 1579, 1580, 1850, and 1945. Food dishes such as "Beqqola" or "Khobiza" made from *Malva sylvestris* L. (common mallow) are well known to Moroccans to this day. Weeds have been used in traditional Moroccan medicine, for example, *Chenopodium ambrosioides* L. (epazote) as an analgesic plant and *Mentha pulegium* L. (pennyroyal) as an anti-cold plant. In addition, many manually pulled weeds were also used as livestock feed, and this practice still continues among some farmers. Other weeds were used as firewood such as *Solymus maculatus* L. (spotted golden thistle), *Scolymus hispanicus* L. (Spanish oyster thistle), and *Silybum marianum* (L.) Gaertn. (milk thistle). Finally, some help to combat soil erosion, such as *Ziziphus lotus* (L.) Lam. (wild jujube) in the Chaouia region of Morocco.

In the early periods of farming, humans had to dedicate a significant portion of their energy to removing weeds from arable lands in order to create conditions conducive to the optimal growth of desired crops. Simultaneously, the idea of managing or controlling weeds began to occupy the minds of early farmers. Addressing these challenges drives weed scientists to develop cutting-edge technologies and innovative methods for sustainable weed management strategies, which can be integrated into current or future agricultural systems (Ward et al. 2014; Chauhan et al. 2017).

In Morocco, herbicide trials began soon after their global introduction, with studies on the effectiveness of 2,4-D and MCPA against wheat weeds being conducted as early as 1947 in Oujda by Rosella (Tanji 2002). In 1950, aerial application of a 2% powder formulation of 2,4-D or MCPA was carried out at a farm in the Merchouch region, followed by additional trials on plots in the Mohammedia region in 1951. With the growing availability of herbicides in Morocco, comprehensive weed control studies in wheat were initiated by key institutions, including the National School of Agriculture of Meknes (ENA), the Directorate of Agricultural Research (DRA), the Hassan II Agronomic and Veterinary Institute (IAV Hassan II), and the National Institute of Agronomic Research (INRA) in 1969 (Tanji 2002). Additional trials were also conducted by phytosanitary companies, plant protection services, and the Regional Offices for Agricultural Development (ORMVA) (Tanji 2002).

6.3 Weed Flora Present in Morocco

Morocco, with around 4,200 plant species, is the richest in North Africa and one of the most diverse in the Mediterranean region (Valdés et al. 2002). Botanical studies on this flora began as early as the eighteenth century with the publication of *Flora Atlantica* by French botanist René Louiche Desfontaines. Further studies continued through the nineteenth and twentieth centuries, such as the *Catalogue des plantes du Maroc* and *Flore de l'Afrique du Nord* (Valdés et al. 2002).

As for the national adventive flora, a total of 851 species have been recorded, including dicotyledons (83.2%) and monocotyledons (16.6%), as well as one non-vascular species, *Equisetum ramosissimum* Desf. (Branched horsetail). This flora represents 20.2% of the specific count, 40.8% of the generic count, and 59.1% of the total vascular plant families in Morocco (Fennane et al. 1999, 2007, 2014). Among the families, 21 contain more than 9 species, representing 85.1% of the total number. Some families, although containing few species, can have a negative impact on crops, especially parasitic plants like the genus *Orobanche*.

In case of weeds associated with cereals, a total of 374 species have been recorded across all the surveyed cereal-growing regions, with a variation ranging from 188 to 315 depending on the region (Table 6.2). Dicotyledons account for 84.5% of the total representing 316 species, while monocotyledons make up to 15.5%, with 58 species. This flora associated with cereals represents 40% of the national adventive flora (Taleb 2000).

In Morocco, studies on weed species in major field crops revealed a low proportion of endemic species (Taleb et al. 1998, 2000). More than 50% of the identified weeds originated from the Mediterranean region, highlighting Morocco's strong botanical linkage within this region. However, species from Asia, Europe, America, and Africa are also present, demonstrating the richness and diversity of the country's weed flora. For example, species of European origin represent 3.3% in cereals and 5.5% in sugar beet, while American species represents 0.8% in cereals and 4.9% in sugar beet, and the lowest number of species have a tropical origin (2.9% in cereal and 1.4% in sugar beet) (Taleb et al. 1998, 2000). This diversity shows that, despite their varied origins, all these species share a common trait: their ability to adapt to disturbances and cultivation techniques, thereby facilitating their spread (Fried 2007).

Table 6.2 Number of weed species recorded in major cereals in different regions of Morocco

Regions	Average annual rainfall (mm)	Number of weed species
Abda	250	270
Chaouia	318	315
Doukkala	384	188
Rhamna	200	204
Haouz	530	258
Loukkos	740	244
Saïs	600	269

Source: Modified from Taleb (2000)

Table 6.3 List of exotic weed species introduced into Morocco

Weed species	Geographic origin
Abutilon theophrasti Medik. (Malvaceae); *Brachiaria eruciformis* (Smith) Griseb. (Poaceae); *Corchorus olitorius* L. (Tiliaceae); *Cyperus difformis* L. (Cyperaceae)	Asia
Ambrosia coronopifolia Torr. & A.Gray (Asteraceae); *Malvastrum coromandelianum* (L.) Garcke (Malvaceae, *Panicum capillare* L. (Poaceae); *Solanum cornutum* Lam. (Solanaceae); *Solanum elaeagnifolium* Cav. (Solanaceae); *Verbesina encelioides* (Cav.) Benth. et Hook. ex Gray (Asteraceae); *Euphorbia heterophylla* L. (Euphorbiaceae); *Ammannia coccinea* Rottb. (Lythraceae); *Euphorbia heterophylla* L. (Euphorbiaceae)	North America
Asclepias curassavica L. (Asclepiadaceae); *Pistia stratiotes* L. (Araceae); *Salpichroa origanifolia* (Lam.) Baillon (Solanaceae)	South America
Heteranthera reniformis Ruiz & Pav (Pontederiaceae)	Central America
Arctotheca calendula (L.) Levyns (Asteraceae)	South Africa
Dactyloctenium aegyptiacum (L.) Willd. (Poaceae)	Africa and Asia
Kochia scoparia (L.) Schrad. (Chenopodiaceae)	Eurasia

Adopted from Bouhache and Taleb (2022)

Morocco has faced the introduction of globally recognized invasive weed species, which have led to significant threats to ecosystems and biodiversity. These species invade both natural and cultivated environments, leading to serious consequences such as crop damage and biodiversity degradation, primarily due to their high propagation capacity. Since the 1980s, 17 new weed species from nine different plant families have been introduced into the country (Tanji and Taleb 1997; Taleb et al. 2005, 2010; Bouhache and Taleb 2013). Majority of the introduced weed species are those originated from North America followed by Asia and South America (Table 6.3).

Among these, two families, Solanaceae and Poaceae, represent half of the newly introduced species. Many of these species originate from the United States, and approximately two-third of them were introduced between 1980 and 1997. Most of these species are considered invasive exotic species, further highlighting the risks they pose to Morocco's biodiversity and agricultural systems.

6.4 Major Weeds, Their Biology, and Characteristics

Approximately 67 dicotyledon and 13 monocotyledon weed species are considered important in cereal crops in Morocco (Table 6.4). Certain species among them are particularly difficult to manage, for example, *Galium* spp., *Convolvulus* spp., *Silybum marianum* (L.) Gaertn. (milk thistle), *Centaurea* spp., *Fumaria* spp., *Emex spinosa* (L.) Campd (devil's thorn), *Lathyrus* spp., and *Malva* spp. Annual species dominate this flora and are considered relatively easy to manage compared to

Table 6.4 Major dicotyledonous and monocotyledon weeds species found in major cereal crops in Morocco

SN	Weed species	Family	Growing period
Dicotyledon			
1	*Anacyclus radiatus* Loisel.	Asteraceae	Annual
2	*Anagallis foemina* Miller	Primulaceae	Annual
3	*Anchusa azurea* Miller	Boraginaceae	Perennial
4	*Astragalus boeticus* L.	Fabaceae	Annual
5	*Beta macrocarpa* Guss.	Chenopodiaceae	Annual
6	*Bellis sylvestris* Cirillo	Asteraceae	Annual
7	*Calendula arvensis* L.	Asteraceae	Annual
8	*Capsella bursa-pastoris* (L.) Medik.	Brassicaceae	Annual
9	*Centaurea diluta* Aiton	Asteraceae	Annual
10	*Chenopodium album* L.	Chenopodiaceae	Annual
11	*Chenopodium murale* L.	Chenopodiaceae	Annual
12	*Chrysanthemum coronarium* L.	Asteraceae	Annual
13	*Chrysanthemum segetum* L.	Asteraceae	Annual
14	*Cichorium endivia* L.	Asteraceae	Annual
15	*Convolvulus althaeoides* L.	Convolvulaceae	Perennial
16	*Convolvulus arvensis* L.	Convolvulaceae	Perennial
17	*Diplotaxis tenuisiliqua* Delile	Brassicaceae	Annual
18	*Echium plantagineum* L.	Boraginaceae	Annual
20	*Emex spinosa* (L.) Campd.	Polygonaceae	Annual
21	*Fumaria agraria* Lag.	Papaveraceae	Annual
22	*Fumaria officinalis*	Papaveraceae	Annual
23	*Fumaria parviflora* Lam.	Papaveraceae	Annual
24	*Galium tricornutum* Dandy	Rubiaceae	Annual
25	*Galium aparine* L.	Rubiaceae	Annual
26	*Galium verrucosum* Huds.	Rubiaceae	Annual
27	*Glaucium corniculatum* (L.) Curtis	Papaveraceae	Annual
28	*Hirschfeldia adpressa* Moench	Brassicaceae	Annual
29	*Lamium amplexicaule* L.	Lamiaceae	Annual
30	*Lathyrus articulatus* L.	Fabaceae	Annual
31	*Lathyrus cicera* L.	Fabaceae	Annual
32	*Launaea nudicaulis* (L.) Hook.f.	Asteraceae	Perennial
33	*Lavatera trimestris* L.	Malvaceae	Annual
34	*Malva parviflora* L.	Malvaceae	Annual
35	*Medicago polymorpha* L.	Fabaceae	Annual
36	*Melilotus sulcatus* Desf.	Fabaceae	Annual
37	*Misopates orontium* (L.) Raf.	Plantaginaceae	Annual
38	*Papaver hybridum* L.	Papaveraceae	Annual
39	*Papaver dubium* L.	Papaveraceae	Annual
40	*Papaver rhoeas* L.	Papaveraceae	Annual

(continued)

Table 6.4 (continued)

SN	Weed species	Family	Growing period
41	*Plantago afra* L.	Plantaginaceae	Annual
42	*Polygonum aviculare* L.	Polygonaceae	Annual
43	*Raphanus raphanistrum* L.	Brassicaceae	Annual
44	*Rhagadiolus stellatus* (L.) Gaertner	Asteraceae	Annual
45	*Ridolfia segetum* Moris	Apiaceae	Annual
46	*Rumex pulcher* L.	Polygonaceae	Annual
47	*Scandix pecten-veneris* L.	Apiaceae	Annual
48	*Scolymus hispanicus* L.	Asteraceae	Annual
49	*Scolymus maculatus* L.	Asteraceae	Perennial
50	*Scorpiurus muricatus* L.	Fabaceae	Annual
51	*Scorpiurus vermiculata* L.	Fabaceae	Annual
52	*Senecio vulgaris* L.	Asteraceae	Annual
53	*Silene gallica* L.	Caryophyllaceae	Annual
54	*Silene vulgaris* (Moench) Garcke	Caryophyllaceae	Perennial
55	*Silybum marianum* (L.) Gaertn.	Asteraceae	Annual
56	*Sinapis alba* L.	Brassicaceae	Annual
57	*Sinapis arvensis* L.	Brassicaceae	Annual
58	*Sonchus asper* (L.) Hill	Asteraceae	Annual
59	*Sonchus oleraceus* L.	Asteraceae	Annual
60	*Torilis nodosa* (L.) Gaertner	Apiaceae	Annual
61	*Urtica urens* L.	Urticaceae	Annual
62	*Vaccaria hispanica* (Mill.) Rauschert	Caryophyllaceae	Annual
63	*Valerianella locusta* (L.) Laterr.	Caprifoliaceae	annual
64	*Vicia benghalensis* L.	Fabaceae	Annual
65	*Vicia hybrida* L.	Fabaceae	Annual
66	*Vicia sativa* L.	Fabaceae	Annual
67	*Ziziphus lotus* (L.) Lam.	Rhamnaceae	Perennial
Monocotyledon			
1	*Arisarum vulgare* O. Targ. Tozz.	Araceae	Perennial
2	*Avena sterilis* L.	Poaceae	Annual
3	*Bromus rigidus* Roth.	Poaceae	Annual
4	*Cynodon dactylon* (L.) Pers.	Poaceae	Perennial
5	*Elymus repens*	Poaceae	Annual
6	*Hordeum murinum* L.	Poaceae	Annual
7	*Lolium rigidum* Gaudin	Poaceae	Annual
8	*Muscari comosum* (L.) Mill.	Asparagaceae	Perennial
9	*Phalaris brachystachys* Link	Poaceae	Annual
10	*Phalaris minor* Retz	Poaceae	Annual
11	*Phalaris paradoxa* L.	Poaceae	Annual
12	*Poa annua* L.	Poaceae	Annual
13	*Vulpia geniculata* (L.) Link.	Poaceae	Annual

perennial species (Bouhache 2020). Among the perennial species that are difficult to control are *Convolvulus arvensis* L. (field bindweed) and *Convolvulus althaeoides* L. (mallow bindweed).

Monocotyledonous weeds are represented only by 13 species. In cereal fields, three classic troublesome species cause yield losses: *Avena sterilis* L. (sterile oat), *Lolium* spp., and *Phalaris* spp. However, *Bromus rigidus* Roth (ripgut brome) is considered a formidable and invasive species in cereal fields in Morocco and even in Mediterranean countries (Bouhache 2021). In addition to these species, others that deserve monitoring include *Hordeum murinum* L. (Rat's barley) and *Vulpia geniculate* (L.) Link (geniculate fescue).

6.4.1 Weed Biology

Understanding the biology and ecology of the main weed species is essential for effective weed management. Taleb (2000) found that the biological type of weeds associated with cereals was dominated by Therophytes (annuals), contributing to 77% of all species infesting cereal fields followed by Geophytes (10.9%), Hemicryptophytes (9.4%), Chamaephytes (1.6%), Nanophanerophytes (0.8%), and Parasites (0.8%). Normally, they germinate in the autumn or winter with the occurrence of the first rain. Perennial weeds make up a quarter of the total number of species. Among the geophytes, which reproduce vegetatively while sexual reproduction is very rare or absent, are *Cynodon dactylon* (L.) Pers. (Bermuda grass), *Arisarum vulgare* Targ.Tozz. (friar's cowl), and *Cyperus rotundus* L. (purple nutsedge). Geophytes and hemicryptophytes adapt to the arid and semi-arid zones of Morocco through their survival vegetative organs (bulbs, rhizomes, stolons). Chamaephytes and nanophanerophytes are very few in number and prefer areas with low mechanization and borders (Taleb 2000). They are represented by *Marrubium vulgare* L. (white horehound), *Phlomis herba-venti* L. (Jerusalem sage), and *Ziziphus lotus*. As parasitic plants, *Thesium humile* Vahl (low thesium) is the only species infesting cereals, and the infestations are very frequent in the Abda region of Morocco.

Several studies highlighted that biophysical, genetics, and management factors significantly affect weed biology and diversity (Basch et al. 2020; Ben-Ghabrit 2019; Bouhache 2020; Klingman et al. 1982). Below are the major factors influencing weed biology and diversity as indicated by Bouhache (2020).

- *Soil type*: The type of soil (texture) has a significant impact on the distribution of weed species, more so than climatic factors such as temperature and rainfall.
- *Life cycle variation*: Annuals complete their life cycle in one year or less, biennials in two years or seasons, and perennials in more than two years/seasons.
- *Seed production capacity and longevity*: Weeds are characterized by their prolific seed production and robust vegetative regeneration capabilities. There is considerable variation in seed output and viability among different weed species. For

Table 6.5 Seed production and longevity of some weed species

Weed species	Number of seeds per plant	Longevity (years)
Bromus sterilis	150–300	1–2
Polygonum convolvulus	100–1000	22
Sinapis arvensis	50–250	35
Chenopodium album	1000–5000	40
Amaranthus retroflexus	>40,000	40
Papaver rhoeas	20,000–40,000	40
Anagallis foemina	250–500	60
Polygonum aviculare	500–1000	60

Source: Modified from Bouhache (2020)

instance, *Bromus sterilis* L. (barren brome) typically produces 150–300 seeds per plant with a short longevity of 1–2 years. In contrast, *Polygonum convolvulus* L. (black-bindweed) yields 100–1000 seeds and can remain viable for 22 years, while *Sinapis arvensis* L. (field mustard) produces 50–250 seeds with a longevity of 35 years. Highly prolific species include *Chenopodium album* L. (fat hen) (1000–5000 seeds, 40 years longevity), *Amaranthus retroflexus* L. (redroot pigweed) (over 40,000 seeds, 40 years longevity), and *Papaver rhoeas* L. (common poppy) (20,000–40,000 seeds, 40 years longevity) (Table 6.5). Species with notable longevity include *Anagallis foemina* Mill. (blue pimpernel) (250–500 seeds, 60 years longevity) and *Polygonum aviculare* L. (common knotgrass) (500–1000 seeds, 60 years longevity).

- *Dispersal mechanisms*: Weeds possess diverse and effective mechanisms for seed dispersal.
- *Seed dormancy and persistence*: Seeds buried in the soil can maintain viability for extended periods, largely supported by dormancy mechanisms, contributing to their long-term persistence in the soil seed bank.
- *Growth plasticity*: Plant growth is highly adaptable to environmental conditions and emergence timing, complicating the assessment of individual weeds. This is the case with invasive weeds which adapt their growth and reproduction according to environmental conditions.
- *Emergence factors*: Weed emergence is staggered and influenced by factors such as germination requirements, soil type, burial depth, and agricultural practices.
- *Seed bank dynamics*: A significant portion of the seeds (less than 10%) germinates each year, creating an inexhaustible seed bank.
- *Weed emergence season*: Weeds can emerge in specific seasons: autumn, winter, spring, summer, and indifferent species that germinate year-round.
- *Growing season*: Autumn and some winter weed emergence directly affects cereal yields, while spring emergence influences harvest timing and quality. For example, when rainfall is abundant in spring, many farmers complain about late-cycle weeds such as *Ridolfia segetum* (L.) Moris (false fennel), which hinder the work of the combine harvester and reduce the quality of the harvested grain.

- *Grass weeds*: Grass weeds share morphological and physiological traits with cereals, making their management in cereal crops difficult.
- *Need for bio-ecological knowledge*: Understanding the bio-ecology of major grass weeds is crucial for effective management and control of competitive species.

Table 6.6 provides a summary of the major grass weeds found in cereal fields across Morocco, their reproductive traits, germination behaviors, and ecological preferences. These species, mostly annuals, primarily propagate through seeds and exhibit remarkable adaptability to diverse environmental conditions. Common examples include *Avena sterilis* L. (sterile oat), *Bromus* spp., *Lolium* spp. (ryegrass), and *Phalaris* spp. (canary grasses). Additionally, species like *Vulpia geniculata* (L.) Link (geniculate fescue) and annual *Poa annua* L. (annual bluegrass) are noted for producing large seed quantities per plant. The weeds' germination and dormancy patterns vary widely, influencing their management in farming systems. Key factors such as optimal germination depth, temperature requirements, and seed viability significantly impact their spread and persistence in agricultural landscapes (Bouhache 2020).

Table 6.6 Major grass weeds in cereal fields and their characteristics

Species	Characteristics
Sterile oat (*Avena sterilis*)	This species includes two subspecies: *Avena sterilis* L. ssp. *macrocarpa* (Moench) Briquet and *A. sterilis* L. ssp. *ludoviciana* (Dur.) Gill & Magné, which are difficult to distinguish at the vegetative stage. The first subspecies is responsible for most infestations. It is an annual plant that reproduces and conserves itself through seeds. One plant can produce between 80 and 240 viable seeds, with three types of seeds per spikelet. The first seed, larger than the others, initiates germination in the following winter. The remaining seeds are dormant and germinate later. Most (87%) seeds fall before harvest, with the remaining being spread by harvesting machines, seed lots, and straw. Seeds buried in the soil can remain viable for 4–7 years, leading to staggered germinations. The germination temperature is optimal at 10 °C, and it occurs between September and February. Germination can occur at depths exceeding 25 cm. Sterile oat is common in non-desert areas of Morocco, preferring fertile, deep soils like clay (tirs), and can also grow in sandy soils with sufficient moisture
Brome (*Bromus* spp.)	Four brome species are found in cereal fields in Morocco: *Bromus rubens* L., *B. madritensis* L., *B. sterilis* L., and *B. rigidus* Roth. The last species (*Bromus rigidus*, or "rigid brome") is the most widespread and harmful. It produces 200–300 seeds per plant and can generate up to 51,256 seeds/m^2 in the Saiss region. Brome seeds are dispersed by agricultural machinery, animals, and contaminated seed lots. They have low innate dormancy and lose viability quickly with increasing burial depth. Seeds can remain viable for 2–3 years under forced dormancy. Brome germination is influenced by seed age, burial depth, light, and soil moisture, and it germinates optimally at a depth of 5 cm. It thrives in subhumid and semi-arid areas, receiving more than 250 mm of rainfall on light, well-drained soils

(continued)

Table 6.6 (continued)

Species	Characteristics
Ryegrass (*Lolium* spp.)	Two species of ryegrass are found in Moroccan cereal fields: *Lolium rigidum* Gaud (annual ryegrass) and *Lolium multiflorum* Lam. (Italian ryegrass). The first species is more abundant and widespread. Rigid ryegrass produces 250 to 3000 seeds per plant, with up to 31,000–45,000 seeds/m^2. Seed dispersal occurs through agricultural machinery, animals, and contaminated seed lots. Ryegrass seeds have innate dormancy and require a period of post-maturation to germinate. Optimal germination occurs at 11 °C for buried seeds and 27 °C for surface seeds. Germination is the best at a depth of 2 cm, and seed viability decreases with increasing depth. Ryegrass is characterized by high genetic variability, enabling it to adapt to various climates, soils, and agricultural practices. It is particularly adapted to dry conditions
Canary grass (*Phalaris* spp.)	Three species of canary grass are common in Moroccan cereal fields: *Phalaris brachystachys* Link. (short-spike canary grass), *Phalaris minor* Retz. (small canary grass), and *Phalaris paradoxa* L. (awned canary grass). These are annual plants that reproduce through seeds. *P. minor* can produce 200–9000 seeds per plant, while *P. paradoxa* can produce 3500–21,000 seeds per plant. Seeds are dispersed by agricultural machinery, contaminated seed lots, and straw. Freshly produced seeds have high germination rates (50–100%) but require a post-maturation phase. Seeds buried at depths of greater than 10 cm do not germinate. The optimal germination temperature is between 8 and 20 °C. These species thrive in subhumid and humid zones, preferring clay and clay-loam soils. *P. minor* can tolerate anaerobic conditions.
Annual bluegrass (*Poa annua* L.)	*Poa annua* L. is a small annual herbaceous plant (2–40 cm in height) that reproduces and conserves itself through seeds. One plant can produce between 500 and 7000 seeds, and even a 0.5 cm tall plant can produce up to 360 seeds. Seed dispersal is done by agricultural machinery, animals, wind, water, and contaminated seed lots. The seeds have very low dormancy and germinate quickly, within 8 days of falling to the ground. The optimal germination temperature is 15–20 °C, and the optimal germination depth is between 0 and 2 cm. The species exhibits high genetic and phenotypic variability, enabling it to adapt to different climatic and edaphic conditions. Annual bluegrass prefers compacted, moist soils but does not tolerate very acidic soils
Rat's barley (*Hordeum murinum*)	Rat's barley refers to a complex of three subspecies: *Hordeum murinum* ssp. *murinum*, *H. murinum* ssp. *H. leporinum*, and *H. murinum* ssp. *glaucum*. The second subspecies is the most widespread in Morocco. It is an annual plant that reproduces through seeds, producing 19–29 seeds per spike, with a viability rate of over 92%. Seed dormancy is short and can be overcome by autumn cooling. Seeds are dispersed by wind, animals, and contaminated plant products. Germination occurs from October to January at temperatures ranging from 8 °C to 33 °C. Seeds tend to germinate near the surface, and deep burial hinders emergence. Rats barley prefer fertile, well-drained soils and are found at altitudes from 30 to 2750 m with precipitation ranging from 250 to 500 mm
Knee-high vulpia (*Vulpia geniculate* (L.) Link)	*Vulpia geniculata* (L.) Link is an annual mesotrophy plant that reproduces through seeds. One plant can produce up to 1000 seeds, or 50,000–500,000 seeds/m^2, with heavy infestations possibly exceeding one million seeds/m^2. Seeds require a post-maturation phase of 2–3 months. Most seeds buried in soil will germinate in the first year, with a small portion (<7%) remaining viable for up to 3 years. Germination occurs throughout the period from October to February at temperatures ranging from 10 °C to 30 °C. Seeds tend to germinate at shallow depths (optimal at 0.5–2 cm) and may even germinate on the soil surface. This species prefers sandy, acidic, silty-clay soils that are lightly disturbed (e.g., no-tilled fields). It can tolerate slight salinity

6.5 Major Challenges of Weed Management in Cereals

Cereal cultivation in Morocco is heavily influenced by the country's climate, with rainfed agriculture being the predominant method in the cereal-growing regions. The main cereal production areas are concentrated in the northern and central parts of the country, particularly in the Gharb, Sais, Doukkala, and Tadla regions. Wheat is typically grown during the winter season, with sowing occurring between October and December, and harvesting takes place in late spring or early summer. Despite its importance, wheat production in Morocco faces several challenges, including irregular rainfall, soil erosion, and the increasing threat of droughts, which can severely impact yields.

6.5.1 Weed Infestations and Challenges

Weeds pose a significant threat to cereal yields, causing losses ranging from 13 to 67%, depending on the region and severity of infestation. Moroccan farmers commonly use herbicides for weed control, as they are both effective and economical. However, improper use of herbicides has led to the emergence of herbicide-resistant weed species, diminishing their effectiveness. Globally, 65 weed species associated with wheat cultivation have evolved herbicide resistance. In Morocco, herbicide resistance has been documented since 2000, with *Lolium rigidum* showing resistance to ACCase inhibitors (fops, dims, and dens) and *Papaver rhoeas* L. (common poppy) to dicot herbicides like 2,4-D and tribenuron-methyl. By 2011, ryegrass also became resistant to ALS-type herbicides, affecting more than 1000 ha of crops (Bouhache 2017). Currently, *Lolium rigidum* is resistant to 13 herbicides in the major agriculture region in Morocco, impacting both wheat and sugar beet crops. Resistance mechanisms vary by herbicide, for example, *Papaver rhoeas* has also shown resistance to multiple herbicides as indicated in the studies of Baye (2014) and Tanji (2009, 2011, 2013, 2024). Furthermore, under pressure from demands of consumers and environmental advocates, several previously used herbicides have been withdrawn from the pesticide market, and the number of new herbicides being introduced is limited, further reducing the range of herbicide solutions available for weed control. This highlights the importance of a reasoned use of herbicides, which takes into account the alternation of herbicides with different modes of action to counteract the resistance phenomenon.

Invasive Weeds Invasive species further threaten Moroccan crops. Five highly invasive species—*Solanum elaeagnifolium* Cav. (silverleaf nightshade), *Verbesina encelioides* Cav.) Benth. & Hook. (golden crownbeard), *Bromus rigidus* Roth (ripgut brome), *Echinochloa phyllopogon* (Stapf) Koss. (late watergrass), and *Oxalis pes-caprae* L. (Bermuda buttercup)—have been extensively studied for their impact on cultivated plants, invasion mechanisms, and control strategies (Ben-Ghabrit 2019). *Bromus rigidus*, found in all cereal-growing regions, is particularly a concern due to its high frequency. Invasive weeds like *Verbesina encelioides* and *Oxalis pes-caprae* not only compete for resources but also exhibit allelopathic effects, suppressing the germination and growth of wheat (Bouhache 2021).

Herbicide Tolerance and Crop Management Cereal crops can suffer herbicide damage, especially under stress conditions, which can impact their herbicide metabolism. Research indicates that wheat cultivars vary in their tolerance to herbicides, with some experiencing severe injury that does not necessarily reduce yield (Lachgar 1996; Tanji et al. 2017; Tanji and Boutfirass 2018). Combining irrigation with weed control and applying fertilizers during planting and post-weed control has been shown to improve crop growth and yield (Hamid 1995; Diab 1996). Effective weed management also enhances the efficacy of fungicides, contributing to increased grain yields.

Monoculture The practice of cereal monoculture exacerbates weed problems, reducing soil fertility and encouraging the proliferation of difficult-to-control species. Farmers often opt for monoculture due to simplified cultivation techniques, but challenges such as legume weed infestations and parasitic plants like *Orobanche* spp. (broomrape), which can destroy up to 100% of faba bean, lentil, and pea crops, remain significant.

Climate Change Effects Climate change is another pressing factor on weed infestation. Increasing atmospheric CO_2, higher temperatures, and altered precipitation patterns affect both crops and weeds. Weeds, with their genetic diversity and physiological plasticity, adapt more rapidly than crops to these changes (Bouhache 2021). In the Tadla region, climate changes over recent decades have made the months of September and October increasingly warmer. As a result, farmers have been sowing crops earlier in September to avoid end-of-cycle drought issues. Consequently, early infestations by summer species, typically very uncommon at this time under normal conditions, have been observed, such as *Portulaca oleracea* L. (common purslane), *Amaranthus albus* L. (common tumbleweed), *A. retroflexus* L. (redroot pigweed), *Solanum elaeagnifolium* Cav. (silverleaf nightshade), *Setaria viridis* (L.) P. Beauv. (green foxtail), and *Cyperus rotundus* L. (purple nutsedge). Additionally, the spring parasitic plant *Cuscuta* spp., which typically germinates after winter, has started appearing in autumn, attacking crops like sugar beets, legumes, and forage plants (Bouhache 2021). To address these challenges, future weed control strategies must adapt to changing climate conditions. Sustainable practices, such as rotating herbicides with different modes of actions, improving crop rotation, and integrating non-chemical control methods, are critical for long-term weed management and environmental protection.

6.6 Major Weed Management Practice

Weed management practices in wheat cultivation in Morocco are influenced by the country's climate, available water resources, and the type of wheat grown. In rainfed areas, farmers optimize sowing periods to capture rainfall, while in irrigated regions, controlled irrigation techniques are used to manage water efficiently and maximize yields. Crop rotation is commonly practiced, with wheat alternated with legumes or other cereals to improve soil fertility and manage pests, diseases, and weeds. In irrigated areas, intensive irrigation practices are adopted to optimize water use, especially during dry seasons. The adoption of improved wheat varieties, particularly those resistant to drought and tolerant to diseases, is encouraged to enhance productivity and reduce climate-related risks. The government policies support the use of modern farming techniques, improved seeds, and irrigation infrastructure to increase wheat yields and resilience in Morocco. However, despite these practices, wheat production remains highly vulnerable to challenges posed by drought and fluctuating rainfall. Weeds, among other crop enemies, pose a significant barrier to increasing yields, highlighting the need for greater attention to weed management.

The main methods for controlling weeds in cereals rely, on the one hand, on non-chemical means through cultural or mechanical techniques and, on the other hand, on chemical means, specifically the use of herbicides. Non-chemical methods include crop rotation, competitive varieties, delayed sowing, seeding density, soil tillage, manual weeding, and animal grazing.

6.6.1 Agronomic Practices

Crop Rotation

Crop rotation plays a critical role in managing weeds and sustaining productivity in Morocco's rainfed drylands, where monocropping of cereals—particularly barley and wheat—has become dominant. Incorporating diverse crops disrupts the weed life cycle and reduces their dominance. It slows down the emergence of herbicide-resistant weeds, particularly in systems that include forage crops. For example, studies have shown that inclusion of alfalfa in the rotation inhibits the germination, regeneration and growth of *Solanum elaeagnifolium* Cav. (silverleaf nightshade) since it contains allelopathic substances (Bouhache and Assali 1995; Bensellam 2003); hence, there has been an interest in integrating the cultivation of alfalfa into the rotation with cereals. Recent study by Ouakrim (2025) found that cereal-cereal rotation has less weed density but has more dominant weed species than in cereal-legume rotation.

Competitive Crops and Varieties Different species and varieties of cereals have distinct abilities to compete with weeds for light, water, and nutrients. Barley shows the greatest ability to suppress weed growth, significantly reducing weed biomass. Soft wheat, durum wheat, and triticale also display strong competitive characteristics, especially early in the growing season. Competitive crops typically have larger leaf area and stronger stem and root growth, enhancing their ability to outcompete weeds.

Crop Density and Seed Rates Increasing plant density improves competition with weeds. Higher planting densities can eliminate or reduce the need for herbicides. The study by Tanji (2017b) showed that a higher density of wheat (400 plants/m^2) effectively reduced weed biomass compared to a lower density (200 plants/m^2).

6.6.2 *Mechanical Weed Control*

- *Soil tillage*: Tillage is crucial for weed control especially in dry conditions. It serves both to prepare the seedbed and manage weeds. Full control of existing weeds at sowing time is essential for integrated crop management (Tanji 2017b).
- *Manual weed control (Hand pulling)*: Common in rain-fed areas, weeds are collected for livestock forage. Hand pulling is done from January to April in cereal crops by small scale farmers to remove weeds such *Sinapis* spp., *Diplotaxis* spp., *Emex spinosa* (L.) Campd (devil's thorn), *Silybum marianum* (L.) Gaertn. (milk thistle), etc. However, this practice tends to decrease due to the increasing cost of agricultural labor (Tanji 2017b).
- *Grazing*: Land left fallow and invaded by weeds is grazed by livestock to utilize the forage. Weeds often have good forage quality, especially at the vegetative stage.

6.6.3 *Chemical Weed Control*

Farmers in Morocco typically use portable sprayers or tractor-mounted sprayers to apply herbicides. However, calibration is often neglected, leading to errors in nozzle settings, application rates, volume, speed, and pressure. Additionally, climatic conditions are frequently overlooked during herbicide application, which can cause crop damage. Proper education on herbicide technology and sprayer adjustments to improve efficiency and weed control is recommended.

Farmers consider several factors while deciding to apply herbicides; the aim is to ensure effective weed control, minimize costs, and reduce environmental impacts. In a study conducted by Hajjaj (2010) in Sidi El Aidi and Jemaa Shaim regions of Morocco, the two main criteria for selecting herbicides by farmers were the cost of herbicides (41%) and the effectiveness of herbicides (30%). However, high herbicide costs, cited by 50% of farmers, and limited access to suitable equipment remain significant barriers. Application of herbicides is justified because of its return on investment, both through yield gains and the quality of cereal grains at harvest and through the reduction of weed seed banks in the soil (Hajjaj 2010).

According to Bouhache (2008), farmers mostly apply weeding practices reflexively, often following neighboring farmers, rather than making informed decisions based on the target weed species and selecting the appropriate herbicides for their own situation. Effective weed management requires consideration of the herbicide dose, timing of application, and proper calibration of herbicide equipment. Ineffective control of cereal weeds can cause the presence of weeds at the end of the cereal cycle (Tanji 2017a). A survey conducted in the irrigated region of Tadla in Morocco (Tanji 2000) revealed that the five main reasons why irrigated wheat fields become infested with weeds are: First, farmers consider weed infestation to be low at the appropriate treatment stage (24% of the plots surveyed). Next, farmers report that the prices of anti-grass herbicides are high (in 15% of plots). Then, water cuts and delayed irrigation hinder weeding (13%). In addition, the treatment stage is exceeded (12%). Finally, in 11% of cases, weeds re-infestation occurs after herbicide treatments.

6.7 Major Herbicides for Cereals

Selective weed control in cereals began worldwide in 1945 with the discovery and commercialization of the herbicides 2,4-D (2,4-dichlorophenoxyacetic acid) and 2,4,5-T (2,4,5-trichlorophenoxyacetic acid), which were selective in foliar application against dicot weeds. Around the same time, IPC (isopropyl-N-phenylcarbamate) was introduced for its ability to selectively control grass species when applied to soil. The 1950s were marked by the development of substituted urea and triazine herbicides that inhibit photosynthesis, while the 1960s were characterized by the introduction of dinitroanilines (Chtaina and Hormatallah 2022). In 1974, glyphosate was commercialized as a non-selective herbicide. During the subsequent decades, several new herbicide chemical families, including imidazolinones, sulfonylureas, pyrimidinyl-oxybenzoates, and triazolo-pyrimidines, were developed (Chtaina and Hormatallah 2022).

In Morocco, herbicide trials began soon after their global introduction, with studies on the effectiveness of 2,4-D and MCPA against wheat weeds being conducted as early as 1947 in Oujda by Rosella (Tanji 2002). In 1950, aerial application of a 2% powder formulation of 2,4-D or MCPA was carried out at a farm in the Merchouch region, followed by additional trials in plots in the Mohammedia region in 1951. With the growing availability of herbicides in Morocco, comprehensive weed control studies in wheat were initiated by key institutions, including the National School of Agriculture of Meknes (ENA), the Directorate of Agricultural Research (DRA), the Hassan II Agronomic and Veterinary Institute (IAV Hassan II), and the National Institute of Agronomic Research (INRA) in 1969 (Tanji 2002). Additional trials were also conducted by phytosanitary companies, plant protection services, and the Regional Offices for Agricultural Development (ORMVA) (Tanji 2002).

Table 6.7 Categories of herbicides used in Morocco and their description

Category	Description
Pre-sowing and pre-emergence herbicides	Widely used for direct-seeded or irrigated cereals. Non-selective herbicides (e.g., paraquat, diquat, glyphosate) are frequently applied to weeds before crop sowing. All pre-emergence herbicides require well-prepared, moist soil for effectiveness. Only four herbicides (chlortoluron, prosulfocarbe, triallate, or a mix of diflufenican + flurtamone + flulufenacet) are approved for use against dicotyledons and grasses in wheat (Bouhache 2020)
Post-emergence herbicides	Fifteen approved herbicides, mainly ALS enzyme inhibitors (Group 2), can be applied from two-leaf stage to early tillering (Bouhache 2020)
Grass herbicides	Commonly used in intensive cereal production and seed production. There are 17 herbicides (FOPs and DEN) that inhibit acetyl-CoA carboxylase (ACCase) and are effective against certain grasses. However, they do not control certain resistant weed species (Bouhache 2020)
Dicotyledon herbicides	A range of 65 approved herbicides is available for managing dicotyledon weeds in wheat, barley, and triticale. Over 50% of these are hormone-based products (2,4-D. MCPA, dicamba) and are effective at advanced growth stages (Bouhache 2020)
Pre-harvest herbicides	No pre-harvest herbicides are approved in Morocco. However, late intervention may be necessary in rainy years to control perennials and facilitate harvesting (Bouhache 2017, 2020)

As of 2022, 97 companies are authorized to market plant protection products in Morocco. The Moroccan agricultural sector uses an average of 17,000 tons of plant protection products annually (Croplife Morocco, 2024/www.croplife.ma). The market distribution consists of 45% insecticides, 40% fungicides, and 15% herbicides in Morocco (Croplife Morocco, 2024/www.croplife.ma). These herbicides can be applied at various stages of crop development, including before sowing, pre-emergence, or post-emergence of the cereal crop (Table 6.7).

6.7.1 Environmental and Health Impacts of Herbicides

While herbicides are a crucial tool for weed management, their use can have negative environmental and health impacts. By reducing species diversity, herbicides diminish habitats for insects, birds, and animals and contaminate soil, water, and air. The environmental risk depends on the toxicity of the herbicide and its persistence in the environment (Chtaina and Hormatallah 2022). Some herbicides, such as metsulfuron used on cereals, have been found to persist in soil, affecting subsequent crops like sugar beets in the Doukkala region. Residual effects of herbicides like aminopyralid and sulfosulfuron can harm sensitive rotational crops and soil microflora (Sayah 1982; Machhouri 1983).

The fate of the applied herbicide depends on its penetration into the plant, its potential metabolism, translocation to the site of action, interaction with the biochemical target, and triggered physiological changes that disrupt growth or destroy the plant. Most systemic herbicides enter the plant through the leaves or both the leaves and roots, where they migrate to reach five sites of action within the cells: auxins, branched-chain amino acids, aromatic amino acids, lipids, and photosynthesis (Bouhache 2017).

With the evolution of herbicide resistance in weeds to various herbicides, the Herbicide Resistance Action Committee (HRAC) recommended classifying the active ingredients of herbicides according to their sites of action, assigning a number to each group. Rotating or changing active ingredients from different chemical families sharing the same number is ineffective in preventing or delaying the appearance of resistance (Bouhache 2017).

6.8 Weed Management Challenges and Opportunity in Conservation Agriculture (CA) in Morocco

The Conservation Agriculture (CA) system in Morocco is gaining recognition as a solution to the environmental and agricultural challenges the country faces, particularly in its semi-arid region. Moroccan agriculture has long been affected by soil degradation due to excessive tillage, overgrazing, and improper management of crop residues, leading to erosion, compaction, and depletion of organic matter. In the face of increasing water and land scarcity and the pressure to meet the food needs of a growing population, sustainable practices like CA offer a promising future (Devkota et al. 2022). This system involves planting crops directly into minimally disturbed soil, keeping crop residues on the surface to reduce water evaporation and soil erosion, thus improving soil structure and conserving moisture. In this regard, Morocco's Ministry of Agriculture launched a national level CA program in 2021, aiming to expand no-till sowing of cereals and legumes across 1.0 million ha by 2030. CA systems, when integrated with appropriate crop rotations and weed management, have great potential to improve the sustainability of Moroccan agriculture and play a key role in strengthening food security and environmental protection in the face of current climate challenges.

The transition to no-tillage systems raises several questions among farmers regarding weed management. No-tillage systems, with minimal soil disturbance, plant residue coverage, and crop rotation, have an impact on weed dynamics (Basch et al. 2020). Indeed, the removal of tillage as a method of weed control means that producers must adjust crop rotations, herbicide use, and other cultural practices as compensation (Labrada 2005; Aibar 2006; Nichols et al. 2015).

Resistance to various herbicides, including glyphosate, is a major issue in CA (Wall et al. 2020). In Morocco, resistance in *Lolium rigidum* and *Papaver rhoeas* has been documented (Bouhache 2017). Hence, weed management in CA needs special consideration, especially during the initial few years of adoption of CA. In Morocco, a shift towards perennial weeds such as *Cynodon dactylon*, *Convolvulus arvensis*, and *Bromus rigidus* has been noted under CA (El Brahli and Mrabet 2000;

Hajjaj 2018). Some weed species thrive in no-tillage systems, often due to their adaptability to less disturbed soils (Dorado 2012). In Spain, *Bromus diandrus* is a major problem; in Aragon, problematic species include *Lolium rigidum* and *Avena* spp. (Aibar 2006). Similarly, Hajjaj (2018) reported that *Vulpia geniculata* (L.) Link. (geniculate fescue) prefers less disturbed soils, and 40% of farmers have reported issues with *Bromus rigidus* under CA.

6.8.1 Case Studies on Crop Rotation and Weed Dynamics Under CA

Case Study 1. From Long-Term CA Experiment, Merchouch

The experiment was established in 2014 with the objective of monitoring the effect of different crop genotypes in CA and conventional agriculture systems in a cereal-legume rotation (Devkota et al. 2021). The experiment was redesigned in 2019 with three different crop rotations as below and weed dynamics were evaluated:

1. Rotation−1: Cereal-cereal where three cereal crops (bread wheat, durum wheat, and barley) were rotated in annual basis
2. Rotation−2: inclusion of food legume (33% legume in cereal rotation): Bread wheat, lentil, durum wheat, bread wheat
3. Rotation−3: inclusion of forage mixture in rotation: Bread wheat–forage mixture (80% vetch + 20% oat), bread wheat–barley mixture

The experiment was designed in a split-plot design with tillage as the main plot and crop rotation as the sub-plot with four replications. Weed samples were collected from 0.25 m^2 area from randomly selected two points and were separated into broad-leaf and narrow-leaf weeds and oven-dried to measure weed dry biomass per m^2.

In all three-rotation systems, the average weed biomass remained higher in the initial two years under CA compared to CT. In 2020, compared to CT, dry weed biomass was higher under CA by 2.5 times in Rotation−1, 1.3 times in Rotation-2–2, and 1.6 times in Rotation−3. Similarly, in 2021, weed biomass under CA was 2.1, 1.8, and 2.6 times higher than in CT under Rotation−1, Rotation−2, and Rotation−3, respectively. In 2023, weed biomass was reduced under CA than under CT in all rotations, and the reduction was higher under rotation with forage mixture (Rotation−3) (Table 6.8). This indicated that using a forage mixture in the rotation can reduce the weed biomass and herbicide application under the CA system, as herbicide was not applied in the forage mixture. As reported by Taji et al. (2017), the adoption of no-tillage combined with an appropriate forage mixture in the wheat-based systems can be a better option for mixed crop-livestock systems in rainfed drylands.

In an in-depth study in 2024 from the same experiment, it was observed that across tillage methods and rotation systems in cereal-based systems, dicot weeds

Table 6.8 Weed biomass (g/m^2) under three different crop rotations practiced under conservation agriculture (CA) and conventional tillage (CT) systems in four different cropping seasons (2019–2023) in Merchouch, Morocco

	Conservation agriculture (CA)			Conventional tillage (CT)		
Year	Rotation–1	Rotation–2	Rotation–3	Rotation–1	Rotation–2	Rotation–3
2019/20	6.03 ± 5.4	14.53 ± 22	2.94 ± 3	2.39 ± 3.7	10.8 ± 11.8	1.76 ± 1.6
2020/21	10.84 ± 8.6	9.04 ± 7.1	9.46 ± 5.3	5.03 ± 6.1	4.95 ± 2.8	3.54 ± 1.6
2021/22	2.47 ± 1.6	2.24 ± 1.1	2.81 ± 3.1	0.96 ± 0.6	6.73 ± 6	0.70 ± 0.5
2022/23	2.55 ± 1.47	4.03 ± 2.65	2.53 ± 1.87	3.33 ± 2.24	4.72 ± 6.93	7.33 ± 7.65
2023/24	39.72 ± 28	74.9 ± 32	49.52 ± 33	61.51 ± 49.1	34.27 ± 23	61.61 ± 47.5

Note: Rotation−1: Cereal-cereal rotation (bread wheat, durum wheat, and barley); Rotation−2: inclusion of food legume (33% legume in cereal rotation): Bread wheat, lentil, durum wheat, bread wheat; Rotation−3: inclusion of forage mixture in rotation: Bread wheat, forage Mixture (75% vetch + 25% oat), bread wheat–barley mixture
Values are average (± standard deviation)

Table 6.9 Weed density (number/m^2) in three different crop rotations grown under conservation and conventional agriculture systems in four different cropping seasons (2022–2023) in Merchouch, Morocco

	Weed density (number/m^2)					
	Conservation agriculture			Conventional tillage		
Rotation	Monocots	Dicots	Total	Monocots	Dicots	Total
Rotation–1	11.0 ± 15	30.5 ± 10	31.5	4.0 ± 7	50.5 ± 25	54.5
Rotation–2	14.0 ± 19	37.5 ± 11	51.5	1.0 ± 2	52 ± 47	53
Rotation–3	4.0 ± 8.6	33.5 ± 7	37.5	0.0	59 ± 15	59

Values are average (± standard deviation). In continuous nine years of conservation

were more dominant than monocot weeds. Compared to the CT, the monocot weed population was higher under CA in all rotation systems (Table 6.9). Nichols et al. (2015) also reported that CT favors annual dicotyledons, while CA encourages perennials and annuals germinating at the soil surface.

Case Study 2

In a field study in Settat province, Chaouia, Morocco, Tanji et al. (2017) reported that with crop rotation combined with timely herbicide application, the weed seed bank and weed diversity can be significantly reduced under a no-tillage system. In a weed seed bank study, they found that after two cropping seasons with no-tillage, the weed seed bank was reduced by 23% in continuous wheat rotation, followed by 68% in wheat after canola and by 72% in wheat after barley + pea mixture. Although weed density and diversity reduced over time under no-tillage, continuous wheat increased *Anisantha rigida* (rigid brome) density and seedbank (Table 6.10). They also have indicated that in mixed crop livestock systems in rainfed drylands of Morocco, grazed weedy fallow can be replaced by a forage mixture for hay with the adoption of a no-tillage system.

Table 6.10 Germinable weed seed density in no-till crops in three sites in Chaouia, Morocco, in September 2012, 2013, and 2014

Rotation	Initial (2012)	Year–1 (2013)	Year–2 (2014)
Wheat–wheat	1890	1009	1445
Barley + forage pea-wheat	1890	719	520
Canola–wheat	1890	567	606

Source: Adopted from Tanji et al. (2017)

Weed Seed Bank in Soil

The weed seed bank is crucial for understanding the past and future of weed issues (Bouhache 2021). No-tillage can alter the qualitative and quantitative aspects of the seed bank, favoring species that do not tolerate burial (Aibar 2006). Studies show variable results regarding the seed bank density between no-tillage and CT (Rahali et al. 2010; Ruisi et al. 2015). A study of the viable seed stock in the 0–10 cm horizon in the Oued Zem region (Morocco) shows quantities ranging from 1520 to 2354 seeds/m^2 depending on the year and associated crop rotations (Tanji et al. 2017). A comparison of the weed infestation situation in a no-tillage field in Chaouia (Morocco) over a period of 18 years revealed that new species recorded as surface flora and those found in the soil seed bank included *Anagallis foemina* Mill. (blue pimpernel), *Capsella bursa-pastoris* (L.) Medik. (shepherd's purse), *Fumaria parviflora* Lam. (fineleaf fumitory), *Galium tricornutum* Dandy (rough corn bedstraw), *Glaucium corniculatum* (L.) Cutis (blackspot hornpoppy), *Polygonum aviculare* L. (common knotgrass), *Scolymus maculatus* L. (spotted golden thistle), *Sonchus oleraceus* L. (common sowthistle), and *Torilis nodosa* (L.) Gaertn. (knotted hedge-parsley) (Hajjaj et al. 2018). A comparison of the potential and actual weed flora in no-tillage and CT fields at Sidi El Aidi and Jemaa Shaim (Morocco) showed that the surface flora (actual) represented only 4.1% and 4.6% of the seed bank in no-tillage and CT, respectively, at the first site. Conversely, it was only 1.9% and 3.9%, respectively, at the second site (Hajjaj et al. 2018). This emergence rate varied from 0 to 25% depending on the species, plot management, and site. High emergence rates were recorded for grasses (Table 6.11). This shows that the weed seed bank can spread and maintain an infestation in a field over several years.

6.9 Conclusions

Weed management in Morocco is a major concern for cereal-based system, where weed competition can lead to yield losses ranging from 13% to 57%. Understanding weed biology and ecology is fundamental to addressing this challenge, as it illuminates the dynamics of persistent and evolving species such as *Lolium rigidum*, *Avena sterilis*, and *Convolvulus arvensis*, which continue to pose major problems in agricultural ecosystems. Herbicides are essential for weed control in cereal-based systems; however, their use faces significant economic hurdles, with 50% of

Table 6.11 Emergence rate (%) of surface flora compared to the viable seed bank in the soil

Weed species	Sidi El Aidi		Jemaa Shaim	
	No-tillage	Convention tillage	No-tillage	Convention tillage
Anagallis foemina Mill.	1.2	1.9	1.7	0.9
Avena sterilis L.	4.5	6.7	–	–
Bromus rigidus Roth	13.7	6.2	–	7.7
Capsella bursa- pastoris (L.) Medik.	1.3	1.4	–	–
Centaurea diluta Aiton	4.5	1.2	–	–
Chenopodium album L.	1.9	2.0	1.4	1.2
Chrysanthemum coronarium L.	–	0.5	–	–
Convolvulus arvensis L.	1.9	–	2.4	–
Emex spinosa (L.) Campd.	3.4	3.7	0	0.6
Fumaria parviflora Lam.	0.7	2.1	–	–
Galium tricornutum (Dandy)	0.3	0.9	–	–
Glaucium corniculatum (L.) Curt.	2.5	3.8	–	–
Lamium amplexicaule L.	–	–	–	0.7
Lolium rigidum Gaud.	12.9	18.8	–	8.8
Papaver rhoeas L.	3.7	5.1	0	–
Phalaris brachystachys Link	15.2	25.0	–	8.3
Polygonum aviculare L.	0.2	0.3	–	–
Scolymus maculatus L.	1.3	0.8	–	–
Sonchus oleraceus L.	0.9	0.8	–	–
Spergula arvensis L.	–	–	1.0	2.6
Torilis nodosa L.	2.9	1.3	–	–

Source: Adopted from Hajjaj et al. (2018)

farmers citing the high cost of herbicides as a major barrier and 41% prioritizing affordability when selecting weed management methods. Furthermore, herbicide resistance is a growing concern. For instance, *Lolium rigidum* has evolved resistance to ACCase-inhibitors herbicides, and *Papaver rhoeas* has shown resistance to critical herbicides like tribenuron-methyl and 2,4-D.

To counter these challenges, Moroccan farmers employ agronomic strategies such as delayed sowing, crop rotation, and adjusting seeding density to enhance crop competitiveness against weeds. While these methods can be effective, their success is often limited by knowledge gaps or resource constraints, which hinder their optimal application. In CA systems, weed populations can be managed more effectively through proper crop rotation and management practices, although shifts in weed species composition can complicate management and necessitate integrated approaches.

Given the complexity of Morocco's weed flora, combined with changing climatic and agricultural conditions, continuous research and adaptive management practices are indispensable. Emphasizing efficient herbicide use, promoting cost-effective solutions, and customizing strategies to local conditions are key to improving agricultural productivity and maintaining environmental sustainability. Additionally, implementing training programs tailored to local conditions is crucial to equip farmers with the knowledge and tools needed for effective and sustainable weed management, particularly in rainfed drylands where CA is practiced.

Acknowledgments This work was supported by One CGIAR Science program Sustainable farming and Scaling for Impact (S4I) and Moroccan Collaborative Grants Program (MCGP).

Annex

Images of Some Species Infesting Wheat Fields (Credit: Badr Hajjaj) (Photos 6.1, 6.2, 6.3, 6.4, 6.5, 6.6, 6.7, 6.8, 6.9, 6.10, 6.11 and 6.12)

Photo 6.1 *Sinapis arvensis* L.

Photo 6.2 *Ridolfia segetum* Moris

Photo 6.3 *Lolium rigidum* Gaudin

Photo 6.4 *Papaver rhoeas* L.

Photo 6.5 *Emex spinosa* (L.) Campd

Photo 6.6 *Cynodon dactylon* (L.) Pers

Photo 6.7 *Convolvulus arvensis* L.

Photo 6.8 *Chrysanthemum coronarium* L.

Photo 6.9 *Centaurea diluta* Aiton

Photo 6.10 *Bromus rigidus* Roth

Photo 6.11 *Bromus rigidus* Roth

Photo 6.12 *Avena sterilis* L.

References

Aibar J (2006) La lutte contre les mauvaises herbes pour les céréales en semis direct, Principaux problèmes. Options Méditerranéennes: Série A. Séminaires Méditerranéens 69:19–26

Andersen RN (1991) The north central weed control conference: origin and evolution. North Central Weed Sci Soc, Champaign, p 206

Basch G, Teixeira F, Duiker SW (2020) Weed management practices and benefits in Conservation Agriculture systems. In: Kassam A (ed) Advances in conservation agriculture, vol 2, pp 105–142. https://doi.org/10.19103/AS.2019.0049.04

Bensellam EH (2003) Etude de l'effet allélopathique de la luzerne (Medicago sativa L.) sur la morelle jaune (Solanum elaeagnifolium Cav.): mise en évidence, isolement et identification des substances allélopathiques. Memoire d'Accès au grade d'Ingénieur en Chef, INRA, Rabat, Maroc

Baye Y (2014) Le ray-grass (*Lolium rigidum* Gaudin) résistant aux herbicides Fops dans la région de Tadla (Maroc): état d'infestation et gestion intégrée. Proc 9ème Congrès de l'AMPP, Rabat, pp 243–253

Ben-Ghabrit S (2019) Dynamique et potentiel d'invasion d'une plante exotique envahissante: *Solanum elaeagnifolium* Cav., une *Solanaceae* qui menace le secteur agricole du Maroc. Thèse de Doctorat, I.A.V. Hassan II, Rabat

Bouhache M (2017) Atouts et faiblesses des herbicides utilisables sur céréales au Maroc. Agriculture du Maghreb (Hors série – Céréaliculture), 100, Décembre 2016/Janvier 2017, 10–16

Bouhache M (2008) Désherbage chimique des céréales d'automne. Agiculture de Maghreb 25:57–62

Bouhache M & Assali F (1995) Effet allélopathique de la luzerne (Medicago sativa L.) sur la germination des graines de la morelle jaune (Solanum elaeagnifolium Cav.). Proceedings du 2ème congrés de l'AMPP, Rabat:171–178

Bouhache M (2020) Bases pratiques de désherbage des céréales d'automne au Maroc. Dar Al Qalam, AMPP, Rabat, p 165

Bouhache M (2021) Malherbologie: Biologie et écologie des adventices des agroécosystèmes. Première édition, Dar Al qalam, Rabat, 374 p

Bouhache M, Taleb A (2013) La laitue d'eau: une plante exotique envahissante des eaux douces marocaines. Agriculture du Maghreb 71(Novembre):100–103

Bouhache M, Taleb A (2022) Gestion des Adventices des Principales Cultures du Maroc. Association Marocaine de Protection des Plantes

Chauhan BS, Matloob A, Mahajan G, Aslam F, Florentine SK, Jha P (2017) Emerging challenges and opportunities for education and research in weed science. Front Plant Sci 8, article 1537:1–13
Chauvel B, Darmency H, Nicolas MJ, Rodriguez A (2018) Gestion durable de la flore adventice des cultures. Editions Quae, Versailles, p 350
Chtaina N, Hormatallah A (2022) Utilisation des herbicides et environnement, Gestion des Adventices des Principales Cultures du Maroc Mohamed Bouhache & Abdelkader Taleb
Cremlyn RJ (1991) Agrochemicals: preparation and mode of action. Wiley, Chichester
De Wet JMJ (1966) The origin of weediness in plants. Proceedings of the Oklahoma Acadamy of Science, pp 14–17
Devkota M, Patil SB, Kumar S, Kehel Z, Wery J (2021) Performance of elite genotypes of barley, chickpea, lentil, and wheat under conservation agriculture in Mediterranean rainfed conditions. Exp Agric 57:126–143
Devkota M, Devkota KP, Kumar S (2022) Conservation agriculture improves agronomic, economic, and soil fertility indicators for a clay soil in a rainfed Mediterranean climate in Morocco. Agric Syst 201:103470
Diab A (1996) Essais sur la conduite technique des blés en irrigué (cas de Tadla). Mémoire de fin d'études, Institut Agronomique et Vétérinaire Hassan II, Rabat, p 117
Dorado J (2012) Reduced and no-tillage. Séminaire sur Gestion des mauvaises herbes dans l'agriculture actuelle. IAMZ, Espagne
El Brahli A, Mrabet R (2000) La Jachère chimique: pour relancer la céréaliculture non-irriguée en milieu semi-aride marocain. Proc Journée nationale sur le désherbage des céréales, Settat, pp 133–145
Fennane et al (1999, 2007, 2014) Flore pratique du Maroc. Institut scientifique, Rabat, Volumes 1, 2 et 3
Ferrahi M, Bekkaoui F (2020) Morocco and the Wheat Initiative, cereal sector in Morocco – April 2020. https://inrameknes.info/wpcontent/uploads/2020/12/2019_Wheat_Initiative_Morroco.pdf
Fried G (2007) Variations spatiale et temporelle des communautés adventices des cultures annuelles en France. Thèse de doctorat. Dijon, INRA-Université de Bourgogne, p 357
Green MB, Hartley GS, West TF (1987) Chemicals for crop improvement and pest management, 3rd edn. Pergamon Press, Oxford, p 207
Hajjaj (2010) Diagnostic désherbage céréales et légumineuses sidi el aidi et jemaa shaim Proceeding du septième congres de l'association marocaine de protection des plantes, Mai 2010, Rabat, Maroc
Hajjaj B, Bouhache M, Mrabet R & Taleb A (2018) Stock semencier des mauvaises herbes dans le sol et évolution de la flore de surface sous système de semis direct. Revue Marocaine de Protection des Plantes, 12:1–17
Hamid A (1995) Effet de l'irrigation d'appoint et du désherbage sur le rendement grain du blé tendre à Moulouya. Mémoire de fin d'études, Ecole Nationale d'Agriculture de Meknès, p 113
Hay JR (1974) Gain to the grower from weed science. Weed Sci 22(5):439–442
Klingman GC, Ashton FM, Noordhoff LJ (1982) Weed science: principles and practices. Wiley, New York, p 449
Labrada R (2005) Gestion des mauvaises herbes pour les pays en développement. FAO, Rome, p 285
Lachgar L (1996) Etude de la réaction de quelques variétés du blé dur et du blé tendre aux principaux herbicides anti-graminées. Mémoire de fin d'études. Institut Agronomique et Vétérinaire Hassan II, Rabat, p 107
Machhouri A (1983) Effet de deux herbicides sur la nitrifiication dans trois sols du plateau de Meknès. Mémoire de fin d'études. Ecole Nationale d'Agriculture de Meknès, p 52
Monaco TJ, Weller SC, Ashton FM (2002) Weed Science: principles and practices, 4th edn. Wiley, New York, p 685
Nichols V, Verhulst N, Cox R, Govaerts B (2015) Weed dynamics and conservation agriculture principles: a review. Field Crops Res 183:56–68. https://doi.org/10.1016/j.fcr.2015.07.012

Ouakrim Z (2025) Ecological resilience and dynamics of weeds: study of the seed bank contrasting crop subsystems in Merchouch. Specialized master thesis. University Mohammed V. Faculte des Science, Rabat

Radicetti E (2012) Ecological weed management. Ph.D. thesis, Tuscia University, Italy, 185 p

Rahali A, Makhlouf M, Benkherbache N (2010) Influence de l'itinéraire technique sur le type et le stock semencier du sol des mauvaises herbes. Cas de la zone semi arid de Sétif. Actes des 4èmes rencontres méditerranéennes du semis direct, Setif, Algérie, pp 90–99

Rao VS (2000) Principles of weed science, 2nd edn. CRC Press, New Hampshire

Ruisi P, Frangipane B, Amato G, Badagliacca G, Di Miceli G, Plaia A, Giambalvo D (2015) Weed seed bank size and composition in a long-term tillage and crop sequence experiment. Weed Res 55:320–328

Sayah M (1982) Effet de deux herbicides sur la microflore du sol. Mémoire de fin d'études, Ecole Nationale d'Agriculture de Meknès, 43

Shekhawat K, Rathore SS, Chauhan BS (2020) Weed management in dry direct-seeded rice: a review on challenges and opportunities for sustainable rice production. Agronomy 10(9):1264. https://doi.org/10.3390/agronomy10091264

Taleb A, Bouhache M, Rzozi SB (1998) Flore adventice des céréales d'automne au Maroc. Actes Inst Agron Vet Hassan II 18(2):121–130

Taleb A, Bouhache M & Rzozi SB (2000) Diversité et importance des mauvaises herbes de la betterave à sucre au Maroc. XIème Colloque International sur la Biologie des Mauvaises Herbes. Dijon, France, pp 159–166

Taleb A et al (2005) Flore nuisible aux monuments historiques du Maroc: cas des villes de Rabat et Salé. Proc. Xème Congrès Société Espagnole de Malherbologie, pp 255–259

Taleb A, Bouhache M, El Mfadli B (2010) New species threatening Moroccan biodiversity: *Verbesina encelioides* (*Asteraceae*). In: Proc 2nd international workshop on invasive plants in Mediterranean type regions of the world. EPPO, Trabzon, pp 156–167

Tanji A (2009) Efficacité des herbicides sur l'ivraie raide (Lolium rigidum). Actes du symposium international AGRUMED «Durabilité des systèmes de culture en zone méditerranéenne, gestion des ressources en eau et en sol», Rabat, pp 380–383

Tanji A (2011) Adventices résistantes aux herbicides au Maroc. Troisième Congrès international sur l'amélioration de la production agricole (APA3), Settat, p 30

Tanji A (2013) Betterave à sucre: La cléthodime est encore efficace sur le ray grass résistant aux herbicides. Agriculture du Maghreb 64:102–104

Tanji A (2024) Weed management in small grain cereals in Morocco, a review. In: Moussadek R, Ouabbou H, El Gharras O, Dahan R, El Mourid M (eds) National Institute of Agricultural Research – INRA Morocco Research for promoting sustainable farming systems in arid and semi-arid areas of Morocco Challenges, achievements and future prospects. ICARDA and INRA, Morocco. First published 2024

Tanji A, Boutfirass M (2018) Effective preemergence herbicides for rigid ryegrass (Lolium rigidum Gaud.) control in irrigated bread wheat (Triticum aestivum L.). J Agric Sci

Tanji A, Taleb A (1997) New weed species recently introduced into Morocco. Weed Res 37(1):27–23

Tanji A, Merimi Z, Laftah M, Ezzaari A, Jabbari A (2012) Gestion intégrée de l'ivraie raide (Lolium rigidum) résistante aux herbicides. Proceedings du huitième Congrès de l'Association Marocaine de Protection des Plantes (AMPP), Rabat, pp 421–433

Tanji A (2017a) Désherbage tardif des céréales, pas de traitement chimique au stade épiaison. Agiculture de Maghreb 102:68–69

Tanji, A. (2017b) Weed Dynamics in No-Till Rainfed Crops in Chaouia, Semi-Arid Morocco. Journal of Agricultural Science. https://doi.org/10.5539/JAS.V9N11P198

Tanji A, Baye B, Boutfirass M (2017) Evaluating postemergence herbicides for broadleaf weed control in irrigated bread wheat (Triticum aestivum L.). J Exp Biol Agric Sci 5(5):614–622. https://doi.org/10.18006/2017.5(5).614.622

Tanji A (2000) Mauvaises herbes du blé et de l'orge dans le périmètre du Tadla. Al Awamia, n° 102, pp 49–57. ISSN:0572–2721

Tanji A (2002) Synthèse de 123 essais de désherbage chimique du blé tendre réalisés au Maroc entre 1948 et 2001. Al Awamia, n° 105, pp 57–83

Valdés B, Rejdali M, Achhal El Kadmiri A, Jury SL, Montserrat JM (2002) Catalogue des plantes vasculaires du nord du Maroc. Volume I et II.: Consejo Superior de Investigaciones Científicas, Madrid

Varanasi A, Jugulam M (2017) Introduction: advancement of weed science as an important discipline of agriculture. In: Biology, physiology and molecular biology. Kansas State University, p 12

Wall P, Thierfelder C, Hobbs P, Hellin J, Govaerts B (2020) Chapter 11: Benefits of Conservation Agriculture to farmers and society. In: Kassam A (ed) Advances in conservation agriculture. Volume 2: Practice and benefits. Burleigh Dodds Science Publishing, Cambridge

Ward SM et al (2014) Agricultural weed research: a critique and proposals. Weed Sci 62:672–678

Went FW, Thimann KV (1937) Phytohormones. The Macmillan Co., New York, p 294

Worldweatheronline (Consulted 10.01.2025). https://www.worldweatheronline.com/

Zimdahl RL (2007) Fundamentals of weed science, 3rd edn. Academic Press, Elsevier, 689, New York

Chapter 7
Biology and Management of Weeds in Nigeria

Oyebanji O. Alagbo, Olumide S. Daramola, and Gabriel O. Adesina

Abstract Nigeria, being the fourth most populous country in the world, has an urgent task to feed its growing population. To achieve sustainable food production, efforts must be accelerated to bridge the existing yield gaps through the adoption of good agronomic practices. Weeds are a major constraint in the agricultural areas of Nigeria. This chapter provides a broad overview of the biology and management of weeds associated with major crops in Nigeria. We discussed the emerging challenges of weeds in Nigeria's cropping systems and management opportunities toward achieving the goal of sustainable food production. Due to increased market demands for staple foods in Nigeria, intensive cropping patterns have substituted the old fallow systems in most arable fields. As a result of intensive cropping activities, most arable fields have witnessed excessive accumulation of weed seeds, leading to the persistence of problematic annual weeds adapting to low-resource farming systems. Thus, most arable fields have heavy weed pressure, with persistent yield losses. The intensity of this challenge, i.e., heavy weed pressure, is different for the different cropping systems in the country. Integrated and site-specific weed management is required to address these challenges. Knowledge of the prevalent weeds and farming practices engaged in different agroecologies is important to understand the appropriate weed management approaches.

Keywords Agroecology · Weed biology · Precision weed management

O. O. Alagbo (✉)
Department of Crop Production and Protection, Faculty of Agriculture, Obafemi Awolowo University, Ile-Ife, Nigeria
e-mail: oyebanjialagbo@oauife.edu.ng

O. S. Daramola
University of Maryland, College Park, MD, USA

G. O. Adesina
Department of Crop and Environmental Protection, Ladoke Akintola University of Technology, Ogbomoso, Nigeria

B. S. Chauhan, K. Jabran (eds.), *Biology and Management of Weeds in Asia and Africa*, Sustainability Sciences in Asia and Africa,
https://doi.org/10.1007/978-981-95-7314-1_7

7.1 Introduction

Nigeria is located in West Africa between latitudes 4° and 14° N, and longitudes 2° and 15° E, with a land area of about 923.768 km^2 and a coastline of 853 km. This is the most populous country in Africa and the sixth largest in the world (>200 million) with an annual population growth rate of 2.4% (World Bank 2024). Nigeria's population is expected to reach 377 million by 2050, thus becoming the world's third most populous country after India and China (UDoEaSAP 2019). Hence, the Nigerian agricultural production system faces the challenge of feeding a fast-growing population. The current agricultural production in Nigeria is inadequate to feed the projected population, and meeting this need is considered a major challenge (Olowe 2021). Out of Nigeria's total arable land of 71 million ha, only 56% (40 million ha) is currently under cultivation, whereas estimates show that about 79 million ha of crop production is required to meet the needs of Nigeria's growing population (Anonymous 2019). The burden of meeting this need is further exacerbated by the threat of climate change and variable weather conditions. Less than 5% of the arable land in Nigeria is under irrigation despite abundant freshwater resources, including two of the largest rivers in Africa, namely, Rivers Niger and Benue (Adelodun and Choi 2018).

Agriculture remains a vital sector of the Nigerian economy as it provides employment and livelihood for 70% of the population, contributing between 22% and 25% of the country's gross domestic product (FAO 2023). Nigeria has a tropical climate, with a rainy season mostly from April to October, and a dry season from November to March. The country also experiences a wide range of rainfall, varying from above 2000 mm in the South to less than 500 mm in the North (Shiru et al. 2020). The climate allows for the cultivation of various crops such as cowpea (*Vigna unguiculata* (L.) Walp.), maize (*Zea mays* L.), pearl millet (*Pennisetum glaucum* L.), rice (*Oryza sativa* L.), and sorghum (*Sorghum bicolor* (L.) Moench)—which are cultivated more in the Northern region, while roots and tubers such as cassava (*Manihot esculenta* L.), yam (*Dioscorea alata* L.), and cocoyam (*Colocasia esculenta* L.), and tree crops such as cocoa (*Theobroma cacao* L.) and oil palm (*Elaeis guineensis* L.) are prominent in the Southern region. Additionally, rice and maize are also important crops grown in the Southern region (Adelodun and Choi 2018; Shiru et al. 2020). According to a survey conducted in 2019, maize, cassava, sorghum, yam, and cowpea are the five major crops among farming households in Nigeria (Fig. 7.1), usually in smallholder farm sizes ranging between 0.5 and 1 ha (Statista 2019).

Nigeria is the world's largest producer of cassava, cowpea, and yam, the second largest producer of okra (*Abelmoschus esculentus* L.), the third largest producer of peanut (*Arachis hypogaea* L.), the fourth largest producer of cocoa, and the fifth largest producer of oil palm. It is also Africa's largest producer of rice and the second largest producer of maize and tomato (*Solanum lycopersicum* L.). Nevertheless, the productivity and yields of these crops are far below the world average (Table 7.1), and the yields for most of the crops are only about one-third or one-half of the yields

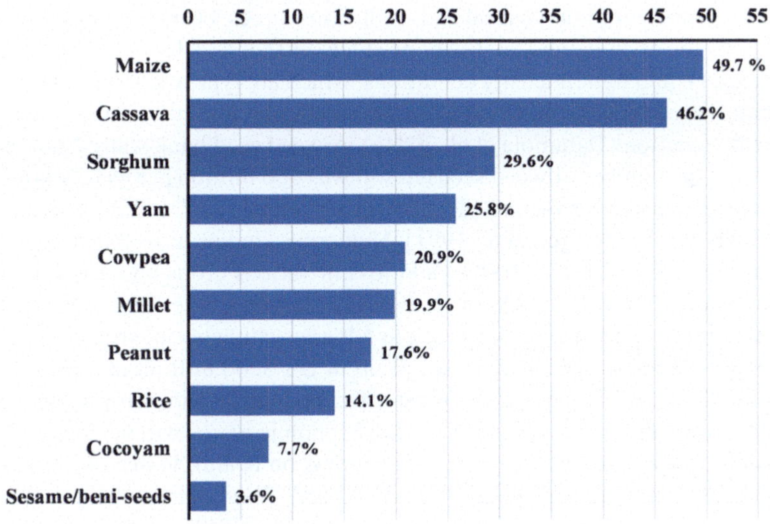

Fig. 7.1 Ten major crops among households in Nigeria. (Data source: Statista 2019)

Table 7.1 Production and yields of major crops in Nigeria

Crop	Area (million ha)	Production (million tons)	Percent of global production (%)	Yield (t ha^{-1})
Cereals				
Maize	5.80	12.70	0.9	2.2
Rice	3.50	1.34	1	6.32
Legumes				
Cowpea	5.97	2.66	–	0.4
Peanut	3.4	4.23	9	1.3
Roots/tubers				–
Cassava	–	–	–	
Yam	–	–	–	–
Vegetables				
Okra	–	–	–	–
Tomato	–	–	–	–

Source: USDA (2023)
– Data not available

demonstrated in research studies (Gianessi 2010). Several biotic and abiotic factors, including pest infestations, decline in soil fertility, and erratic rainfall, in addition to poor agronomic management, are responsible for the low crop yields in Nigeria, but weeds are considered the most harmful biotic constraint (Gianessi 2010). Weeds are perceived by most smallholder farmers as the greatest yield-limiting constraint (Vissoh et al. 2004).

Weeds compete for both natural and applied resources, besides being responsible for reducing the quantity and quality of produce (Gianessi 2010). The high temperature, humidity, and light intensity conditions, which are typical of Nigeria's climate, favor rapid weed seed germination and excessive weed growth, resulting in significant yield reduction (Daramola et al. 2021a). A weed seed bank study from Nigeria showed that the number of weed seeds in soil was 200 million per ha (Chikoye et al. 1997), and weedy experimental plots produced between 17 and 30 tons of weed fresh weight per ha (Adigun et al. 1991). Numerous studies have shown that weeds can cause between 35% and 100% yield loss in various crops under weedy conditions in Nigeria (Table 7.2). Although there are limited studies on economic losses due to weed infestation in various crops in Nigeria, estimates in Africa showed that parasitic weeds alone can cause losses of up to US $200 million in rice and US$1 billion in maize, and Nigeria is the most affected country from crop losses caused by weeds (Rodenburg et al. 2016). The total economic losses will be greater if other weeds are included and indirect effects of weeds on health, losses of biodiversity, nutrient depletion, and crop quality are considered. Weed management is therefore essential to improve crop production in Nigeria and will play an important role in determining whether Nigeria meets the future food production requirements. In this chapter, we therefore provide a broad overview of the biology and management of weeds in Nigeria, briefly explain the major crops of Nigeria, the weeds associated with them, and the methods of their management. We also discuss the current and emerging challenges of weed management in Nigeria's cropping systems and opportunities to overcome these challenges to improve crop production, as well as an outlook on developments in weed science research in Nigeria. This chapter aims to compile existing literature and identify current research gaps and prospects for future research on sustainable weed management in Nigerian cropping systems.

Table 7.2 Yield losses due to weeds in major crops in Nigeria

Crop	Yield loss (%)	Sources
Cereals		
Maize	51–100	Akobundu and Ekeleme (2000)
Millet	46–57	Choudhary and Lagoke (1981)
Rice	48–100	Kolo et al. (2021)
Sorghum	40–80	Lagoke et al. (1990)
Legumes		
Cowpea	64–78	Adigun et al. (2014)
Peanut	75–85	Ademabayoje et al. (2020)
Roots/tubers		
Cassava	25–100	Hauser (2017)
Yam	76–78	Unamma and Akobundu (1989)
Vegetables		
Okra	51–59	Daramola et al. (2020a)
Pepper	55–78	Adigun et al. (2018a) and Daramola et al. (2021b)
Tomato	59–76	Daramola et al. (2021c)

7.2 Biology of Prominent Arable Weeds in Nigeria

Small-scale production systems in sub-Saharan African countries are highly heterogeneous. Smallholder farmers often cultivate crops in mixtures to optimize limited land resources. During the last three decades, the old traditional "fallow" practice has diminished due to intensive farming. "Fallowing" is a well-known strategy to improve soil structure and maintain low weed pressure over the years. Smallholder farmers now engage in continuous cropping patterns due to increased market demands for staple foods. Factors such as intensive and mixed cropping, crop rotation, and soil disturbance have caused shifts in weed communities in Nigeria. Hence, the weed flora in Nigeria's arable fields varies across different agroecological conditions (Ekeleme et al. 2019). Under intensive land use in the savanna, predominant weed genera such as *Commelina, Digitaria, Cynodon,* and *Striga hermonthica* (Del) Benth. continue to ravage cereal-based cultivated fields (Jafun and Abdul 2004). Environmental variables such as soil pH, texture, organic carbon, ion exchange, rainfall, altitude, and temperature explain 34–39% of the total variation in weed species composition across different agroecological conditions (Chikoye and Ekeleme 2001; Ekeleme et al. 2019), whereas other factors such as cropping history, weed control practices, tillage, and related land use systems play a major role in the distribution of weed species (Ekeleme et al. 2019).

7.2.1 Seedbank Characteristics

In Nigeria, high weed pressure has become a major menace in continuous cropping systems, leading to increased cost of weeding with characterized yield reduction in low-resource arable fields. In the rainforest agroecology, weed seed population in continuous cropping systems was estimated between 18,555 and 19,725 seeds m^{-2} compared to 838 and 4697 seeds m^{-2} in fallowed fields (Ukwoma-Eke and Ekeleme 2024). Consequent to intensive land use, persistent, problematic annuals have evolved and adapted to low-resource farming systems. For example, perennial and annual grasses such as *Digitaria horizontalis* Willd., *Eleusine indica* Gaertn, *Pueraria phaseoloides* (Roxb.) Benth., and *Cynodon dactylon* (L.) Pers. have become more abundant within the seed bank of a continuously cultivated maize/cassava cropping system in the rainforest agroecology (Ukwoma-Eke and Ekeleme 2024). Irrespective of agroecology, seeds of *Ageratum conyzoides* (L.) L., sedges, and *D. horizontalis* were reported to be dominant species in the soil seedbank (Chikoye and Ekeleme 2001). Sedges are characterized to generate large seedlings through seeds and rhizomes with little or no seed dormancy (Chikoye and Ekeleme 2001)—this explains their frequent persistence and high density across all agroecologies in Nigeria.

7.2.2 Weed Seed Germination

Most weed seeds in the tropics require optimal light and moisture for germination (Ukwoma-Eke and Ekeleme 2024). In tropical regions, most viable weed seeds remain dormant within the soil during long drought (dry season) due to a lack of rainfall. Seeds naturally emerge at the beginning of wet seasons under optimal light conditions (Campbell and Grice 2000). The wet season also coincides with the emergence periods of most arable crops. Notwithstanding, certain problem weed seeds could deviate from this rule. More studies are needed to understand the dynamics of weed seed germination under different situations in the tropics. According to Takim and Fadayomi (2013), the period between 3 and 10 weeks after planting was observed as the peak period of weed seedling emergence for maize-cowpea intercrop and sole maize or cowpea in the Southern Guinea savanna agroecology.

7.2.3 Weed Seed Dispersal

Depending on seed characteristics, weed seeds are often transported to new areas through wind, flood, human, or animal interventions (Campbell and Grice 2000). Under a changing climatic scenario in Nigeria, increased flooding, application of animal manure for soil improvement, and migration of cattle herders (in search of pastures) into wet arable areas during the dry season are potential means of weed seed spread into low-resource farmlands. For instance, heavy infestation of *Mimosa invisa* (Mart.) and *Sacciolepis africana* (C.E.Hubb. & Snowden) has been observed along cattle routes and flooded wetlands in a rainforest agroecology of Nigeria (Dania Ogbe and Bamidele 2006; Alagbo and Akinyemiju 2018). Farmers claimed that the recent invasion of *M. invisa* (Dania Ogbe and Bamidele 2006) and *S. africana* (Oyebanji Alagbo—personal communication) could be linked to the movement of cattle and heavy flooding in wetland areas, as these weeds were not observed in the area 2–3 decades ago. Also, animal manures commonly used as fertilizer in the northern region can equally harbor viable weed seeds. Understanding what weed seeds can survive passage or become viable through animals' digestive tracts is important to establish the risk of weed invasion. In addition, seedbank studies on weed seeds within flood debris are required to understand potential weed seed spread along flood right-of-ways across different agroecologies in Nigeria.

7.2.4 Soil Characteristics

The flora dynamics of arable weed communities in cereal-based cropping systems have been associated with soil properties such as organic carbon content and soil nitrogen levels in the northern savanna ecology of Nigeria (Weber et al. 1995). For example, a high incidence of *Eleusine* species was associated with low organic carbon content and cation exchange capacity (Weber et al. 1995). In the derived savanna and humid rainforest, the prevalence of *Chromolaeta odorata* (L) R.M. Kings & Robinson and *Euphorbia heterophylla* L. was associated with soils with high pH and available phosphorus (P), whereas *A. conyzoides*, was associated with high clay and silt content (Chikoye and Ekeleme 2001). Soil properties such as magnesium, nitrogen, soil texture, and pH levels have helped predict the spread of *S. hermonthica*—in addition to other climatic variables such as rainfall, temperature, and latitude (Aliyu et al. 2023).

Generally, *Imperata cylindrica* (L.) Raeusch is found in low to moderate soil fertility areas. This makes *I. cylindrica* endemic in many low-resource arable fields in Nigeria. The infestation level of *I. cylindrica* in the tropics ranged from 9% to 79% depending on soil properties and soil moisture conditions. For example, available P explained 42% of this grass's severity variation (Chikoye et al. 1999). Moreover, *I. cylindrica* was confirmed to occur in high abundance in sandy soils, compared to low abundance observed in clay soils. It was opined that *I. cylindrica* growth on heavy clay soils could be restricted, as evidenced by the production of slender rhizomes and lack of node buds (Chikoye and Ekeleme 2001). Increased acidity was reported to break the dormancy of invasive species like *C. odorata* and *Tithonia diversifolia* (Hemsl.) A.Gray (Uwalaka 2023). The latter implies that areas with increased soil acidity could create a competitive advantage for invasive species (with a high affinity for acidic soil) against native weed seeds with similar requirements for germination. For example, over the years, *T. diversifolia* invasion has been restricted to the derived savanna agroecology in Nigeria, probably due to soil acidity (See Table 7.3).

In summary, the association between soil type and the distribution of arable weed communities in Nigeria suggests that the prevalence of certain weed species could help farmers predict soil properties in different localities and vice versa.

7.2.5 Crop-Weed Competition

Weed infestation during the early crop growth stages is a serious issue in Nigeria. Under continuous cropping systems, higher weed density often occurs at the early growth stages of the crop due to a shortage of labor and a lack of tractors for tillage operations during land preparation and the high cost of input, causing prolonged delays in weeding intervention by smallholder farmers. These challenges further cumulate into heavy weed biomass, resulting in significant crop yield losses (Takim 2012). Moreso, several studies in Nigeria have reported a decline in yield components such as plant height, number of leaves, and plant weight in arable crops due to high weed density and prolonged weed interference (Takim 2012; Adigun et al. 2014). Depending on crop morphology and planting pattern, yield losses due to weed competition range between 50% and 100% (see Table 7.2). Besides competition for growth resources, problematic weed species with heavy seed shed

Table 7.3 Prominent weeds of major arable crops and their agroecology in Nigeria

Prominent Weeds	Weed class	Life cycle	Major arable crops							Agroecology					‡Rankings
			Maize	Rice	Okra	Tomatoes	Cowpea	Peanut	Cassava	HR	DS	SGS	NGS	SS	
Euphorbia heterophylla L.	B	A	√	√	√	√	√	√	√	†	†	†	†	†	1
Digitaria horizontalis Willd.	G	A	√	√	√		√		√	†	†	†	†		2
Commelina benghalensis L.	B	A	√	√	√		√		√	†	†	†		†	2
Cynodon dactylon (L.) Pers.	G	P	√	√		√	√			†		†	†	†	3
Panicum maximum Jacq.	G	P	√	√			√		√	†	†	†	†	†	3
Ageratum conyzoides (L.) L.	B	A		√	√		√		√	†	†	†	†		3
Tridax procumbens (L.) L.	B	A	√		√	√				†	†	†			4
Amaranthus spinosus L.	B	A	√	√		√				†	†	†	†	†	4
Brachiaria deflexa (Schumach.) C.E.Hubb. ex Robyns	G	A	√					√	√		†	†		†	4
Rottboellia cochinchinensis (Lour.) Clayton	G	A	√	√					√	†		†	†	†	4
Eleusine indica Gaertn	G	A		√		√		√		†	†		†	†	4
Tithonia diversifolia (Hemsl.) A.Gray	B	P	√		√						†				5
Imperata cylindrica (L.) Raeusch.	G	P					√		√	†	†	†		†	5
Cyperus esculentus L.	S	P				√	√						†	†	5
Pueraria phaseoloides (Roxb.) Benth.	B	A				√			√	†					5
Striga hermonthica (Del.) Benth.		A	√	√											5
Sources			Alagbo et al. (2025)	Alagbo *et al.* (2022)	Olasantan (2001) Ayodele et al. (2024) and Daniya (2013)	Awodoyin et al. (2007) and Na-Allah et al. (2017)	Abaidoo and Fontem (2014) and Imoloame and Osunlola (2017)	Amosun and Aduramigha-Modupe (2016) Daniya (2013)	Chikoye Lum and Udensi (2010) Olorunmaiye et al. (2013) Ekeleme et al.(2019)						

HR Humid rainforest, *DS* Derived Savanna, *SGS* Southern Guinea Savanna, *NGS* Northern Guinea Savanna, *SS* Sudan Savanna, *B* Broadleaves, *G* Grasses, *A* Annual weeds, *P* Perennial weeds, † Weeds are present in agroecology, √ Weeds are prevalent in crop ‡ Ranking was based on frequency of occurrence in agroecology and prevalence in crops (annually) exhibit aggressive growth, capable of interrupting crop harvest, e.g., *Amaranthus spinosus* L., *Ageratum conyzoides*, *Commelina benghalensis* L., *Digitaria* spp., *E. heterophylla*, *Portulaca oleracea* L., and *Tridax. procumbens* L. (Awodoyin and Olubode 2009; Daniya et al. 2023). Prevalent weeds often result in poor market value of perishable crops like vegetables and contamination of harvested grains with weed seeds at harvest when left uncontrolled (Adeyemi et al. 2022).

7.2.6 Herbicide Resistance

Herbicide-resistant biotypes of arable weeds in Nigeria have evolved in other countries. For instance, *Rottboellia cochinchinensis* (Lour.) Clayton biotypes were reported to show resistance to fluazifop-P-butyl in Costa Rica (Castillo-Matamoros et al. 2016) and multiple-level resistance to haloxyfop-R-methyl and sethoxydim in Bolivia (Avila et al. 2007). *Euphorbia heterophylla* biotypes have shown cross-resistance to sulfonylurea and imidazolinone groups (Gelmini et al. 2005) and multiple resistance to fomesafen, imazamox, and glyphosate (Palma-Bautista et al. 2020). Glyphosate-resistant *E. indica* biotypes have been confirmed in Indonesia (Kurniadie et al. 2023). Moreso, *E. indica*–resistant biotypes to paraquat have equally evolved due to several years of repeated paraquat use (Hooda and Chauhan 2023). Also, *E. indica* resistant to oxadiazon—applied preemergence has been established (McElroy et al. 2017). In Nigeria, farmers and several field experiments have reported herbicide-tolerant weed biotypes—probably due to repeated application of similar herbicidal compounds in low-resource arable fields (Imoloame 2020). Unfortunately, there are no known herbicide resistance studies yet in Nigeria.

7.2.7 Weed Allelopathy

Prevalent arable weeds of Nigeria are among weeds reported to have some allelochemical compounds capable of inhibiting germination and growth of several plant species in the surrounding environment through the production of leachates and root exudates from active plants or the release of chemicals from decomposing residues (Erida et al. 2023). For example, *n*-hexane extract of *I. cylindrica* provided effective control of *A. spinosus* (Erida et al. 2023). *Ageratum conyzoides* exerts an allelopathic effect on rice by releasing water-soluble phytochemicals (Negi et al. 2020). Also, exudates of *A. conyzoides*, *Chromolaena odorata*, and *Cyperus esculentus* have been observed to inhibit the germination of tomato seeds in Nigeria Adeleke and Onyebuchi (2022). However, it was opined that the release of allelopathic properties of certain plant extracts may depend on timing and some environmental

Table 7.4 Allelopathic properties and mode of action of some prevalent arable weeds in Nigeria

Weeds	Allelopathic extracts	Mode of action	Sources
Imperata cylindica	Rhizome extracts (*n*-hexane, ethyl acetate and methanol)	Herbicidal control of *Amaranthus spinosus*	Erida et al. (2023)
Rottboellia cochinchinensis	Soil-incorporated powder	Growth inhibition of *B. pilosa, E. crus-galli* and *Lactuca sativa* under dry condition	Bundit et al. (2015)
Amaranthus spp.	Whole plant extract	Inhibition of vegetable and *Conyza bonariensis* (L) Cronquist germination	Prinsloo and Du Plooy (2018)
Ageratum conyzoides	Leaf root and inflorescence extracts	Inhibition of growth and germination of *Parthenium hysterophorus* L.	Javaid et al. (2020)
Tithonia diversifolia	A 10% aqueous leaf extract	Inhibition of weed growth	Ajayi et al. (2017)

factors. For instance, Bundit et al. (2015) established that the allelopathic activity of soil-incorporated *R. cochinchinensis* extract achieved better inhibition of *Bidens pilosa* L., *Echinochloa crus-galli* (L.) P.Beauv, and *Lactuca sativa* L. growth under dry conditions than wet. Table 7.4 shows the allelopathic properties and mode of action of some arable weeds in Nigeria.

7.3 Critical Period of Weed Control in Nigeria

In Nigeria, weeds are better controlled when crops and weeds are at early growth stages—given that seeding or transplanting is carried out within a short window after land preparation (before weed seedling emergence). Unfortunately, farmers' weeding schedules often lag behind the critical period of weed control due to a shortage of labor and the unavailability of herbicides. Timely weed control can be realized when smallholder farmers can access the required inputs ahead of time.

The critical period of weed control (CPWC) in some legumes (soybean and cowpea) or vegetables depends on crop cultivar and cultural practices (Choudhary and Lagoke 1981; Adigun et al. 2014; Daramola et al. 2021c). For example, cultivation of late-maturing cultivars or practice of wide-row spacing might require a longer weed-free period due to a delay in canopy closure. In root tubers like cassava, a weed-free period between 2 and 12 weeks after planting is necessary, as this period corresponds with early canopy closure and initiation of storage roots (Ekeleme et al. 2020) (Table 7.5).

Table 7.5 Critical period of weed control in selected arable crops

Crop	CPWC (weeks after sowing WAS)	Source
Soybean	Late-maturing cultivar 3–6 WAS with 50 and 75 cm spacing or 3–9 WAS with 100 cm row spacing	Daramola (2020)
Cowpea	3–8 WAS for 60 cm row spacing 3–12 WAS for 90 cm row spacing	Osipitan et al. (2016)
Groundnut	3–6 WAS	Osunleti et al. (2022)
Okra	3–6 WAS	Dada and Fayinminnu (2010)
Tomatoes	3–6 WAS	Adigun (2005)
Cassava/maize intercrop	4–8 WAS	Melifonwu (1994)
Cassava mono crop	4–12 WAS (note: 2–12 weeks is important for hoe weeding)	Melifonwu (1994) and Ekeleme et al. (2020)
Maize	3–6 WAS	Alagbo et al. (2025)
Upland rice	3–9 WAS	Kolo et al. (2020)

7.4 Weed Management Practices in Nigeria

7.4.1 Cultural Weed Control

7.4.1.1 Fallow, Crop Rotation, and Intercropping

Traditionally, smallholder farmers in Nigeria have enhanced soil productivity and minimized weed infestations by practicing crop rotation, intercropping, and allowing fallow periods of up to 10 years in shifting cultivation systems (Akobundu and Ekeleme 2002). The fallow period impacts the weed community by depleting and altering the composition of the weed seedbank. To facilitate the intensification of cropping systems, natural fallow is often replaced or complemented with managed weed-suppressing species of cover crops and legumes (Ekeleme et al. 2019). Both natural and planted fallows have been reported to reduce weed infestation by shading weeds surviving after crop harvest and by depleting the weed seedbank through loss of weed seed viability, predation, and non-recruitment of new seeds (Akobundu and Ekeleme 2002; Ekeleme et al. 2019). The weed seedbank was reduced by 35% after 6 years of natural fallow and by 53–86% after 6 years of planted herbaceous legume fallow such as *Pueraria phaseoloides* (Roxb.) Benth. and woody legumes, *Acacia auriculiformis* Benth., *Leucaena leucocephala* (Lam.) de Wit, and *Senna siamea* (Lam.) H.S.Irwin & Barneby, compared to the continuous maize/cassava system (Ekeleme et al. 2019).

Crop rotation and intercropping are also important cultural practices used by smallholder farmers in Nigeria to mitigate the weed problem. Weeds and other pests can quickly adapt to management practices when a single crop is grown in consecutive years, whereas the changes in management practices due to crop rotation or intercropping help to disrupt the buildup of crop-specific weeds (Akobundu and

Ekeleme 2002). Temporal diversification through crop rotation and spatial diversification through intercropping have been demonstrated to lower weed densities and biomass in Nigerian cropping systems (Vissoh et al. 2004; Falade et al. 2023). For instance, *Striga* infestation in maize was reduced by 35–76% after 1–2 years of cowpea-maize and soybean-maize rotations compared with continuous maize cultivation (Kureh et al. 2006). Similarly, the rotation of vegetables and arable crops with upland rice improved the management of herbicide-tolerant weeds in the lowland ecosystem compared with continuous upland rice cultivation (Gbanguba et al. 2011).

Intercropping suppresses weed growth by offering early ground cover, either through a high plant population or by incorporating a fast-growing component crop. Intercropping root crops like cassava with maize, vegetables (okra, pepper, maize, melon), or legumes (cowpea, groundnut, and soybean) has been shown to contribute to improved weed suppression compared to cassava monoculture (Olorunmaiye 2010). However, careful selection of crop species, planting date, and planting pattern is essential to prevent severe competition for resources between crops in intercropping systems (Olorunmaiye 2010).

7.4.1.2 Soil Fertility Management

Fertilizer management can influence both crop and weed growth. The rate of fertilizer applied, especially nitrogen, has been shown to affect weed infestation and crop performance in tomato (Adigun et al. 2018b), pepper (Adigun et al. 2018a), okra (Adigun et al. 2018c), upland rice (Kolo et al. 2020, 2021), and maize (Usman et al. 2001, 2002) in Nigerian farming systems. Most studies indicated that weeds have a lesser impact on well-fertilized crops than unfertilized crops due to more vigorous crop growth that enhances weed suppression. For instance, weed density was reduced by 11–27% and weed biomass by 20–23% in okra fertilized with NPK (20:10:10) and organic cassava peel compost compared with unfertilized crops. Contrarily, prevalent weed species such as *Cyperus rotundus* L., *Cyperus esculentus* L., *I. cylindrica*, and *Cynodon dactylon* were more competitive than vegetable crops such as tomato and okra at higher levels of fertilizer application (Olabode et al. 2010a; Kolo et al. 2021).

The timing of fertilizer application influences the competition dynamics between crops and weeds. Early fertilizer applications, particularly following preemergence herbicide use, have been found to improve the suppression of late-emerging weeds (Adigun et al. 2018b). In maize, applying 120 kg N, 60 kg P, and 60 kg K in two equal split doses—at planting and 6 weeks after planting—provided more effective weed control than applications made at 2 and 6 weeks after planting (Imoloame et al. 2021). Similarly, split applications of cattle dung at 4 and 8 weeks after planting resulted in better weed suppression compared to pre-planting applications in NERICA 2 and CG14 upland rice cultivars (Dada et al. 2017). However, in upland rice grown in Nigeria's tropical forest-savanna transition zone, nitrogen applied at

90 kg ha^{-1} in two equal splits at 3 and 6 weeks after planting increased weed density and biomass, and only benefited yields when it was combined with effective weed control (Daramola et al. 2020a).

7.4.1.3 Crop-Competitive Cultivars

Crop cultivars that suppress weeds, maintain high yields under weedy conditions, and adapt well to the local environmental conditions are of high importance to resource-poor farmers in Nigeria (Yusuf et al. 2019b). Crop cultivars exhibiting high seedling vigor, spreading growth habit, high specific leaf area and biomass accumulation, increased plant height, rapid growth rates, early canopy closure, and cereals with high tillering ability have a capacity to suppress weeds through enhanced competition for growth resources (Yusuf et al. 2019a, b; Garko et al. 2021). Table 7.6 shows selected crops and cultivars with weed-competitive attributes. Nonetheless, there have been limited studies assessing the weed-competitive abilities of crop varieties in Nigeria. In legume crops like cowpea, groundnut, and soybean, fast-growing cultivars with a spreading habit are generally more weed competitive than erect types (Futuless and Bake 2010). However, the response of vegetables like okra cultivars to weed competition in Nigeria has shown inconsistency (Makinde et al. 2020, 2021).

Generally, research indicates the potential of different crops to develop weed-competitive cultivars; however, there may be trade-offs, as traits that enhance weed competitiveness could negatively impact yield. For instance, while African rice (*Oryza glaberrima*) is competitive with weeds, it has lower yield potential due to lodging and grain shattering (Ekeleme et al. 2009). Additionally, the lack of herbicide-tolerant crop cultivars that allow postemergence application has limited the use of herbicides for weed control in Nigeria's cropping system.

Table 7.6 Some weed-competitive cultivars and their characteristics in Nigeria

Crops	Cultivars	Characteristics	Sources
Rice	OS6	Increased plant height	Akobundu and Ahissou (1985)
Cassava	TMS 30572	Larger canopy diameter	Ola et al. (2021)
Cowpea	SAMPEA 16	Increased plant height, produced more root and shoot biomass	Yusuf et al. (2019a, b)
Groundnut	SAMNUT 22, SAMNUT 23, SAMNUT 24, and Angba-Chido	More branching and rapid canopy closure	Salifu (2016)
Soybean	TGX 1835-10E	–	Ene et al. (2019)

7.4.1.4 Row Spacing and Plant Population

In Nigeria, cultural practices that minimize weed interference, weed seedbank buildup, seedling recruitment, and seed production are crucial for weed management because farmers have limited access to herbicides. Crop competitiveness against weeds can be enhanced by increasing crop population densities. High planting densities and narrow row spacing often lead to fast canopy closure (Adigun et al. 2017a). The use of high plant population and narrow row spacing to enhance rapid canopy closure and greater light interception has been observed to improve weed suppression and increase yield in various crops in Nigeria's cropping systems including cowpea (Adigun et al. 2014, 2017a), peanut (Adigun et al. 2017b; Ojelade et al. 2018), soybean (Daramola et al. 2020a, 2021a), rice (Orluchukwu and Molokwu 2023), maize (Udensi and Omovbude 2018), okra (Osadebe et al. 2022), pepper (Mustapha et al. 2021), and tomato (Adigun et al. 1994). For example, reducing row spacing from 75 cm to 50 cm reduced weed density and biomass in cowpeas, peanuts, and soybeans (Amosun and Aduramigha-Modupe 2016; Daniya 2013; Na-Allah et al. 2017). Similarly, a slight decrease in intra-row spacing from 20 cm to 15 cm resulted in a 35% reduction in weed density in peanuts (Adigun et al. 2017b). Makinde et al. (2022) found that a spatial arrangement of 100 × 25 cm in okra reduced weed density and biomass compared to arrangements of 100 × 50 cm or 100 × 75 cm due to rapid canopy closure.

Nonetheless, reducing row or plant spacing to enhance weed suppression often leads to a higher seeding rate, which can increase production costs. However, research has shown that the weed control benefits associated with narrow spacing often result in increased crop yield and higher net returns (Daramola et al. 2019, 2021a). In addition to improved weed suppression, narrow row spacing has been found to reduce the need for herbicide use in weed management when compared to wider spacing in crops such as soybean (Daramola et al. 2020a) and cowpea (Osipitan et al. 2013). Despite these benefits, most studies indicate that while narrow spacing and higher seeding rates improve weed suppression, they do not eliminate the need for herbicides to protect crop yield (Adigun et al. 2017b). Additionally, not all weed species respond similarly to changes in row spacing and seeding rates. For instance, in soybeans, a narrow row spacing of 50 cm significantly reduced the density and biomass of grass weed species such as *D. horizontalis*, *P. maximum*, *C. dactylon*, and *Paspalum scrobiculatum* compared to 75 cm and 100 cm spacing, while broadleaf weed species like *C. benghalensis*, *Gomphrena celosioides*, and *A. spinosus* were unaffected (Daramola et al. 2021a). This can be attributed to the fact that broadleaf weeds tend to be more shade-tolerant than grass weed species (Rowan and Athena 2006).

7.4.2 *Mechanical Weed Control*

7.4.2.1 Tillage and Land Preparation

In a typical tropical climate, weeds in fallow fields could develop rapidly up to 2 m high due to higher temperature, humidity, and rainfall. Although these fallow weeds are beneficial as soil cover, they often result in heavy seed rain and biomass on soil surfaces, creating difficulties during land preparation. On small-scale farms, land preparation is typically done manually using hand hoes, whereas larger farms commonly employ mechanized methods such as disc plows or harrows. Traditionally, the weeds are slashed manually with a cutlass and burnt to enhance the ease of manual tillage operations (Fasinmirin and Reichert 2011).

Mechanical weed control includes land preparation before planting, involving tillage and leveling and in-crop interventions for weed removal. Surface weed seeds and heavy biomass are buried mechanically with one or two deep plowing operations, depending on the competitive potentials of dominant fallow weed species with crops at early growth stages (Akobundu and Ekeleme 2002). Due to limited mechanization in Nigeria, smallholder farmers have consistently practiced zero or minimum tillage for many years (Ajah 2014). Such no-tillage practice involves preplant treatment with herbicides to suppress fallow weeds, followed by direct seeding of crops, leaving the dead biomass as cover for weed suppression (Fasinmirin and Reichert 2011). Unfortunately, no-till arable fields tend to accumulate a significant amount of weed seeds in the topsoil over time, resulting in the prevalence of dominant weed species with low initial dormancy and high light requirement for germination (Chauhan et al. 2006; Chauhan and Johnson 2008). In contrast to no-till, minimum tillage practices involving hoeing and ridging or conventional tillage (plowing and harrowing) have been shown to reduce weed infestation by burying weed seeds that are present on the soil surface (Kayode and Ademiluyi 2004; Nwagwu et al. 2022).

To achieve optimal crop yield benefits in Nigeria, proper tillage operation must be followed with timely seeding to retard early weed infestation. Importantly, considering the risk of heavy weed seed rain on soil surfaces as a result of intensive practice of no-till, routine breaks of no-till practice with minimum or deep plowing (at least once in 3 years) are important to enhance the burial of surface weed seeds (Nwagwu et al. 2022). However, there is limited research on how different tillage practices impact weed seed dynamics and weed control in Nigeria's cropping systems.

7.4.2.2 Hand Weeding Practices

Hand weeding practices, including hand pulling, hand slashing with a cutlass, and hand hoeing, are the most predominant methods of weed control on small-scale crop farms in Nigeria, yet this method is inefficient, time-consuming, expensive, labor-intensive (Adigun et al. 2018a), and requires 250–780 man h ha^{-1} (Akobundu

1987). In addition, the availability of labor for weeding is often uncertain, particularly during the critical period of weed control, resulting in delayed weeding, or weeding after the crops have suffered irreversible damage from weed competition (Ojelade et al. 2018; Daramola et al. 2020a). On subsistence farms, weeding is primarily done by women from the farm household, with children often participating as well. In contrast, larger farms may hire external labor for hand weeding, which can result in higher costs compared to herbicide use (Akobundu 1987). When adequate labor is available, hand weeding can effectively control shallow-rooted annual weeds and prevent seed production. However, deep-rooted perennial weeds like *I. cylindrica* and *Oryza longistaminata* A.Chev.& Roehr. (Akobundu 1987; Hamma and Ibrahim 2013) are difficult to control, as these are capable of rapid regrowth from rhizomes.

Research has demonstrated that two to three timely hand weedings within the first 60–90 days after sowing are essential for maximizing yields in crops such as rice (Kolo et al. 2021), soybean (Daramola et al. 2020a), cowpea (Adigun et al. 2014), peanut (Adigun et al. 2017b), tomato (Adigun et al. 1994, 2018b), pepper (Adigun et al. 2018a), and okra (Daramola et al. 2021c) in Nigeria cropping systems. The effectiveness of these weedings depends on the level of weed infestation and the dominant weed species present. In soybeans, Daramola et al. (2021a) found that early-season hand weeding effectively eliminated weed species like *Commelina benghalensis* and *Gomphrena celosioides*. However, controlling late-emerging weeds such as *Axonopus compressus*, *C. dactylon*, *D. horizontalis*, *E. indica*, and *P. scrobiculatum* required a longer period of hand weeding, extending up to 9 weeks after sowing. Nevertheless, hand weeding during the reproductive stage of crop growth can result in yield loss in cowpea and soybean due to mechanical damage (Adigun et al. 2014; Daramola et al. 2020a).

7.4.3 Chemical Weed Control

The use of herbicides is an important labor-saving alternative to hand weeding. Farmers in Nigeria use various herbicides for weed control depending on crop selectivity, weed community composition, environmental conditions, and economic benefits (Lagoke et al. 2014; Adeyemi et al. 2021). While significant research has been conducted on herbicide efficacy in Nigeria, the adoption of herbicide technology by smallholder farmers remains low due to a lack of effective extension linkages to disseminate herbicide knowledge (Akinrinola and Fagbola 2021). To achieve effective and safe use of herbicides, appropriate application rates and timing are important factors to be considered in relation to the crop type, weed growth stage, and weather conditions (Daramola et al. 2021a). Moreover, effective and safe use of herbicides requires application skills, technical know-how, and additional costs to growers (including the cost of herbicides, application equipment, and protective kits). These requirements are often not met by smallholder farmers in Nigeria due to high levels of illiteracy, inadequate knowledge of proper herbicide use, lack of financial means to purchase products and protection equipment, and limited access to information (Daramola et al. 2021a). As a result, herbicide selection, application rates, and timing frequently deviate from label recommendations, leading to inefficient weed control, crop injury, yield reductions, and environmental contamination (Akinrinola and Fagbola 2021). Additionally, Nigerian farmers have limited access to markets, which are often characterized by a limited variety of products and inconsistent supplies. The incorrect use of herbicides, caused by the aforementioned problems, may accelerate the evolution of herbicide resistance in weeds.

Subject to proper application, research has shown that herbicides can effectively suppress weeds, increase yield, and offer economic advantages by decreasing the time required for weeding, enabling farmers to adopt more efficient time and labor-saving practices (Daramola et al. 2019, 2021a). While the effectiveness of herbicides could vary (depending on weed flora and the pattern of weed emergence), the majority of these studies indicate that preemergence herbicide application was profitable than two hand weedings (Daramola et al. 2019, 2020a, 2021b, c). Surveys conducted across different agroecological zones in Nigeria also showed that most farmers using herbicides observed improved weed control, higher crop yields, and reduced labor and production costs (Akinrinola and Fagbola 2021).

Moreover, the negative impact of herbicides on non-target organisms has been encountered (Akinrinola and Fagbola 2021). Herbicide-induced reduction in metabolism of nodule bacteria, nodulation, and N-fixation has been reported in legume crops in Nigeria, including soybean and cowpea (Olabode et al. 2010a, b; Abaidoo and Fontem 2014). For example, imazaquin at 0.1–0.4 kg ai ha^{-1} and pendimethalin at 1.0–8.0 kg ai ha^{-1} reduced nodulation, nitrogen fixation, vesicular-arbuscular mycorrhizal fungi colonization, and species diversity and richness in cowpea and soybean (Abaidoo and Fontem 2014). Similarly, the colonization of arbuscular mycorrhizal fungi and the population of soil macro-fauna, such as earthworms and bacteria, were decreased with the increasing rates of atrazine in maize (Akinrinola and Fagbola 2021).

Commonly used herbicides in Nigerian cropping systems are presented in Table 7.6. Preemergence control of annual grasses and small-seeded broadleaf weeds in Nigeria is achieved mainly with very long-chain fatty-acid inhibitors from the chloroacetamide chemical family (e.g., alachlor, butachlor, metolachlor, pretilachlor, and *S*-metolachlor) and mitosis inhibitors from the dinitroaniline chemical family (e.g., pendimethalin and trifluralin). Pre-plant weed control in Nigeria is achieved mainly with paraquat and glyphosate. Their broad-spectrum activity, affordability, and widespread acceptance have made them popular choices as pre-plant herbicides, particularly in no-till production systems. In legume crops like soybean, postemergence graminicides such as clethodim and fluazifop have been studied for controlling grass weeds (Alonge 2000), while acifluorfen and bentazon have been tested for managing broadleaf weeds (Aluko et al. 2023). Postemergence application of acifluorfen and bentazon at 4 weeks after planting provided effective weed control and achieved yields comparable to two hand weedings, with good crop tolerance in the soybean (Aluko et al. 2023). While these postemergence herbicides provide effective weed control, they are rarely used by farmers (Table 7.7).

Table 7.7 Common herbicides used in Nigerian cropping systems, application rates, and timing

Herbicides	Rates (kg a.i. ha^{-1})	Timing	Reference
Soybean			
Butachlor	1.5–2.0	PRE	Daramola et al. (2020a, 2021a)
Metolachlor	1.5–2.5	PRE	Aluko et al. (2023)
S-metolachlor	0.6–1.9	PRE	Ezebuiro et al. (2021)
Metolachlor + prometryn	1.5–2.0	PRE	Ojelade et al. (2018)
Clethodim	0.04–0.1	POST	Aluko et al. (2023)
Oxyfluorfen	0.9–1.4	POST	Aluko et al. (2023)
Quizalofop	0.01–0.2		Aluko et al. (2023)
Pendimethalin	1.0–2.5	PRE	Abaidoo and Fontem (2014)
Trifluralin	0.5–0.8	PRE	Aluko et al. (2023)
Diuron	0.5–1.5	PRE	Imoloame (2014)
Imazethapyr	0.10	PRE, POST	Omisore and Olofintoye (2008)
Fluazifop	0.25–3.0	POST	Omisore and Olofintoye (2008) and Fayinminnu et al. (2010)
Imazaquin	0.05–0.72	PRE, POST	Alonge (2000) and Abaidoo and Fontem (2014)
Acifluorfen	0.2–04	POST	Fayinminnu et al. (2010)
Bentazon	0.5–1.0	POST	Fayinminnu et al. (2010)
Oflinuron	2.0–2.5	POST	Agahiu (2020)
2,4-D	2.0–3.0	POST	Agahiu (2020)
Linuron	2.0–2.5	POST	Agahiu (2020)
Peanut			
Acetochlor	1.6–2.4	PRE	Eni et al. (2021)
Butachlor	2.0–2.5	PRE	Ademabayoje et al. (2020)
Metolachlor	1.5–2.5	PRE	Ademabayoje et al. (2020)
Prometryn	2.0	PRE	Adigun et al. (2016)
Metolachlor + prometryn	1.5–2.4	PRE	Ademabayoje et al. (2020) and Adigun et al. (2017b)
Pendimethalin	1.5–2.0	PRE	Olayinka and Etejere (2015) and Lawal et al. (2023)
Imazethapyr	0.10	PRE	Lawal et al. (2023)
Cowpea			
Pendimethalin	1.5–2.0	PRE	Abaidoo and Fontem (2014)
Atrazine	2.0		Ugbe et al. (2016)
S-metolachor	2.0		Ugbe et al. (2016)
Imazaquin	0.05–0.4		Abaidoo and Fontem (2014)
Haloxyfop	0.05–0.6	PRE	Ibrahim et al. (2024)
Diuron	0.6	PRE	Imoloame and Osunlola (2017)
Metolachlor	1.5	PRE	Adigun et al. (2020)

(continued)

Table 7.7 (continued)

Herbicides	Rates (kg a.i. ha^{-1})	Timing	Reference
S-metolachlor	0.6–1.9	PRE	Nzeako et al. (2019)
Metolachlor + prometryn	1.2–2.0	PRE	Adigun et al. (2017a, b)
Metolachlor + metobromuron	2.0–3.0	PRE	Ugbe et al. (2016)
Tomato			
Butachlor	1.5	PRE	Yusuf et al. (2019a, b)
Metolachlor	1.0–3.0	PRE	Yusuf et al. (2019a, b)
Pendimethalin	1.0	PRE	Oladejo et al. (2021)
Prometryn	1.5	PRE	Yusuf et al. (2019a, b)
Pepper			
Butachlor	1.0–2.0	PRE	Na-Allah et al. (2017)
Diphenamid	1.5–3.0	PRE	Adigun et al. (1991)
Pendimethalin	1.0–2.0	PRE	Mustapha et al. (2021)
Prometryn	1.0–2.0	PRE	Adigun et al. (1991) and Daramola et al. (2021b)
Metribuzin	1.0–1.5	PRE	Adigun et al. (1991)
Metolachlor	0.25–2.0	PRE	Adigun et al. (1991) and Daramola et al. (2021c)
Metobromuron	1.25	PRE	Adigun et al. (1991) and Daramola et al. (2021c)
Okra			
Butachlor	1.0–2.0	PRE	Kolo et al. (2021) and Obiazi et al. (2024)
Pendimethalin	0.33–2.0	PRE	Obiazi et al. (2024)
Metolachlor	1.0–2.5	PRE	Daramola et al. (2020a)
Prometryn	1.0–2.0	PRE	Daramola et al. (2020a)

PRE Preemergence, *POST* Postemergence

7.5 Parasitic Weeds and Their Management in Nigeria

Parasitic weeds, especially witchweed (*Striga* spp.) and *Alectra vogelii* Benth., pose a significant challenge to agricultural productivity in savanna arable fields of Nigeria, with yield losses ranging from 10% to 98% (Olusoji 2012; Alagbo et al. 2023). Despite various control efforts, heavy infestation of parasitic weeds still ravages the savanna arable fields due to poor management practices (Alagbo et al. 2023). Such infestation significantly reduces crop quality and quantity—leading to complete crop loss in extreme cases.

Chemical control measures such as seed coatings with glyphosate, imidazolinones, and sulfonylureas have proven effective in controlling parasitic weeds like *Striga*. Soil-incorporated herbicides also reduce seed germination and attachment for weeds like *Orobanche* spp. (Kabambe et al. 2008; Chikoye et al. 2010a, b). Non-chemical control measures such as breeding for resistance, biological control, and cultural practices like hand pulling and trap cropping have been attempted in Nigeria (Chikoye et al. 2010b; Aly et al. 2014; Kebede and Bogale 2018). Among the later, host plant resistance and rotations with trap crops are considered the most cost-effective approaches for smallholder farmers. Suitable cultivars of cotton, soybean, peanut, and some legume cover crops (Emechebe 1991; Isah and Lagoke 2010) have shown promise in inducing suicidal germination of *Striga* seeds, preventing their attachment to main crops. Application of nitrogen fertilizer in low-fertility soils has also been proven effective in reducing *Striga* infestation (Adagba et al. 2002, 2003), while the use of natural products (allelopathy) to inhibit seed germination and growth has some potential (Aly et al. 2014; Kebede and Bogale 2018). Research programs that engage farmers' inclusivity are needed to integrate research findings with the indigenous knowledge for parasitic weed control in Nigeria.

7.6 Precision Weed Management Technologies

Precision weed management is an important component of precision agriculture with a special focus on providing alternatives or reducing herbicide loads in the environment. These technologies combine computer vision and decision support algorithms to make precise and accurate weeding applications (Gerhards 2019). Future opportunities for precision weed management in Nigeria include the deployment of artificial intelligence (AI) for weed-crop detection, and the development of smart spraying systems to reduce excessive herbicide use.

7.6.1 Artificial Intelligence (AI) for Weed-Crop Detection

Considering the specificity of weed communities in different agroecologies and cropping systems, creating a knowledge base for weed identification becomes necessary to enhance farmers' decisions on site-specific weed management. In Nigeria, depending on specific tasks, neural networks and other machine learning models have been optimized, with precise accuracy for plant detection and spraying applications in real time (Olaniyi et al. 2020, 2021; Nnadozie et al. 2023). In cassava, Nnadozie et al. (2023) deployed YOLOv5 convolutional neural network models for real-time detection of cassava and weeds using large image datasets from UAV imagery. Two versions of YOLOv5 (YOLOv5n and YOLOv5s) were trained with 1700 cassava images and further evaluated and deployed for cassava detection (with UAV imagery) at different growth stages, weed interference, and lighting scenarios. The result showed that YOLOv5s yielded the best accuracy, with fewer computer memory constraints. In practical terms, YOLOv5s would be more precise for cassava counting, especially where computing memory and storage are major constraints (Nnadozie et al. 2023). Such smart counting estimations could help identify cassava selectivity to different herbicide programs in varying agroecologies using UAV imagery. Additionally, with robotic automation, such information could be useful for mapping spot spraying of cassava fields (excluding the crop detected).

Olaniyi et al. (2023) deployed a machine learning approach for intelligent maize and weed recognition using acquired weeds and maize image datasets from 18 arable fields in southern savanna agroecology. These data are necessary to model decision support algorithms for the site-specific application of herbicides in real time.

7.6.2 Smart Spraying Systems

In addressing the challenge of suboptimal or over-usage of herbicides by smallholder farmers in Nigeria, studies on smart weeding technologies have emerged. For example, decision algorithms such as fuzzy logic (Olaniyi et al. 2020) and a single-shot multi-box detector (SSD) network models (Olaniyi et al. 2021) have proven high precision and accuracy in detecting weeds for selective herbicide spraying in lowland rice. Also, with the recent emergence of Nigerian agricultural drone companies (e.g., the Nigerian flying labs company https://flyinglabs.org/nigeria and Integrated aerial precision https://www.linkedin.com/in/femiadekoya), some commercial farms have embraced drone technology for smart spraying operations—in place of knapsack spraying applicators.

7.7 Conclusions

Considering the intensive land use for agricultural production in Nigeria, integrated weed management approaches may offer long-term solutions in addressing prevalent weeds threatening sustainable food production in low-resource arable fields. Successful weed management programs must be designed to target problem weeds in different agroecologies. For example, applying herbicide mixtures or temporal rotation with different herbicide compounds is necessary to combat herbicide-resistance problems. Use of smart sprayers is recommended to have better spray application and hence save on herbicides. Since herbicides cannot provide sufficient weed control in all situations, herbicide programs should be supported with cultural and deep tillage practices. For instance, adopting a spatiotemporal model of intercropping cum seasonal rotation of different crops has great potential to reduce weed seed population and disrupt the emergence of problematic weed species associated with a particular cropping system. The practice of minimum or deep tillage (at least once in 3 years) could help promote the burial of accumulated weed seeds on soil surfaces.

7.8 Future Outlook

More studies are needed to evaluate postemergence herbicides in Nigeria, as previous studies focused more on preemergence herbicides. Also, there are no known herbicide-resistant studies yet in Nigeria; hence, it is necessary to find out if herbicide-resistant weed biotypes exist in Nigeria. Long-term studies on the effects of tillage practices on weed communities and crop productivity across various agro-ecological zones are needed. Such knowledge could help guide farmers in developing management strategies for problematic weed species associated with specific tillage methods. More studies are needed to explore the weed-competitive potential of crop varieties in Nigeria.

References

Abaidoo DR, Fontem LA (2014) Response of weeds and soil microorganisms to imazaquin and pendimethalin in cowpea and soybean. Crop Prot 65:168–172

Adagba MA, Lagoke STO, Imolehin ED (2002) Nitrogen effect on the incidence of *Striga hermonthica* (Del.) Benth in upland rice. Acta Agron Hung 50(2):145–150. https://doi.org/10.1556/AAgr.50.2002.2.4

Adagba MA, Lagoke STO, Ndahi WB (2003) Effect of nitrogen fertilizer on the response of resistant and susceptible upland rice varieties to Striga hermonthica infestation. Niger J Weed Sci 16:47–54

Adelodun B, Choi KS (2018) A review of the evaluation of irrigation practice in Nigeria: past, present and future prospects. Afr J Agric Res 13(40):2087–2097

Ademabayoje O, Adigun J, Adeyemi O, Daramola O, Ajiboye G (2020) Efficacy and economics of integrated weed management in groundnut (Arachis hypogea L.). Acta Agric Slov 116(1):137–144

Adeyemi OR, Eni E, Lagoke TO, Ebenezer O (2021) Enhancing weed control efficacy of two pre-emergence herbicides in maize/jackbeans and maize/groundnut intercrop in derived savanna agro-ecology. J Agric Sci Belgrade 66(4):335–350

Adeyemi OR, Bashiruddin AA, Adigun JA, Adejuyigbe CO, Osunleti SO (2022) Fruit quality and marketability of Okra (Abelmoschus esculentus (L.) Moench) as influenced by biochar rates and weeding regime. Int J Pest Manag 25:1–9

Adigun JA (2005) Critical period of weed interference in rainfed and irrigated tomatoes in the Nigerian savanna. Agric Trop Subtrop 38(2):73–80

Adigun JA, Lagoke STO, Karikari SK (1991) Chemical weed control in irrigated sweet pepper (*Capsicum annuum* L.). Int J Pest Manag 37(2):155–158

Adigun JA, Lagoke ST, Kumar V, Erinle ID (1994) Effect of intra-row spacing, nitrogen level and period of weed interference on growth and yield of transplanted tomato (*Lycopersicom esculentum* Mill) in the Nigerian savanna. Samaru J Agric Res 11:31–42

Adigun J, Osipitan AO, Lagoke ST, Adeyemi RO, Afolami SO (2014) Growth and yield performance of cowpea (Vigna unguiculata (L.) Walp) as influenced by row-spacing and period of weed interference in South-West Nigeria. J Agric Sci 6(4):188

Adigun JA, Adeyemi OR, Olorunmaiye P, Daramola O, Babatunde A (2016) Influence of inter-row spacing and weed control methods in groundnut (Arachis hypogeal (L.)). J Agric Sci Environ 16(1):86–95

Adigun JA, Daramola OS, Adeyemi OR, Olorunmaiye PM, Osipitan OA, Idiodemise O (2017a) Effects of row-spacing and period of weed interference on growth and yield of Cowpea (*Vigna unguiculata* (L.) Walp). Niger J Ecol 16(2):88–100

Adigun JA, Adeyemi OR, Daramola OS, Odueme PU, Fadeyi OJ (2017b) Growth and yield performance of groundnut (*Arachis hypogeae* L.) as affected by row-spacing and weed control methods in the Nigerian forest–savannah transition zone. Niger J Ecol 16(1):22–35

Adigun JA, Daramola OS, Adeyemi OR, Olorunmaiye PM (2018a) Response of transplanted Chilli Pepper (*Capsicum frutescens* L.) to nitrogen application and weed management in the Nigerian Forest Savanna transition Zone. Niger J Ecol 17(2):1–4

Adigun JA, Daramola OS, Adeyemi OR, Olorunmaiye PM, Osipitan OA (2018b) Nitrogen and weed management in transplanted tomato in the Nigerian forest-savanna transition zone. Ann Agrar Sci 16(3):281–285

Adigun J, Daramola O, Adeyemi O, Ogungbesan A, Olorunmaiye P, Osipitan O (2018c) Impact of nitrogen levels and weed control methods on growth and yield of okra (*Abelmoschus esculentus* (L.) Moench) in the Nigerian Forest-savanna transition zone. J Exp Agric Int 20(2):1

Adigun JA, Adeyemi OR, Daramola OS, Olorunmaiye PM (2020) Response of cowpea to inter-row spacing and weed competition. Agric Trop Subtrop 53(2):73–79

Agahiu AE (2020) Assessment of some herbicides for the control of purple nutsedge (*Cyperus rotundus*) in soybean (*Glycine max* (L.) Merrill). GSC Biol Pharm Sci 12(1):229–234

Ajah J (2014) Factors limiting small-scale farmers' access and use of tractors for agricultural mechaniza tion in Abuja, North Central Zone, Nigeria. Eur J Sustain Dev 3(1):115–124. https://doi.org/10.14207/ejsd.2014.v3n1p115

Ajayi OA, Akinola MO, Rasheed OA (2017) Allelopathic potentials of aqueous extracts of *Tithonia Diversifolia* (Hemsley) A. Gray in biological control of weeds in cowpea cropping system. Int J Agric Econ Dev 5(1):11

Akinrinola T, Fagbola O (2021) Fertilizers, mycorrhizal inoculation and atrazine interactions on weed biomass and yield of maize. Int J Agric Environ Food Sci 5(4):477–487

Akobundu IO (1987) Weed science in the tropics—principles and practices. Wiley, Chichester, p 522

Akobundu IO, Ahissou A (1985) Effect of interrow spacing and weeding frequency on the performance of selected rice cultivars on hydromorphic soils of West Africa. Crop Prot 4(1):71–76

Akobundu IO, Ekeleme F (2000) Effect of method of *Imperata cylindrical* management on maize grain yield in the derived savanna of South Western Nigeria. Weed Res 40(4):335–334

Akobundu IO, Ekeleme F (2002) Weed seedbank characteristics of arable fields under different fallow management systems in the humid tropical zone of southeastern Nigeria. Agrofor Syst 54(2):161

Alagbo OO, Akinyemiju OA (2018) Performance of different herbicides in transplanted rice in a rainfed forest zone of Nigeria. Niger J Weed Sci 31:55–69

Alagbo OO, Akinyemiju OA, Chauhan BS (2022) Weed management in rainfed upland rice fields under varied agro-ecologies in Nigeria. Rice Sci 29(4):328–339. https://doi.org/10.1016/j.rsci.2021.11.004

Alagbo OO, Akinyemiju OA, Chauhan BS (2023) Management of *Striga hermonthica* (Del.) Benth in Nigerian savanna upland rice fields – current challenges and approaches. Int J Pest Manag:1–10. https://doi.org/10.1080/09670874.2023.2264802

Alagbo OO, Gabriel OA, Akinyemiju OA (2025) Weed management in maize under fiscal pressure: innovative and affordable options for Nigeria. Niger J Weed Sci 37:12–30

Aliyu KT, Lado A, Hussaini MA, Kamara AY, Musa AS, Dawaki MU, Suleiman MS, Bello TT, Fagge AA, Isa HM, Ibrahim H (2023) Spatial modelling indicates Striga seedbank density dependence on rainfall and soil traits in the savannas of northern Nigeria. Weed Res 63:88–101. https://doi.org/10.1111/wre.12569

Alonge SO (2000) Effect of imazaquin applications on the growth, leaf chlorophyll and yield of soybean in the Guinea Savanna of Nigeria. J Environ Sci Health B 35(3):321–336

Aluko OA, Amosun JO, Ayodele OP, Udemba IO, Olasoji JO (2023) Screening of herbicides for weed control in soybeans (*Glycine max*) in derived Savanna Agroecology of Nigeria. Moor J Agric Res 24(1):56–71

Aly R, Dubey NK, Yahyaa M, Abu-Nassar J, Ibdah M (2014) Gene silencing of CCD7 and CCD8 in Phelipanche aegyptiaca by tobacco rattle virus system retarded the parasite development on the host. Plant Signal Behav 9:e29376. https://doi.org/10.4161/psb.29376

Amosun JO, Aduramigha-Modupe V (2016) Influence of groundnut populations on weed suppression in cassava-groundnut systems. J Agric Sci 8(5):72–78

Anonymous (2019) Hunger relief in Africa: Nigeria action against hunger. Available online at https://www.actionagainsthunger.org/location/africa/nigeria/. Accessed 29 June 2024

Avila W, Bolaños A, Valverde BE (2007) Characterization of the cross-resistance mechanism to herbicides inhibiting acetyl coenzyme-A carboxylase in itchgrass (*Rottboellia cochinchinensis*) biotypes from Bolivia. Crop Prot 26(3):342–348. https://doi.org/10.1016/j.cropro.2006.03.019

Awodoyin RO, Olubode OS (2009) On-field assessment of critical period of weed interference in okra [*Abelmoschus esculentus* (L.) moench] field in Ibadan, a rainforest-savanna transition eco-zone of Nigeria. Asian J Food Agro-Ind pages 288–296

Bundit A, Thongjoo C, Chompoo J, Pornprom T (2015) Allelopathic activity of itchgrass (*Rottboellia cochinchinensis*) and its phytotoxicity in soil. Thai J Agric Sci 48(2):73–80

Campbell SD, Grice AC (2000) Weed biology: a foundation for weed management. Trop Grassl 34:271–279

Castillo-Matamoros R, Brenes-Angulo A, Herrera-Murillo F, Gómez-Alpízar L (2016) Base molecular para resistência a fluazifop-P-butyl em capim-camalote (*Rottboellia cochinchinensis*) da Costa Rica. Planta Daninha 34(1):143–150. https://doi.org/10.1590/S0100-83582016340100015

Chauhan BS, Johnson DE (2008) Germination ecology of goosegrass (Eleusine indica): an important grass weed of rainfed rice. Weed Sci 56(5):699–706

Chauhan BS, Gill GS, Preston C (2006) Tillage system effects on weed ecology, herbicide activity and persistence: a review. Aust J Exp Agric 46(12):1557–1570

Chikoye D, Ekeleme F (2001) Weed flora and soil seedbanks in fields dominated by *Imperata cylindrica* in the moist savannah of West Africa. Weed Res 41(6):475–490. https://doi.org/10.1046/j.1365-3180.2001.00251.x

Chikoye D, Ekeleme F, Akobundu I (1997) Weed composition and population dynamics in intensified smallholder farms in West Africa. In: Brighton crop protection conference. British Crop Protection Council, Famham, Surrey, pp 161–166

Chikoye D, Ekeleme F, Ambe JT (1999) Survey of distribution and farmers' perceptions of speargrass [*Imperata cylindrica* (L.) Raeuschel] in cassava-based systems in West Africa. Int J Pest Manag 45(4):305–311. https://doi.org/10.1080/096708799227725

Chikoye D, Lum AF, Udensi UE (2010a) Efficacy of a new glyphosate formulation for weed control in maize in Southwest Nigeria. Crop Prot 29(9):947–952. https://doi.org/10.1016/j.cropro.2010.06.011

Chikoye D, Fontem LA, Menkir A (2010b) Seed coating herbicide tolerant maize hybrids with imazapyr for *Striga hermonthica* (Del.) Benth control in the West African savanna. J Food Agric Environ 9(1):416–421

Choudhary AH, Lagoke STO (1981) Weed control in pearl millet in the savanna zone of Nigeria. Int J Pest Manag 27(4):465–471

Dada OA, Fayinminnu OO (2010) Period of weed control in okra [*Abelmoschus esculentus* (L.) Moench] as influenced by varying rates of cattle dung and weeding regimes. Not Bot Horti Agrobot Cluj-Napoca 38(1):149–154

Dada OA, Oladiran EM, Togun AO (2017) Optimizing grain yield in upland rice field by synchronizing time of cattle dung application with period of weed removal in Southwestern Nigeria. Trop Agric 94(2):141

Dania Ogbe FM, Bamidele JF (2006) Incidence and spread of an invasive weed, Mimosa invisa Mart in Benin City Metropolis, Nigeria. Int J Bot 2(3):336–339. https://doi.org/10.3923/ijb.2006.336.339

Daniya E (2013) Response of weeds and pod yield of some okra cultivars to weed control treatments. Savannah J Agric 8(1):105–112

Daniya E, Mohammed AB, Onyenuforo CC (2023) Weed control and productivity of maize (*Zea mays* l.) interseeded with Jointvetch (*Aeschynomene histrix* Poir.) at different sowing arrangements. J Agric Environ 19(2):89–96

Daramola SO (2020) Timing of weed management and yield penalty due to delayed weed management in soybean. Planta Daninha 38:e020236046

Daramola OS, Adeyemi OR, Adigun JA, Adejuyigbe CO (2019) Economics of row spacing and integrated weed management in soybean (*Glycine max* L.). J Agric Sci Belgrade 64(3):265–278

Daramola OS, Adigun JA, Olorunmaiye PM (2020a) Challenges of weed management in rice for food security in Africa: a review. Agric Trop Subtrop 53(3):107–115

Daramola OS, Adeyemi OR, Adigun JA, Adejuyigbe CO (2021a) Influence of row spacing and weed control methods on weed population dynamics in soybean (Glycine max L.). Int J Pest Manag 68(1):43–58

Daramola OS, Adigun JA, Adeyemi OR (2021b) Efficacy and economics of integrated weed management in chilli pepper (*Capsicum frutescens* L.). J Crop Improv 35(1):38–50

Daramola OS, Adigun JA, Olorunmaiye PM, Adeyemi OR (2021c) Efficacy and economic comparison of weed management in tomato. Int J Veg Sci 27(4):352–363

Division UDoEaSAP (2019) World population prospects: comprehensive tables. United Nations, New York

Ekeleme F, Kamara AY, Oikeh SO, Omoigui LO, Amaza P, Abdoulaye T, Chikoye D (2009) Response of upland rice cultivars to weed competition in the savannas of West Africa. Crop Prot 28(1):90–96

Ekeleme F, Atser G, Dixon A, Hauser S, Chikoye D, Olorunmaiye PM, Sokoya G, Alfred J, Okwusi MC, Korieocha DS (2019) Assessment of weeds of cassava and farmers management practices in Nigeria. Tropicultura 37(2):2295–8010. https://doi.org/10.25518/2295-8010.586

Ekeleme F, Dixon A, Atser G, Hauser S, Chikoye D, Olorunmaiye PM, Olojede A, Korie S, Weller S (2020) Screening preemergence herbicides for weed control in cassava. Weed Technol 34(5):735–747

Emechebe AM (1991) Pan-African Striga Control Network (PASCON): feasibility and relevance. A Consultancy Report submitted to FAO Regional Office for Africa. 59 p

Ene C, Anyim A, Chukwudi U, Okechukwu E, Ikeogu U (2019) Growth and yield response of selected improved soybean (*Glycine max* [L.] Merrill) varieties to varying weeding regimes under a tropical condition. J Cent Eur Agric 20(1):157–178

Eni E, Adeyemi OR, Lagoke ST, Asiribo OE (2021) Enhancing the weed control efficacy of two pre-emergence herbicides in maize/jack bean and maize/groundnut intercrops in derived savanna. J Agric Sci Belgrade 66(4):335–350

Erida G, Ichsan CN, Hafsah S, Husna R, Marliah A, Nurahmi E, Hayati E (2023) Herbicidal effects of n-hexane, ethyl acetate and methanol extracts of cogon grass (*Imperata cylindica* l.) and their phytotoxicity on spiny amaranth (*Amaranthus spinosus* l.) growth. IOP Conf Ser Earth Environ Sci 1183(1):012106. https://doi.org/10.1088/1755-1315/1183/1/012106

Ezebuiro P, Yakubu A, Musa M (2021) Effect of pre-emergence herbicides on the growth and yield of soya bean [*Glycine max* (L.) Merrill] in Sudan savanna. FUDMA J Agric Agric Technol 7(1):16–26

Falade AA, Lagoke ST, Adigun JA, Pitan OR, Osunleti SO (2023) Growth and yield of maize (*Zea mays* L.) as influenced by cropping pattern and weed control treatments in the forest–savanna agro-ecological zone of Southwest Nigeria. Acta Fytotech Zootech 26(2). https://doi.org/10.15414/afz.2023.26.02.155-162

Fasinmirin JT, Reichert JM (2011) Conservation tillage for cassava (*Manihot esculenta* Crantz) production in the tropics. Soil Tillage Res 113(1):1–10

Fayinminnu OO, Adesiyan SO, Sosanya SO (2010) Effect of paraquat as post emergence herbicide on yield of cowpea (*Vigna unguiculata* (L) walp). J Agric Soc Res 10(2). https://doi.org/10.4314/jasr.v10i2.67590

Food and Agriculture Organization of the United Nations (2023) India at a glance. Available online at https://www.fao.org/india/fao-in-india/india-at-a-glance/en

Futuless KN, Bake ID (2010) Evaluation of yield and yield attributes of some cowpea (*Vigna unguiculata* (L.) Walp) varieties in Northern Guinea Savanna. J Am Sci 6(8):508–511

Garko MS, Yawale MA, Mohammed IB, Yahaya SU, Ishaya DB, Mahadi MA (2021) Effects of poultry manure fertilization and weed control method on leaf chlorophyll and other indices of three maize (*Zea mays*) cultivars in Kano Sudan Savanna, Nigeria. J Curr Opin Crop Sci 2(1):74–85

Gbanguba AU, Ismaila U, Kolo MG, Umar A (2011) Effect of cassava/legumes intercrop before rice on weed dynamics and rice grain yield at Badeggi, Nigeria. Afr J Plant Sci 5(4):264–267

Gelmini GA, Victória Filho R, Novo MDCDSS, Adoryan ML (2005) Resistence of *Euphorbia heterophylla* L. to ALS-inhibiting herbicides in soybean. Sci Agric 62(5):452–457. https://doi.org/10.1590/s0103-90162005000500007

Gerhards R (2019) Precision weed management systems. In: Stafford J (ed) Managing soil health for sustainable agriculture fundamentals, vol 1. Burleigh dodds Science Publishing, pp 399–416

Gianessi LP (2010) Solving Africa's weed problem: increasing crop production & improving the lives of women. Asp Appl Biol 96:23

Hamma I, Ibrahim U (2013) Weed management techniques of horticultural crops in Nigeria. Am Eurasian J Agric Environ Sci 13(3):362–366

Hauser EF (2017) Weed control in cassava cropping systems. In: Hershey HC (ed) Achieving sustainable cultivation of cassava, vol 2: genetics, breeding, pests, and diseases, Burleigh Dodds series in agricultural science, vol 21. Burleigh Dodds Science Publishing Limited, Cambridge, pp 1–26

Hooda VS, Chauhan BS (2023) Herbicide response and germination behavior of two goosegrass (*Eleusine indica*) populations in the Australian environment. Weed Sci 71(6):584–593. https://doi.org/10.1017/wsc.2023.51

Ibrahim AM, Sanusi J, Aderemi GA, Kerau M (2024) Effect of groundnut (*Arachis hypogea* l.) varieties as influence by sowing date and weed control methods in semi-arid environment Nigeria. Br J Multidiscip Adv Stud 5(2):45–56

Imoloame EO (2014) The effects of different weed control methods on weed infestation, growth and yield of soybeans (*Glycine max* (L) Merril) in Southern Guinea Savanna of Nigeria. Agrosearch 14(2):129–143

Imoloame EO (2020) Weed control and productivity of maize (L.) in Malete, Kwara State of Nigeria. Agric Trop Subtrop 53(2):63–71

Imoloame EO, Osunlola OS (2017) Effect of method of weed control and insecticide (Lambda cyhalothrin) rates on the productivity of cowpea (*Vigna unguiculata* (L.) Walp.) in a Southern Guinea savanna of Nigeria. Rev Bras Herbic 16(3):216–229

Imoloame EO, Alagbe E, Lawal OO (2021) Influence of fertilizer application timing and reduced herbicide dosage on weed infestation and maize grain yield. Indian J Weed Sci 53(4):374–380

Isah K, Lagoke STOM (2010) Effect of Striga Hermonthica on host crop genotypes planted in rotation with trap crop varieties in the southern Guinea savanna of Nigeria. Niger J Weed Sci 23:25–44

Jafun FB, Abdul SD (2004) Predominant weeds of some cereal crops in the scrub Savannah region of Nigeria. Pertanika J Trop Agric Sci 27(1):57–64

Javaid N, Javaid A, Shah MH, Khan IH, Waleed SM (2020) Herbicidal activity of Ageratum conyzoides against parthenium. J Weed Sci Res 27(2):137–146. https://doi.org/10.28941/pjwsr.v26i2.823

Kabambe VH, Kanampiu F, Ngwira A (2008) Imazapyr (herbicide) seed dressing increases yield, suppresses Striga asiatica and has seed depletion role in maize (Zea mays L.) in Malawi. Afr J Biotechnol 7:3293–3298

Kayode J, Ademiluyi B (2004) Effect of tillage methods on weed control and maize performance in southwestern Nigeria location. J Sustain Agric 23(3):39–45

Kebede M, Bogale A (2018) Economically important parasitic weeds and their management practices in crops. J Environ Earth Sci 8(12):104–115

Kolo E, Adigun JA, Adeyemi OR, Daramola OS, Bodunde JG (2020) Effect of nitrogen levels and weed management methods on weed abundance and yield of upland rice (*Oryza sativa* L.). J Agric Sci Belgrade 65(2):137–150

Kolo E, Adigun JA, Daramola OS, Adeyemi OR, Olorunmaiye PM (2021) Nitrogen effect on weed management in rice in a tropical forest-savanna transition zone of Nigeria. J Crop Improv 35(2):169–183

Kureh I, Kamara AY, Tarfa BD (2006) Influence of cereal-legume rotation on Striga control and maize grain yield in farmers' fields in the Northern Guinea savanna of Nigeria. J Agric Rural Dev Trop Subtrop 107(1):41–54

Kurniadie D, Widianto R, Umiyati U, Widayat D, Nasahi C, Budiawan A (2023) Management of *Eleusine indica* (L.) Gaertn resistance to glyphosate herbicide in Indonesia. Agronomy 13(6):1–10. https://doi.org/10.3390/agronomy13061649

Lagoke STO, Shebayan JAY, Adeosun JO, Iwafor ENO, Olukosi J, Adu A, Olufajo O, Emechebe A, Zaria N, Adeoti A, Onyibe J, Chobe S (1990) Survey of Striga control methods in maize, sorghum and cowpea in the Nigerian savanna. In: Unpublished paper presented at the pan African Striga control network conference held at IITA-Ibadan, pp 1–65

Lagoke ST, Eni E, Adigun JA, Phillip BB (2014) Influence of intercropped groundnut (*Arachis hypogaea* L.) on the performance of weed control treatments in maize production. Int J Agron Agric Res 5(6):1–8

Lawal A, Olayinka B, Ayinla A, Bulala FA, Muktar S, Abdulra'uf L (2023) Effects of expired and non-expired Pendimethalin and hand weeding on the bio-productivity and seed quality of groundnut (*Arachis Hypogea* L.). Sci J Univ Zakho 11(1):104–111

Makinde EA, Adeyemi OR, Odeyemi OM, Salau AW, Abiodun OL (2020) Comparison of weed suppression ability and yield of two okra cultivars and with different planting density. Trop Subtrop Agroecosystems 23(2). https://doi.org/10.56369/tsaes.2927

Makinde EA, Adeyemi OR, Odeyemi OM, Salau AW, Abiodun OL (2021) Planting density on weed suppression and yield of okra. Int J Veg Sci 27(3):260–267

Makinde EA, Odeyemi OM, Abiodun OL, Ayeni OS (2022) Plant spatial arrangement on growth, weed control, and yield of okra. Int J Veg Sci 28(1):3–14

McElroy JS, Head WB, Wehtje GR, Spak D (2017) Identification of Goosegrass (*Eleusine indica*) biotypes resistant to Preemergence-applied Oxadiazon. Weed Technol 31(5):675–681. https://doi.org/10.1017/wet.2017.41

Melifonwu AA (1994) Weed and their control in cassava. Afr Crop Sci J 2(4):519–530

Mustapha AB, Felix T, Tashiwa MC, Gworgwor NA (2021) Effect of plant population density and methods of weed control on the yield of pepper (Capsicum annum L.) in Northeastern Nigeria. J Appl Sci Environ Manag 25(2):261–267

Na-Allah MS, Mukhtar AA, Mahadi MA, Tanimu MU, Muhammad A (2017) Performance of cowpea (*Vigna unguiculata* (L.) Walp) under irrigation as influenced by weed management methods and intra-row spacing. J Agric Sci Pract 2(4):58–65

Negi B, Bargali S, Bargali K, Khatri K (2020) Allelopathic interference of *Ageratum conyzoides* L. against rice varieties. Curr Agric Res J 8(2):69–76. https://doi.org/10.12944/carj.8.2.01

Nnadozie EC, Iloanusi ON, Ani O, Yu K (2023) Detecting cassava plants under different field conditions using UAV-based RGB images and deep learning models. Remote Sens 15:2322

Nwagwu F, Athanasius M, Michael G, Obok E, Ibrahim A (2022) Weed control and cowpea yield under different tillage systems. J Appl Biol Biotechnol 10(4):110–115

Nzeako EN, Udensi UE, Omovbude S, Orluchukwu JA (2019) Maize and weed response to legume cover short fallow and fertilizer in Port Harcourt South Eastern Nigeria. J Appl Life Sci Int 20(2):1

Obiazi CC, Okogu GO, Ugbo O, Alama SI, Uwabor CS, Akpomedaye JF, Otite FO (2024) Combined use of butachlor and cowpea for weed management in okra production. Int J Phytol Res 4(2):08–12

Ojelade OB, Lagoke ST, Adigun JA, Babalola OA, Daramola OS, Osipitan OA (2018) Intra-row spacing and weed control influence growth and yield of groundnut (*Arachis hypogea* L.). Adv Agric Sci 6(4):1–11

Ola OT, Ogedengbe OO, Adesina GO, Raji TM, Akinyemiju OA (2021) Effect of weed control methods and canopy characteristics on weed biomass, yield and yield components of two contrasting cassava varieties in the rainforest zone of Nigeria. J Nat Sci Res 12(14):45–54

Olabode OS, Adesina GO, Ajibola AT (2010a) Seasonal effects on the critical period for weed removal and okra performance on *Tithonia diversifolia* (Helmsl) A. Gray infested field. Acad J Plant Sci 3(4):156–160

Olabode OS, Adesina GO, Babajide PA (2010b) Weed control efficiency of reduced atrazine doses and its effect on soil organisms in maize (*Zea mays* L.) fields of south western Nigeria. J Trop Agric 48(2):52–54

Oladejo LF, Oni OO, Awogbade AL, Makinde IA, Ogunleti DO, Adediran BO, Jokanola OO, Okunade RF, Adedeji JA (2021) Effect of different methods of weed control on the growth and yield of tomato (*Lycopersicum esculentum* mill.). In: Proceedings of the 39th annual conference of Horticultural society of Nigeria (Hortson)

Olaniyi OM, Daniya E, Kolo JG, Bala JA, Olanrewaju AE, Mikail OO (2020) A computer vision-based weed control system for lowland rice precision farming. Int J Adv Appl Sci 9(1):51–61. https://doi.org/10.11591/ijaas.v9i1.pp51-61

Olaniyi OM, Daniya E, Abdullahi IM, Bala JA, Olanrewaju EA (2021) Weed recognition system for low-land rice precision farming using deep learning approach. In: Tawfik M, Anass C, Ibtissam EH (eds) Advances in intelligent systems and computing, vol 1193. Springer, pp 385–402. http://www.springer.com/series/11156

Olaniyi OM, Salaudeen MT, Daniya E, Abdullahi IM, Folorunso TA, Bala JA, Nuhu BK, Adedigba AP, Oluwole BI, Bankole AO, Macarthy OM (2023) Development of maize plant dataset for intelligent recognition and weed control. Data Brief 47. https://doi.org/10.1016/j.dib.2023.109030

Olayinka BU, Etejere EO (2015) Growth analysis and yield of two varieties of groundnut (*Arachis hypogaea* L.) as influenced by different weed control methods. Indian J Plant Physiol 20:130–136

Olorunmaiye PM (2010) Weed control potential of five legume cover crops in maize/cassava intercrop in a Southern Guinea savanna ecosystem of Nigeria. Aust J Crop Sci 4(5):324–329

Olowe V (2021) Africa 2100: how to feed Nigeria in 2100 with 800 million inhabitants. Org Agric 11(2):199–208

Olusoji OO (2012) Agronomic performance of improved cowpea varieties under natural infestation with *Alectra vogelii* (Benth.) in the Northern Guinea Savannah of Nigeria. Agricultura Tropica Et Subtropicav 45(2):66–71

Omisore JK, Olofintoye JA (2008) Evaluation of chemical and mechanical weed control methods in soyabean (*Glycine max* L. Merr) production. Niger Agric J 39:82–87

Orluchukwu JA, Molokwu CI (2023) Effects of spacing on growth, yield and weed suppression of rice variety in south-south, Nigeria. Int J Appl Sci Res 6(3):40–47

Osadebe VO, Obi OJ, Dauda N, Uzoigwe AD (2022) Effect of plant spacing and weed control treatments on growth and yield of okra (*Abelmoschus Esculentus* (L.) Moench) in derived savanna, Nigeria. J Agricu Environ Sci Res 22(1):1–10

Osipitan OA, Adigun JA, Lagoke ST, Afolami SO (2013) Effect of inter-row spacing and weed control methods on growth and yield of cowpea (*Vigna unguiculata* L. Walp) in South Western Nigeria. Niger J Plant Prot 27:97–111

Osipitan OA, Adigun JA, Kolawole RO (2016) Row spacing determines critical period of weed control in crop: cowpea (*Vigna unguiculata*) as a case study. Azarian J Agric 3:90–96

Osunleti S, Ajani A, Olaogun O, Osunleti T, Olatunde E (2022) Assessing the critical period of weed interference in groundnut Arachis hypogaea L. in Ogun State, south western Nigeria. Acta Fytotechn Zootech 25(3):219–225

Palma-Bautista C, Rojano-Delgado AM, Vázquez-García JG, Yanniccari M, De Prado R (2020) Resistance to Fomesafen, Imazamox and glyphosate in *Euphorbia heterophylla* from Brazil. Agronomy 10(10 October). https://doi.org/10.3390/agronomy10101573

Prinsloo G, Du Plooy CP (2018) The allelopathic effects of Amaranthus on seed germination, growth and development of vegetables. Biol Agric Hortic 34(4):268–279. https://doi.org/10.1080/01448765.2018.1482785

Rodenburg J, Demont M, Zwart SJ, Bastiaans L (2016) Parasitic weed incidence and related economic losses in rice in Africa. Agric Ecosyst Environ 235:306–317. https://doi.org/10.1016/j.agee.2016.10.020

Rowan FS, Athena DM (2006) 2Is C4 photosynthesis less phenotypically plastic than C3 photosynthesis? J Exp Bot 57(2):303–317

Salifu YI (2016) Yield response of groundnut (*Arachis hypogaea*) to weeding regime and plant spacing. J Biol Agric Healthc 6(6):65–73

Shiru MS, Shahid S, Dewan A, Chung ES, Alias N, Ahmed K, Hassan QK (2020) Projection of meteorological droughts in Nigeria during growing seasons under climate change scenarios. Sci Rep 10(1):10107

Statista (2019) Five major crops among farming households in Nigeria. https://www.statista.com/statistic Nigeria: major crops | Statista. Accessed 2 Sept 2024

Takim F (2012) Weed competition in maize (*Zea mays* L.) as a function of the timing of hand-hoeing weed control in the southern Guinea savanna zone of Nigeria. Acta Agron Hung 60(3):257–264

Takim FO, Fadayomi O (2013) Pattern of emergence of weeds in an arable field in the Southern Guinea savanna ecology of Nigeria. Niger J Weed Sci 26:53–63

Udensi EU, Omovbude S (2018) Influence of plant spacing on weed suppression and maize performance in the humid forest agro-ecology of south eastern Nigeria. Niger Agric J 49(1):32–39

Ugbe LA, Ndaeyo NU, Enyong JF (2016) Efficacy of selected herbicides on weed control, cowpea (*Vigna Uniguiculata* L. Walp) performance and economic returns in Akamkpa, southeastern Nigeria. Int J Res Agric For 3(5):19–27

Ukwoma-Eke O, Ekeleme F (2024) Seedbank composition and periodicity of seedling emergence at two locations in Southeast Nigeria. Niger Agric J 55(1):54–60

Unamma RPA, Akobundu IO (1989) Effects of tropical weeds on yield in white yam (*Dioscorea rotundata* Poir). Weed Res 29(1):1–6. https://doi.org/10.1111/j.1365-3180.1989.tb00833.x

United States Department of Agriculture, Foreign Agricultural Service. Nigeria | USDA Foreign Agricultural Service. 2023, https://www.fas.usda.gov/regions/nigeria. Accessed May 3 2025

Usman A, Elemo KA, Bala A, Umar A (2001) Effect of weed interference and nitrogen on yields of a maize/rice intercrop. Int J Pest Manag 47(4):241–246

Usman A, Elemo KA, Misari SM, Lagoke ST, Adigun JA (2002) Nitrogen and weed management in maize intercropped with upland rice. J Sustain Agric 21(1):5–16

Uwalaka NO (2023) Comparative germination study between some native and invasive weed species in Nigeria. Biologia 78(5):1349–1354. https://doi.org/10.1007/s11756-022-01180-x

Vissoh PV, Gbehounou G, Ahanchede A, Kuyper TW, Röling NG (2004) Weeds as agricultural constraint to farmers in Benin: results of a diagnostic study. NJAS Wageningen J Life Sci 52:305–329

Weber G, Elemo K, Lagoke STO (1995) Weed communities in intensified cereal-based cropping systems of the northern Guinea savanna. Weed Res 35(3):167–178. https://doi.org/10.1111/j.1365-3180.1995.tb02030.x

World Bank Data (2024) GDP per capita (current US$) - Nigeria | Data. data.worldbank.org. Accessed 25 Sept 2024

Yusuf MJ, Musa UT, Ifeoluwa AO (2019a) Evaluation of maize varieties under varying fertilizer application rates in Owode-Egba area of OGUN state Nigeria. J Agric Vet Sci 12:2532

Yusuf MA, Uzoma AO, Mamudu, AY, & Salaudeen MT (2019b) Growth of cowpea (Vigna unguiculata (L.) Walp.) and weed infestation as affected by cultivar and N-sources in Minna, Niger State. In: Proceedings of the 43rd Annual Conference of the Soil Science Society of Nigeria. Soil Science Society of Nigeria, pp 45–53

Chapter 8
Biology and Management of Weeds in Pakistan

Amar Matloob, Farhena Aslam, and Tahir Hussain Awan

Abstract Weed-related challenges in Pakistan include high financial loss due to weed infestation in economically important crops, invasion of managed and natural ecosystems by alien invasive weeds, floral shifts in weed communities, and the evolution of herbicide-resistant biotypes of major weeds. Labor shortages during the critical period of weed removal, limited automation, and fault lines in implementing and maintaining weed control are significant concerns. Conventional agronomic practices and weed management approaches are insufficient to combat the weed menace and sustain crop productivity. Research on weed ecology and related management aspects, although scanty in Pakistan, is still crucial for a better understanding of how weeds grow, spread, and impact agroecosystems. In this context, improved knowledge of weed ecology and biology (weed germination, seedling emergence, growth, reproductive potential, and seed dormancy) seems indispensable for developing sustainable weed management practices. This will assist in the development of ecologically based weed management approaches, especially in response to evolving cropping systems and related crop management practices under changing climatic patterns. Weeds respond to environmental and edaphic factors under diverse agroclimatic conditions of Pakistan in a complex manner and prompt an in-depth exploration of weed eco-biology considering environmental influences, biological traits, and adaptive management practices. As the country continues to face challenges like climate change, water scarcity, and increasing herbicide resistance, strategic research and innovations will be the key to mitigating the adverse impacts of weeds on food security and the country's economy. There is an urgent need to reorient weed research in Pakistan toward a more holistic framework

A. Matloob (✉)
Department of Climate Change, MNS University of Agriculture, Multan, Pakistan
e-mail: amar.matloob@mnsuam.edu.pk

F. Aslam
Department of Agronomy, University of Agriculture, Faisalabad, Pakistan

T. H. Awan
Rice Research Institute, Kala Shah Kaku, Punjab, Pakistan

B. S. Chauhan, K. Jabran (eds.), *Biology and Management of Weeds in Asia and Africa*, Sustainability Sciences in Asia and Africa,
https://doi.org/10.1007/978-981-95-7314-1_8

that emphasizes ecological understanding, long-term monitoring, and integrated weed management principles. Besides conventional weed management approaches, innovations in weed management like drone-enabled mapping and precision spraying, artificial intelligence (AI)–driven robotics, use of thermal or laser weeders, biodegradable carrier-based nanoherbicides, herbicide-resistant crops, RNAi-mediated spray-induced gene silencing, and harvest weed seed control need to be considered for technological adoption and diffusion at the farm level.

Keywords Agroecology · Chemical weed control · Eco-friendly weed management · Field crops · Weed-crop competition · Weed dynamics

8.1 Introduction

There is an accruing interest in enhancing agricultural production in a sustainable manner while limiting the environmental impacts of agricultural practices. The escalating demand for food and other ecosystem services, with a concurrent increase in population and unprecedented urbanization, is putting huge pressure on agricultural production systems with already dwindling natural resources and declining total factor productivity. Weed management is crucial to sustain crop production and ensure food security in the backdrop of climate change and ever-increasing population (Matloob et al. 2020). Weeds are the plants that compete with human interest and pose multifaceted issues ranging from reduction in crop yields and deterioration of produce quality to posing ecological and environmental challenges, incurring economic impacts in addition to those involved in controlling them. Weeds interfere with crops through competitive and allelopathic interactions (Aslam et al. 2017). Besides weed-related problems in the managed ecosystems, the threat of alien invasive weeds has endangered the sustainability of forests, pastures, rangelands, and watersheds, posing serious ecological implications (Qureshi et al. 2014; Bajwa et al. 2016a; Shabbir et al. 2023).

In Pakistan, weed-related yield losses mount up to 36 million tons in important crops like wheat, rice, corn, sorghum, barley, sugarcane, cotton, pulses, oil seeds, millets, vegetables, and fruits (Mushtaq and Chema 2007). Previous estimates indicated that weeds were causing a loss of US$ 3 billion annually (Mushtaq and Chema 2007; Marwat et al. 2008; Zafar et al. 2010), which was significantly greater than losses caused by pathogens and insect pests (Naseer-ud-Din et al. 2011). The average crop yield loss due to weeds is estimated as 11.5% in Pakistan, as against the global average of 9.5% (Alam 2003). In monetary terms, these losses have magnified over time, and currently, weed-inflicted losses in five important field crops are US$ 6.46 billion (Table 8.1). Quite often, weeds are ignored by farmers and go unchecked, primarily because their damage is not visually apparent early in the season.

Table 8.1 Agronomic and economic impacts of weeds on major crops of Pakistan

Crop	Total production (million metric tons)[a]	Average yield loss due to weeds (%)	Average yield loss due to weeds (million metric tons)	Market price (PKR/kg)	Monetary loss due to weeds	
					Billion PKR	Billion US$
Cotton	1.74	32	0.56	213	118.60	0.43
Corn	9.80	33	3.23	75	242.55	0.88
Rice	9.90	34	3.37	110	370.26	1.34
Sugarcane	87.60	30	26.28	10	262.8	0.95
Wheat	31.40	26	8.16	97	791.91	2.86
Total			41.6		1786.12	6.46

Modified after Matloob et al. (2020)
[a]Pakistan Economic Survey 2023–24, Economic Advisors Wing, Finance Division, Government of Pakistan (https://www.finance.gov.pk/survey/chapter_24/2_agriculture.pdf)

Depending upon agroclimatic conditions and socio-economic scenarios, the management of weeds is carried out by different approaches, including manual pulling/hoeing, mechanical, cultural, biological, and chemical control (Khan et al. 2016). However, mechanical and manual weeding are costly and time-consuming, and crop mimicry can occasionally limit their efficacy and render them less effective against weeds, especially during the critical period of weed-crop competition. Use of herbicides for weed control has been an integral component of contemporary agriculture and has increased manifold in recent years. It offers selective, timely, and efficient weed control, resulting in enhanced productivity and ease in other farm operations. In Pakistan, herbicides have been substituted for manual weeding (the most labor-intensive task among farm operations) owing to a 5–10 times increase in wages. Now, herbicides are being used on 50% of the wheat acreage. A study found a 13%–29% wheat yield advantage as compared to farms where herbicides were not used (Khan et al. 2011). The application cost of pre-emergence herbicides in rice is 38%–46% less than that of one manual weeding. Nevertheless, herbicides drive an agroecosystem toward declining species diversity, besides causing changes in floristic composition and behavior of weed populations (Mahajan et al. 2011; Matloob et al. 2015a; Qi et al. 2020). Repeated use of herbicides with similar modes of action may exert selection pressure and lead toward the evolution of herbicide-resistant biotypes of economically important weeds in major crops, hampering the productivity of cereal-based cropping systems (Abbas et al. 2017). Irrational use of herbicides and their improper application technique can also cause crop phytotoxicity along with harmful implications for the ecosystem and human health.

A better understanding of interactions of weeds with their environment and influence of weed control measures on species abundance, richness, survival, and potential for resistance development could yield crucial information that can help devise sustainable weed management programs (Matloob et al. 2020). For effective and sustainable weed management, it is essential to know the habits and characteristics of the unwanted plants in croplands and the ways they interfere with the crop (Tanveer 2008). Eco-biology of the weeds in terms of life cycle, adaptability to climatic conditions, seed dormancy, root system, viability and longevity of the reproductive organs, reproductive process, seed retention/dispersal modes, productivity, and interference mechanisms can be decisive aspects for strategic weed management in cropping systems (Chauhan 2012; Chauhan et al. 2012; Ali et al. 2017). Contrarily, neglecting the eco-biology of weeds while managing weeds could result in the evolution of resistant biotypes of weeds, a shift in weed populations to difficult-to-control weeds, dominance of minor weeds, limitation regarding choice of crops and loss of agronomic usefulness of fields, etc. This chapter aims to provide an in-depth exploration of weed ecology in Pakistan, highlighting environmental influences, biological traits, and practical management approaches, enriched by case studies and relevant examples, while establishing the knowledge gaps.

8.2 Weed Biology

8.2.1 Crop-Weed Association

Several weeds are associated with a particular crop or cropping sequence. Timing of emergence, floristic composition, abundance, persistence, competitive, and reproductive ability of weeds are significantly affected by the crop species, habitat conditions, and level of disturbance. In agricultural systems, weeds with consistent periodicity of germination and those which mimic the crop life cycle are more prevalent and successful and are often regarded as segetal weeds (Tanveer 2008). Nevertheless, field-to-field variations regarding the level of weed infestation are common owing to differences in soil type, management practices (crop rotation, sowing methods, irrigation and fertilization techniques, weed control methods), and socio-economic factors (Matloob et al. 2015a, 2020). Certain crop establishment methods also favor certain types and groups of weeds, for example, conservation tillage practices and aerobic rice culture, with weed species and category-specific effects influencing the final outcome of crop-weed interactions (Chauhan and Johnson 2009, 2010; Matloob et al. 2015a, b, 2020).

The common weeds of wheat in Pakistan causing huge economic losses are *Phalaris minor* Retz., *Avena fatua* L., *Alopercurus myosuroides* Huds., *Chenopodium album* L., *C. murale* L., *Convolvulus arvensis* L., *Rumex dentatus* L. and *Cirsium arvense* (L.) Scop. The weeds with relatively less economic importance include *Lolium temulentum* L., *Coronopus didymus* L., *Carthamus oxycantha* Bieb., *Emex spinosa* (L.) Campd., *Euphorbia helioscopia* L., *Galium aparine* L., *Lathyrus aphaca* L., *Medicago polymorpha* L., *Anagallis arvensis* L., *Fumaria indica* L., *Sisymbrium irio* L., *Sonchus arvensis* L., *S. asper* (L.) Hill, *Vicia sativa* L. *Raphanus raphanistrum* L., and *Silybum marianum* (L.) Gaertn (Tanveer 2008; Ashiq and Aslam 2014). Nevertheless, any of these weeds may pose a serious management challenge under any condition, and their relative importance may vary on temporal and spatial scales across fields and regions (Ali et al. 2017).

The significant weeds of cotton crop in Pakistan are *Cyperus rotundus* L., *Cynodon dactylon* L. (Pers.), *Echinochloa colona* (L.) Link, *Dactyloctenium aegyptium* (L.) Beauv., *Leptochloa chinensis* (L.) Nees, *Dichanthium annulatum* (Forssk.) Stapf, *Trianthema portulacastrum* L., *Digera arvensis* Forsk., *Cucumis melo* var. agrestis, and *Tribulus terrestris* L. (Rajput et al. 2008). The less common weeds in this crop are *E. glabrescens* Munro ex Hook. f., *C. pluricaulis* Choisy, *Portulaca oleracea* L., *Euphorbia hirta* L., *C. difformis* L., *Eleusine indica* (L.) Gaertn, *Paspalum distichum* L., *Xanthium strumarium* L., *Corchorus antichorus* Raeusch, *C. depressus* (L.) Stocks, and *Cleome viscoa* L. (Tanveer et al. 2003; Tanveer 2008; Manalil et al. 2017).

The most common weeds in conventional puddled transplanted rice system are *Echinochloa crus-galli* (L.) P. Beauv., *Echinochloa colona, Echinochloa glabrescens, Eleusine indica, P. distichum, Cyperus difformis, Cyperus maritimus, Cyperus iria, Ipomoea aquatica* Forssk., *Sagittaria guyanensis* H.B.K., *Nymphaea stellata* B.P., *Phyla nodiflora* (L.) Green, *Sphenoclea zeylanica* Gaertn, and *Marsilea minuta* (Tanveer 2008; Farooq et al. 2011; Ashiq and Aslam 2014; Matloob et al. 2015a, b). On the other side, the grain yield losses in the direct seeded rice (DSR) production system vary from 48% to 80% and major weeds are *Cyperus rotundus, C. iria, C. difformis, Echinochloa colona, Echinochloa crus-galli. Eleusine indica, L. chinensis, Eclipta prostrata* L., *T. potulacastrum, D. aegyptium*, and *P. oleracea* (Mann et al. 2007; Hussain et al. 2008; Khaliq et al. 2014; Matloob et al. 2015a, b).

Major weeds infesting the autumn corn crop are *C. dactylon, C. rotundus, D. aegyptium, T. portulacastrum, D. arvensis, T. terrestris, C. melo, Sorghum halepense* (L.) Pers., and *E. colona* (Riaz et al. 2007; Khatam et al. 2013). In cornfields, phyto-sociological attributes of weeds such as frequency, relative frequency, density, relative density, and importance values varied to a much greater extent in response to edaphic factors such as potassium concentration, organic matter content, and sandy nature of the sampled soil (Ahmad et al. 2016). However, the weed flora of spring-planted corn differs and consists of *Amaranthis viridis* L., *C. album, C. arvensis, C. didymus, T. portulacastrum, C. dactylon, Setaria viridis* (L.) P. Beauv., and *P. minor* (Tanveer 2008; Ashiq and Aslam 2014).

Both the perennial and summer and winter annual weeds belonging to the grass, broadleaf, and sedge categories infest sugarcane fields. Weed flora varies in sugarcane crop grown in different provinces of Pakistan. For instance, in Sindh province, the most prevalent and dominant weed species of sugarcane crop are *Desmostachya bipinnata, C. dactylon, T. portulacastrum, T. terrestris, Alhagi maurorum* Medik., *C. rotundus, D. annulatum, C. arvensis*, and *Achyranthes aspera* L. (Qureshi 2004). However, in KPK province, major weeds of sugarcane fields are *C. dactylon, Brachiaria eruciformis* (Sm.) Griesb., *C. rotundus, S. halepense, C. dactylon*, and *Digera* spp. are the most abundant weeds (Afridi et al. 2015). In Punjab, *Conyza bonariensis* (L.) Cronq., *C. trilocularis* (L.) Mant., *Mukia maderaspatana* (L.) Roem., *Polygonum plebeium* R. Br., *P. barbatum* L., *S. oleraceus* L., *Spergula arvensis* L., *L. sativus* L., *Stelleria media* (L.) Vill., and *S. irio* L. are also present in sugarcane fields (Tanveer 2008; Ashiq and Aslam 2014).

8.2.2 Weed Seed Dormancy and Seed Bank Dynamics

Weeds produce seeds that remain dormant even under conditions favorable for germination as a part of a survival strategy and thus remain buried in soil for extended periods. Timing of weed seedling emergence and persistence of weed seed banks in soil is, in part, also governed by dormancy (Nakabayashi and Leubner-Metzger 2021). Dormancy is a highly adaptive trait and describes seed dispersal in time. Seeds perceive environmental signals like temperature, moisture, light, oxygen, nutrients, germination inhibitors, and even smoke, and adjust their dormancy status accordingly (Qaderi 2023). Eradication of weed populations from an infested area is almost impossible, as most of the weed species pause their life cycle by entering into seed dormancy after maturity (Ali et al. 2012; Munawar et al. 2015). Moreover, seed withdrawals occur when seeds are removed from the seed bank through processes like predation, germination, and senescence. However, deposits occur when new seeds are added through seed rain during dispersal. Weed seeds possess unique features of dormancy induction, and dormancy can also be broken without completing germination (Finch-Savage and Footitt 2017). Such induction and release of dormancy is modulated by dormancy proteins and phytohormones synthesized during the dynamic processes of seed maturation and storage (Wang et al. 2017; Zhang et al. 2021). These factors, in combination with ecological and edaphic ones, regulate weed seed dormancy in agricultural fields (Ikram et al. 2014).

In most weed species, seasonal patterns of seed dormancy have been documented, and factors such as temperature, light duration, rainfall, and humidity during seed development influence seed dormancy (Javaid et al. 2010). Seed dormancy has been reported for several weed species, including *Zaleya pentandra* (L.) C. Jeffrey, *Rhynchosia capitata* (RC), *Euphorbia dracunculoides* Lam., *Convolvulus arvensis* L., *A. fatua* L., *P. minor* Retz., *L. aphaca* L., *R. crispus* L., *F. polymorpha* L., *Parthenium hysterophorus* L., and *Astragalus* spp. in Pakistan (Hassan et al. 2004; Javaid et al. 2010; Ali et al. 2011, 2012; Ikram et al. 2014; Munawar et al. 2015).

Various methods of dormancy breaking involving hot water, dry heat, cold treatment (stratification), and scarification for different durations have been investigated under laboratory studies to break seed dormancy in weed species (Ali et al. 2012; Munawar et al. 2015). In addition to this, different chemicals, like hydrochloric acid (HCl), potassium nitrate (KNO_3), sulfuric acid (H_2SO_4), and thiourea [$SC(NH_2)_2$] have also been extensively studied for their possible role as seed dormancy-breaking agents in various weed species (Ali et al. 2012; Ikram et al. 2014; Munawar et al. 2015).

Estimation of weed seed dormancy and its characterization is helpful in predicting the timing of weed seedling emergence in response to environmental cues, weed survival rate, and success of weed management interventions under field conditions, as dormant weed seeds escape the application of herbicides and emerge as flushes over the years. Nevertheless, the prediction of weed emergence timing and pattern is a complex phenomenon requiring the knowledge of trait evolution and weed ecology in the backdrop of changing climatic scenarios and altered cropping practices (Nakabayashi and Leubner-Metzger 2021). Realistic simulation of eco-physiological processes and population-based models can be effective in integrating dormancy responses and persistence across generations and diverse agroecosystems (Fernández Farnocchia et al. 2021). Improved dormancy-breaking methods under field conditions, such as alternating temperature caused by tillage practices, positioning weed seeds near the soil surface, increasing exposure to light and soil wetting as well as drying, could be helpful for efficient weed management. Researchers should also investigate trends in seed dormancy of important noxious weed species in terms of future climate change scenarios.

The term "seed bank" describes the collection of viable weed seeds/vegetative propagules present in the soil of a field, which are the source of future weed infestations (Tanveer 2008; Chauhan and Johnson 2010; Zamljen et al. 2024). It comprises both newly shed seeds (seed rain) and those that have persisted in the soil over time. The seed banks are highly dynamic and respond to the changes in inputs and losses (Forcella 2003). Weeds escaping weed control or late-emerging flushes after the application of post-emergence herbicides contribute toward the indigenous seed bank. Seeds of new weed species may be introduced via dissemination by external sources, and upon successful establishment and completion of the "seed to seed cycle," can result in permanent establishment in that particular field (especially in the absence of effective control measures). Distribution of weed seeds is highly heterogeneous within seed banks, and weed seeds occur in both vertical and horizontal planes, and hydrological events can redistribute weed seeds horizontally (Tanveer 2008).

The size and composition of the weed seed bank vary in time and space and respond to changes in agroclimatic conditions. Mahmood et al. (2012) found significant differences regarding weed seed bank size sampled from 0 to 8 and 8 to 15 cm soil depths in wheat fields. The size of the seed bank was highest for *Cyperus rotundus* followed by *Convolvulus arvensis*, *F. parviflora* Lam., and *A. fatua*. Ali et al. (2022) noted a remarkable increase of 61% in the weed seed bank under untreated control plots of wheat after one season, which substantiates the need for effective weed management. In this study, the initial seed bank before sowing was 38,624 weed seeds m^{-2}, which increased to 62,186 seeds m^{-2} at harvesting in untreated control plots. In another study, it was found that the majority of weed seeds emerged from the upper 5 cm soil surface (Khan et al. 2012). While appraising the status of weed seed banks in Chitral, Pakistan, Luqman et al. (2021) found 31 weed species (23 broadleaf, 8 grasses, and 1 sedge) belonging to 17 distinct families. Additionally, out of these 31 weeds, 24 were annual and the rest were perennials.

Seasonal variations are also evident, and the germinable seed bank was fourfold higher in the spring cropping season compared to the autumn cropping season (Zamljen et al. 2024). The persistent seed bank was consistently higher (117,000 seeds m^{-2}) as compared to the germinable seed bank (27,000 seeds m^{-2}) in this study. It is well established that the density and floristic composition of weed seeds in the seed bank are directly related to the initial weed population in the field

(Teasdale et al. 2004; Brainard et al. 2008). Estimates indicate that only 1–9% of the viable seed produced in a given year or season develop into seedlings, and out of total deposits to the seed bank, 95% come from annual weeds. The few dominant weeds that adapt themselves to the cropping system and disturbances account for 70–90% of the total seed bank (Tanveer 2008; Hosseini et al. 2014). Nevertheless, noxious weeds like *C. rotundus* emerging throughout the year can contribute to the seed bank enormously and continuously, as compared to weeds like *Solanum nigrum* L., depicting a limited period of emergence and limited annual addition to its seed bank.

Weed seed bank reflects the physical history of failures and successes of past and current soil and weed management practices, and knowledge of its size and composition can help to not only anticipate weed infestation in terms of qualitative and quantitative differences but also ameliorate the same. For example, the largest weed seed bank was noticed under an organic corn production system as opposed to a conventional system (Barberi et al. 1998), and such differences in weed seed banks among corn production systems were attributed to variation in weed control efficacy rather than due to the tillage. Moreover, the relative importance of troublesome weeds (*A. retroflexus*, *C. album*, *D. sanguinalis*, and *S. nigrum*) of corn in the cropping system in this study varied according to the corn management system opted. Contrarily, Maqsood et al. (2018) reported that tillage systems were more decisive in modulating weed seed density than weed management practices in corn. The study by Zamljen et al. (2024) reported greater diversity and evenness in both persistent and germinable seed banks of weeds under conservation and no-till than the conventional tillage system. Depleting the seed bank by reducing both persistent and germinable seeds can be an effective strategy to limit potential weed infestation in field crops. The behavior of the weed seed bank in response to biotic as well as abiotic factors and crop husbandry needs more explicit studies, especially under divergent cropping systems currently prevailing in various parts of the country, to develop rational weed management practices.

8.2.3 Weed Seed Ecology

A better understanding of weed seed ecology could lead toward effective prevention, management, and even control of weeds. In this pursuit, a lot of work has been undertaken in Pakistan and across the globe. Eco-biology and related management aspects of economically important weeds like *Avena* spp. (Bajwa et al. 2017), *Conyza* spp. (Bajwa et al. 2016b), *C. rotundus* (Peerzada 2017), *E. colona* (Peerzada et al. 2016), *E. crus-galli* (Bajwa et al. 2015), *P. hysterophorus* (Shabbir et al. 2012, 2023; Tanveer et al. 2015; Bajwa et al. 2016a), *P. minor* (Shad and Siddiqui 1996; Yasin et al. 2011), *S. marianum* (Khan et al. 2009), *S. halepense* (Peerzada et al. 2023), and invasive weeds in general (Nasim and Shabbir 2012) have been extensively reviewed. Weed species vary considerably regarding their germination timing and requirements, including soil temperature, water availability, light, soil pH, and nutrient levels, leading to different emergence times and ecological niches.

Studies on the seed germination ecology of weeds like *Asphodelus tenuifolius* Cav. (Tanveer et al. 2014), *Cleome viscosa* (Javaid et al. 2020), *Convolvulus arvensis* (Tanveer et al. 2013a), *Cucumis melo* var. *agrestis* (Tanveer et al. 2012), *Emex* spp. (Javaid and Tanveer 2014), *Lathyrus aphaca* (Aamir et al. 2019), *Leptochloa chinensis* (Safdar et al. 2024), *Lolium perenne* (Javaid et al. 2022), *Rhynchosia capitata* (Ali et al. 2012), and *Trianthema portulacastrum* (Tanveer et al. 2013b) are available. The significant findings of these studies are summarized in Table 8.2.

Germination behavior of weeds can be categorized into quasi-simultaneous (the majority of weed seeds lying on or in the upper layer of soil germinate in a brief period) or intermittent (seeds germinate at irregular intervals over weeks, months, or years, and weed flushes are also common) patterns (Salisbury 1961). Weeds exhibit an intermittent pattern as a survival strategy, where seeds with varying degrees of dormancy germinate over extended periods (Qaderi 2023). Numerous factors such as light, temperature, moisture, soil reaction, salinity, surface mulch, and seed burial depth influence the intricate process of weed seed germination (Matloob et al. 2015a, b, 2020). For *S. halepense*, *C. rotundus*, *T. portulacastrum*, and *D. aegyptium*, germination commenced earlier under reduced light as compared to natural conditions. However, across various light intensities, the final germination percentage was higher under natural sunlight (Shad et al. 1989, Table 8.3). Nevertheless, this study reported numerical values regarding germination percentage without statistical significance, although statistical differences among means were reported for other parameters such as plant height and shoot fresh weight of the aforementioned weed species.

Seeds of "cool-season" weeds tend to germinate at much lower temperatures than do the seeds of "warm-season" species (Tanveer 2008). A day/night temperature of 20/10 °C promoted germination of scarified seeds of *Cressia cretica* (Khan 1991; Table 8.4). Soil temperatures are critical for seed germination and subsequent seedling growth, especially in areas where soils freeze, which can affect the survival of vegetative structures of perennial weeds. Germination of *A. fatua* and *Sinapis arvensis* L. was greatly suppressed due to limited oxygen availability in compacted soils. Compared to the compacted soil, cultivation enhanced the germination of aforementioned weeds by six- and twofold, respectively (Tanveer 2008). Germination of many weeds is inversely related to the burial depth of the seed in the soil from about 1 cm downward. Maximum emergence of grass and broadleaf weeds was recorded from 0 to 10 cm layers (Hassan et al. 2003; Table 8.5). A progressive increase in soil seed reserves and species richness was recorded with an increase in annual rainfall (Naeem et al. 1993; Table 8.6).

Altering the seed distribution within the soil profile and increasing the frequency of disturbances can change the germination dynamics of weeds in the field. Under adequate soil moisture and temperature regimes, soil cultivation may stimulate 0.5–6% of weed seeds present in soil to germinate (Tanveer 2008). A better knowledge of factors influencing weed seed germination will help rationalize the management options to suppress weed seed germination or stimulate the same so that weed seedlings can be killed later on by manipulating tillage practices or non-selective herbicide/s (Chauhan 2012).

Table 8.2 Key findings and ecological optima of different weeds reported from weed seed germination ecology undertaken in Pakistan

Test species	Temperature (°C)	Light/dark conditions (h)	Salt stress (NaCl mM)	Osmotic potential (MPa)	pH	Burial depth (cm)	Significance	Reference
Asphodelus tenuifolius	Highest (80%) germination at 15 °C and lowest (32%) at 25 °C	87% germination under 10 h photoperiod and 78% germination under complete darkness	88% germination at 0 mM NaCl, while only 8% seeds germinated at 150 mM NaCl	Decrease in germination from 87% to 8% as drought stress increased from 0 to −0.6 MPa	Highest germination (98%) at pH 7	79% germination from seed placed on the surface, while only 3% germination when placed at 4 cm	Major weeds of winter crops (chickpea, lentil, and wheat) in rainfed areas	Tanveer et al. (2014)
Cleome viscosa	Highest (81%) germination at 30 °C and no germination at 15 °C	Not studied	83% germination at 0 mM NaCl, while only 24% seeds germinated at 250 mM NaCl, with complete inhibition at 300 mM	Reduction from 87.5% at 0 MPa to 27.5% at −0.8 MPa	Highest germination (79%) at pH 7 and complete inhibition at pH 10	87.5% germination from the surface as against 55% and 30% at 3 and 4 cm burial depths	Fast-growing herb found in disturbed sites, gardens, paddies, pastures, and riverbanks	Javaid et al. (2020)

Convulvulus arvensis	Highest germination (58.33 and 48.33) at 20 and 25 °C, respectively, with no germination at 45 °C	Not studied	Seed germination response varied as a function of seed size, with large seeds showing 50% of the maximum germination at 235.3 mM	50% inhibition of maximum germination occurred at −0.4 MPa	Highest germination (47.50%) at pH 7, which decreased either below or above this pH	Regardless of seed size, burial depth negatively affected the germination	Problematic weed of wheat and other winter cereals especially under irrigated conditions	Tanveer et al. (2013a)
Cucumis melo var. *agrestis*	Temperatures (25–40 °C) had a nonsignificant effect on seed germination	Not studied	Maximum seed germination (80%) with distilled water and minimum (15%) at 150 mM NaCl	Reduction from 70% at 0 MPa to 2.5% at −0.6 MPa	Not studied	Highest (62.5%) germination at 1 cm depth, below and above this depth, germination was reduced	Branched prostrate weed of summer crops (sorghum, pearl millet, cotton, and corn)	Tanveer et al. (2012)
Emex spp.	Highest germination in *E. australis* and *E. spinosa* occurred at 20/12 °C (light/dark), no germination at 35/25 and 10/5 °C	Maximum germination under 10 h photoperiod in both species. Seeds were still able to germinate in complete darkness	Germination at 50 mM and 200 mM was 87% and 21%, respectively. Complete inhibition at 250 mM	For *E. spinosa*, germination decreased from 98% to 10% from 0 to −0.6 MPa For *E. australis*, germination decreased from 84% to 0% from 0 to −0.6 MPa	No significant differences in germination at pH 6–8	Similar emergence for the seed placed on the surface and buried at 2 cm. Further increase in burial depth reduced germination	Invasive problematic weeds of winter crops on light-textured shallow soils	Javaid and Tanveer (2014)

(continued)

Table 8.2 (continued)

Test species	Temperature (°C)	Light/dark conditions (h)	Salt stress (NaCl mM)	Osmotic potential (MPa)	pH	Burial depth (cm)	Significance	Reference
Lathyrus aphaca	Highest germination (91.6%) at 15/10 °C (light/dark)	Germination was stimulated by 10/14 and 12/12 light/dark periods as compared to complete darkness	92% germination at 0 mM NaCl, while 42% seeds germinated at 250 mM NaCl	Germination (91.6%) at 0 MPa was reduced to 45% and 11.6% at −0.6 and −0.8 MPa	Highest germination at neutral pH, which decreased afterward at 9.0	Highest germination at 2 cm burial depth, which decreased as the depth increased beyond this value, with no germination at 12 cm	An important broadleaf weed of marginal lands in a rainfed cropping system	Aamir et al. (2019)
Leptochloa chinensis	Highest germination (92%) at 35 °C, which decreased at 40 and 45 °C	Not studied	No germination at 200 mM NaCl, with 50% inhibition of the maximum germination occurring at 102 mM NaCl	Not studied	Highest germination (77–78%) at pH 6–7, which decreased at higher pH	Emergence was greatly reduced as burial depth increased from 11 to 5 cm, beyond which no seed emerged. Likewise, emergence also declined as the flooding depth was increased from 0 to 4 cm	Difficult-to-control weed of aerobic rice cultures with widespread tolerance to ALS inhibitors	Safdar et al. (2024)

Lolium perrene	Highest germination (92%) at 25 °C, which decreased at 35 °C	Not studied	The highest germination was at 0 mM NaCl, while only 15% seeds germinated at 250 mM NaCl. 50% inhibition of the maximum germination occurring at 170.6 mM NaCl	Germination was not affected by osmotic potentials up to −0.4 MPa but inhibited completely at 0.8 or −1.0 MPa	No effect of pH range 5–10 on seed germination	90% germination from 1 cm burial depth, with no seed germinating from surface or 6–7 cm depths	An important pasture and forage weed	Javaid et al. (2022)
Rhynchosia capitata	Highest (97.5%) seed germination at 30 °C and lowest (17.5%) at 45 °C	Seeds germinated in both light and dark conditions	Highest germination (90%) at 0 mM NaCl, while only 10% seeds germinated at 200 mM NaCl. Complete inhibition at 250 mM. 50% inhibition of the maximum germination occurring at 137.6 mM NaCl	Germination (100%) at 0 MPa was reduced to 15% at −0.6 MPa	Seed germination was statistically alike at pH range 6–8 with a reduction observed at below or above this range	Seedling emergence decreased with an increase in burial depth. Some seedlings still emerged from 10 cm	An emerging summer annual weed in Southern Punjab	Ali et al. (2012)

(continued)

Table 8.2 (continued)

Test species	Temperature (°C)	Light/dark conditions (h)	Salt stress (NaCl mM)	Osmotic potential (MPa)	pH	Burial depth (cm)	Significance	Reference
Trianthema portulacastrum	85% germination at 35 °C and lowest (71.5%) at 45 °C	Not studied	More than 80% germination up to 150 mM and only 10% germination at 250 mM. 50% inhibition of the maximum germination occurring at 193.8 mM NaCl	Germination (100%) at 0 MPa was reduced to 20% at −0.8 MPa	Maximum germination at alkaline pH (8–10)	Germination for surface located seeds was high with reduction realized as burial depth increased. A seed burial depth of 2.7 cm reduced emergence by 50%	A major broadleaf weed of the summer season capable of causing complete crop failure	Tanveer et al. (2013b)

Table 8.3 Percent germination of different weed species in response to different light intensities

Light intensities	Weed species							
	Sorghum halepense		*Cyperus rotundus*		*Trianthema portulacastrum*		*Dactyloctenium aegyptium*	
	6 DAS	Final	6 DAS	Final	6 DAS	Final	6 DAS	Final
Natural light	0	62	0	98	0	48	0	72
2000 Lux	16	56	44	68	4	58	0	36
1000 Lux	12	26	36	70	10	62	6	40
500 Lux	16	46	40	90	10	64	4	48
Below 500 Lux	36	–	30	–	50	–	52	–

Shad et al. (1989)
DAS days after sowing

Table 8.4 Germination of scarified and non-scarified seeds of *Cressa cretica* at different temperatures and salinity levels

	Scarified seeds				Non-scarified seeds			
NaCl salinity (%)	0	1	3	5	0	1	3	5
Day/night temperature (°C)								
20/10	89 ± 6.5	78 ± 4.5	39 ± 8.1	16 ± 4.2	25 ± 2.0	18 ± 1.8	8 ± 1.4	4 ± 0.6
25/10	60 ± 4.3	32 ± 3.1	20 ± 1.2	9 ± 2.4	15 ± 1.8	9 ± 1.1	5 ± 1.2	3 ± 1.1
25/15	32 ± 1.2	30 ± 4.2	16 ± 1.6	6 ± 1.0	10 ± 1.9	7 ± 0.8	4 ± 0.6	2 ± 0.5
30/20	28 ± 2.8	20 ± 1.8	4 ± 1.1	0 ± 0.0	8 ± 4.0	6 ± 4.00	3 ± 0.8	0 ± 0.0

Khan (1991)
± Standard error of mean

Table 8.5 Effect of soil depth on weed seed germination

Soil depth (cm)	Germination (number per pot)	
	Grassy weeds	Broadleaf weeds
0–10	3.30[a]	7.20[a]
10–20	2.10[a, b]	3.55[a, b]
20–30	1.10[b]	0.90[b]

Hassan et al. (2003)
[a]SD p≤0.05 is 1.811 for grassy weeds
[b]LSD p≤0.05 is 4.790 for broadleaf weeds

Table 8.6 Relation of annual rainfall to weeds in wheat fields of the rainfed area of Punjab, Pakistan

Annual rainfall (mm)	Number of weed species	Total number of seedlings
150–300	10	128
300–500	14	276
500–1000	23	1587
1000–1500	27	2756

Naeem et al. (1993)

In order to develop sustainable and site-specific weed management strategies, integration between weed biology and management practices needs to be considered and tailored to the specific cropping systems. In addition to weed seed ecology studies conducted under controlled conditions, exclusive studies pertaining to seed bank dynamics under different cropping systems, dispersal mechanisms and vectors involved, along with their relative importance, traits conferring persistence among economically significant weeds, and management implications based on ecological knowledge are lacking in the Pakistani context. Traditional short-term studies on germination ecology should be extended to comparative germination ecology across multiple populations and their subsequent generations under different environmental stressors. In Pakistan, weed research is frequently misconstrued and limited to herbicide efficacy trials conducted under field conditions (Matloob et al. 2020). This narrow perspective undermines the broader scope of the discipline and does not focus on the real breadth of weed science, which is basically based on the ecology and biology of weeds. A holistic approach to the biology of weeds, their species interactions, and ecological processes is critical for the development of sustainable and integrated plans for weed management in the wake of increasing challenges like herbicide resistance and climate change.

8.2.4 *Weed Phenology*

Phenology is the study of growth and development events (emergence, days to flowering, flowering period, seed retention and shedding period, time lapse from emergence to maturity, etc.) with reference to calendar or thermal time. For any particular weed species, these attributes are confined either to calendar time (Julian date) or growing degree days (GDDs). The GDDs reflect the summation of the number of degrees Celsius when the day's average temperature was above the base temperature. Below the base temperature, seeds remain quiescent, and when thermal time in terms of GDDs accumulates above the base temperature (differential species thresholds), seed germination and subsequent development proceed.

Phenology deals with functional rhythms of the plants or their development dynamics, where various phenological events are matched or fine-tuned to the environment (representing both biotic and abiotic conditions) with emphasis on the inherent schedule of events. Phenotype influences phenology, while phenology regulates the developmental plasticity of weed species (Ghersa and Holt 1995). In wheat fields, 95% of weeds were at the vegetative phase in January, while during March, 61% and 26% of weeds were at the flowering and fruiting stages, respectively (Hussain et al. 2009). Observed variations in flowering response are often ascribed to differences in seasonal temperature ranges, and in most studies, the blooming period commenced from the onset of spring and lasted till the end of summer (Asim et al. 2016; Haq et al. 2023). For example, a study revealed that July was the peak flowering season, wherein 80 plant species (26.1%) had flowering, and it was followed by June, during which 77 species (25.1%) showed flowering. Peak fruiting for 78 species was noticed during September (Rahman et al. 2018).

The success of weed management largely depends on how and when weeds grow, and synchronizing management timing with "weed happening time" is crucial. In this backdrop, herbicide labels often relate application timing to the weed growth stage or size. Moreover, the success of harvest weed seed control is also based on the time when weeds are ready to shed their seeds, and this varies across weeds due to innate species traits, year, cropping systems, and environments. Differences in seed retention could translate into variable levels of harvest weed seed control (Barker 2022). Piskackova and Leon (2022) reinforced the idea of using weed emergence timing and phenological development for effective decision-making. This approach could complement the emergence and critical period of weed-crop competition models to appraise the time of implementing and maintaining weed control and assess crop risks beyond the economic thresholds of weed competition.

8.3 Weed Dynamics and Associated Challenges in Major Crops

8.3.1 Wheat and Rice

Weeds can become serious limiting factors, especially where tillage intensity is reduced. Under conservation tillage systems, early weed seedling emergence from fresh seed banks developed near the soil surface, lack of disturbance to the root system of perennial weeds, herbicide interception by crop residues, and changes in pattern and timing of weed emergence are important challenges (Matloob et al. 2015a, b). Moreover, aspects such as mechanical damage to the roots of weed plants, seedling uprooting by implements, exposure of weed seeds to the sunlight, and predation are also compromised due to the absence of soil disturbance under conservation tillage systems. Floristic changes in weed community dynamics in response to tillage practices, crop establishment techniques, and irrigation regimes have been documented in resource-conserving rice-wheat cropping systems (Matloob et al. 2015a, b; Farooq and Nawaz 2014; Nawaz and Farooq 2016).

Fields of DSR are dominated by diverse and difficult-to-control weeds, while the wheat fields established under zero tillage (ZT) as a resource conservation technique are dominated by perennial grasses (Kumar and Ladha 2011; Matloob et al. 2015b; Bhatt et al. 2016; Nawaz and Farooq 2016). New, difficult-to-control grass weeds, such as *A. myosuroides* and *L. temulentum*, do infest the wheat fields as well. Before the Green Revolution, *C. oxycantha* was the dominant weed of wheat fields. Nevertheless, increased inputs of water, nutrients, and tillage coupled with enhanced cropping intensity have almost eliminated this weed from irrigated areas. Likewise, *A. fatua* is also being replaced by *A. myosuroides* on heavy soils where rice is grown in the Kallar tract of Punjab, Pakistan.

In the backdrop of anticipated changes in cropping systems and associated rise in climatic variables like temperature and CO_2, weeds like *P. minor* will continue to dominate and cause greater interference in wheat fields (Matloob et al. 2020). Moreover, climate change is expected to alter the geographic distribution of weeds, inducing range shifts, transforming minor weeds to highly relevant species, increasing herbicide volatilization, and reducing rain-safe periods for timely and effective herbicide application (Matloob et al. 2020).

A switch from puddled transplanted rice (PTR) to DSR has altered weed pressure as well as weed spectrum. Terrestrial weeds that are typically not found in PTR fields, like *T. portulacastrum* and *Alternenathera philoxeroides* (Mart.) Griesb. dominated DSR systems in Pakistan (Farooq et al. 2011; Matloob et al. 2015a, b). Under conventional tillage (CONT), these weeds posed severe competition to DSR, especially early in the season and may result in crop failure by complete crop shading (Khaliq et al. 2010). These authors reported prolonged competition by grasses such as *Echinochloa* spp., *D. aegyptium*, and *E. indica*, and sedges (*C. rotundus* and *C. iria*) that lasted the whole growing season. Matloob (2014) reported that CONT favored the dominance of broadleaf weeds like *T. portulacastrum* in DSR, while grass weeds, especially *D. aegyptium*, were favored in DSR established under ZT (Fig. 8.1). Tillage practices opted for rice crop had a profound influence on weed growth in the subsequent wheat crop. For example, the density of *C. album* was significantly lower in wheat crops sown after PTR (Farooq and Nawaz 2014). Nevertheless, wheat grown on fields vacated by DSR had lower densities of *P. minor* and *R. dentatus*.

A major breakthrough in conservation tillage was the development of modern wheat drills capable of wheat sowing in rice residues without land preparation or CONT, enabling the ZT as the most promising resource conservation technology in the rice-wheat cropping systems (Erenstein and Laxmi 2008; Nawaz and Farooq 2016). Nawaz et al. (2017) advocated that DSR followed by ZT wheat can sustain the productivity and profitability of the rice-wheat cropping system as a resource-conserving approach. Under such ZT systems, less weed density was reported compared with wheat sown under CONT (Mann et al. 2008; Usman et al. 2012). However, annual grasses and perennial weed density increase in subsequent years (Nawaz and Farooq 2016), making available herbicide options redundant. In addition to its effects on weed growth, ZT can influence herbicide efficacy due to retention by surface residues (Kumar and Goh 2000). This aspect is seldom considered in most of the local studies available. Chauhan and Opeńa (2012) evaluated the effect of tillage systems (ZT and CONT) and herbicides on weed emergence, weed growth, and grain yield in DSR. The efficacy of herbicides (oxadiazon followed by fenoxaprop + ethoxysulfuron and oxadiazon followed by penoxsulam + cyhalofop) was lower in the ZT than in the CONT system. In Pakistan, years-long research and field experiences resulted in the development of successful weed control packages for the conventional sown rice and wheat crops (Matloob et al. 2020, 2022). However, with modifications in growing environments and establishment techniques such as water-saving methods of rice cultivation, adoption of permanent raised beds, and ZT systems, the existing weed control practices became ineffective or less important.

Fig. 8.1 Influence of (**a**) CONT—conventional tillage and (**b**) ZT—zero tillage on dynamics of weed emergence in direct seeded rice fields in Pakistan. Dominance of broadleaf weeds like *T. portulacastrum* in DSR was favored under CONT, while, grassy weeds especially *D. aegyptium* were favored in DSR established under ZT

Chemical weed control gained popularity owing to its apparent benefits. In contemporary agriculture, herbicide usage has increased, and now herbicides are considered a vital element to secure crops against weeds. Nevertheless, sole reliance on herbicides for controlling weeds needs to be discouraged due to the associated trade-off of weed flora shifts, evolution of herbicide-resistant biotypes, and seed bank buildup (Matloob et al. 2020; Heap 2023). Lack of a critical moisture regime at the time of application may render available PTR herbicides less suitable for weed control in DSR systems (Khaliq and Matloob 2012). Contrary to PTR, herbicide options in DSR are very limited. Pendimethalin, as a major pre-emergence herbicide in DSR, is less effective against sedges. The available post-emergence herbicide options are also limited to penoxsulam and bispyribac sodium, either alone or in combination with bensulfuron methyl. These herbicides also provide poor control of sedges and grasses such as *D. aegyptium* (Mann et al. 2007; Khaliq et al. 2012a, b, 2014). Matloob (2014) reported the tolerance of weeds like *C. rotundus* and *D. aegyptium* to the application of pre- (pendimethalin) and post-emergence (bispyribac sodium) herbicides in Pakistan. Weeds like *L. chinensis*, *D. aegyptium*, and *E. indica* showed tolerance to the herbicides like bispyribac sodium and penoxsulam applied in DSR (Khaliq et al. 2014; Matloob 2014). Therefore, weed control in DSR is a real challenge for growers and is still a major obstacle in its large-scale adoption (Jabran and Chauhan 2015).

Likewise, ACCase inhibitors such as fenoxaprop and pinoxaden controlled annual as well as perennial grass weeds poorly in ZT wheat (Matloob et al. 2020). Endemic rice and wheat herbicides targeting acetolactate synthase (ALS) enzyme represent major chemical weed control options, and their use is increasingly constrained by the evolution of herbicide resistance (Heap 2023). Currently, sulfonylurea and triazolopyrimidine sulfonanilide herbicides targeting ALS enzyme, along with auxin-type herbicides such as fluroxypyr, are ineffective against weeds like *C. arvensis*, which inflict severe damage in terms of resource competition and cause crop lodging (Fig. 8.2). Moreover, the timing of post-emergence herbicide application often coincides with the period of dense fog or smog, resulting in acute crop phytotoxicity. Hence, tackling complex challenges under such systems necessitates a holistic approach integrating principles of weed control with the wisdom of conservation agriculture (Farooq and Nawaz 2014; Bhatt et al. 2016). A better understanding of the cause-and-effect relationship pertaining to weed ecology in response to altered soil moisture regimes, stand establishment techniques, and soil physicochemical properties is needed to combat the weed menace and devise eco-efficient weed management practices (Matloob et al. 2020).

Fig. 8.2 *Convolvulus arvensis* inflicts severe damage in terms of resource competition and is difficult-to-control weed in irrigated wheat in Pakistan

8.3.2 Cotton

In cotton, farmers use intensive weed control by chemical and mechanical means. Early-season chemical weed control is confined to two herbicides, namely, s-metolachlor and pendimethalin, which are mostly used in Pakistan as a pre-emergence application for controlling weeds in cotton. Pendimethalin is poor against sedges (*C. rotundus*) and broadleaf weeds (*Digera* spp., *C. pluricaulis*, and *M. maderaspatana*). However, s-metolachlor 960EC at 2.5 L ha^{-1} gave efficient (93.8%) weed control and the highest seed cotton yields (2396 kg ha^{-1}) (Jarwar et al. 2005). For cotton crops sown under the bed-furrow system, which has dominated the flat sowing in recent years, increased proliferation of weeds such as *C. rotundus* has been observed due to the porous mass of pulverized soil, fine tilth, less compaction, and frequent irrigation. Moreover, conventional tillage practices involving cultivator and rotavator tend to fragment its rhizomes and tubers into smaller pieces and mix evenly within the upper soil profile, thereby promoting further growth of this weed. Manual weeding, although practiced by small-scale farmers, is inefficient in controlling this weed and increasingly constrained by a lack of labor and rising wages. Usually, shielded direct application of ethoxysulfuron is done to control sedges in cotton. For this, too, cotton plants should be of a certain age (45–60 days old).

The recent introduction of glyphosate-resistant cotton varieties has its own pros and cons, thus needs careful consideration and assessment of current interventions along with their future implications. Farmers of the cotton-wheat system using triple gene (Cry1Ac-Cry2Ab-EPSPS) cotton varieties are spraying glyphosate at least 1–2 times in the same summer season, and this frequency is further increased, especially under the rainy season of the monsoon. This might pose economic and ecological concerns. It is quite clear that over-dependence on glyphosate in herbicide-resistant cotton (as the sole weed control method) has caused the evolution of resistance in 43 weeds across the globe, and out of those, 16 weed species are reported from Australia (Iqbal et al. 2019). The widespread adoption of glyphosate-tolerant cotton in Australia from 2000/2001 growing season onward has caused evolution of numerous glyphosate-resistant weed biotypes (Werth et al. 2012, 2013; Heap and Duke 2018). In the rainy season, grasses like *E. colona* and *D. aegyptium* grow at a much faster pace and deprive the crop of essential inputs, impairing the quantity and quality of seed cotton. To control these weeds, aryloxyphenoxypropionate herbicides such as quizalofop and haloxyfop as ACCase inhibitors are largely used. However, weeds often grew bigger and set seeds while farmers were waiting for fields to dry and become ready for post-emergence application of grass killers.

8.3.3 Corn

Grain yield losses due to mixed weed flora in corn ranged from 35% to 50% (Maqsood et al. 1999; Maqbool et al. 2006, 2016). However, corn crop infestation by *Parthenium hysterophorus* alone, as an invasive weed, may cause a reduction of up to 50% in its grain yield (Safdar et al. 2015, 2016). Chemical weed control is the most commonly used method in Pakistan due to the availability of a broad range of herbicidal products in the country that give efficient control of corn weeds. The most commonly used selective herbicides for weed control in cornfields are atrazine, s-metolachlor, nicosulfuron, triasulfuron, metribuzin, mesotrione, and propisochlor that are used in different formulations and combinations (Matloob et al. 2020). Pre-emergence formulation of atrazine and s-metolachlor is used for early-season weed control (Ashiq and Aslam 2014; Matloob et al. 2020). Mix formulation of atrazine with mesotrione is currently the major post-emergence herbicide being used to control weeds in cornfields of Punjab, Pakistan. Recently, a formulation containing isoxaflutole and theincabrazone was launched in the market; however, its limited application window and high cost compared to the other herbicide options have limited its large-scale application. Moreover, under autumn planting occurring during the hot months of July and August, improper use of this product may cause acute crop stress. Formulations containing a combination of pendimethalin with acetochlor, or atrazine with halosulfuron may also cause phytotoxicity to the main crop.

8.3.4 Sugarcane

Sugarcane is an important industrial crop that is cultivated in two seasons (spring and autumn) in Pakistan. Being a long-duration crop, the critical weed-crop competition period in sugarcane is 90–120 days after sowing (Afghan 1996), which is longer than other field crops. Hot and humid conditions, high tillage intensity, high soil fertility, and frequent irrigation are conducive to the luxurious growth of diverse weed flora comprising annuals, biennials, and perennial plants. Tall perennial grasses like *S. halepense* offer severe competition and decrease cane density, inflicting cane and sugar yield losses. Weeds like *C. rotundus* and *C. dactylon* are found almost throughout the sugarcane growing season (Matloob et al. 2020). Incidence of non-crop weeds like *P. hysterophorus* is also increasing in sugarcane fields (Bajwa et al. 2016a; Shabbir et al. 2012, 2023). Considering recent climatic changes and associated uncertainties, prolonged drought spells and heat waves will favor C_4 and parasitic weeds like *Striga hermonthica* (Delile) Benth. and *Cuscuta* spp. (Matloob et al. 2020). Chemical weed control is prevalent; nevertheless, during hot summer months, triazines and sulfonylurea herbicides also pose phytotoxicity to the main crop, and repeated flushes of weeds are encountered, especially in the rainy season.

8.4 Non-chemical Weed Management

Besides herbicides, mulches can also be used for weed control in conservation tillage systems. Weed suppression is one of the several benefits of applying mulches in the agricultural fields (Jabran 2019, 2020). The mulches originating from the residues of previous crops probably suppress the weeds in two ways. First, these physically suppress the germination and growth of weeds. Second, these affect the weeds by exuding allelochemicals after decomposition. Plastic (black polyethylene) mulch, on the other hand, blocks the sunlight to prevent weed seed germination while the growth processes of already emerged weeds, including photosynthesis and respiration, are severely disturbed. Some studies indicate the success of both the straw and plastic mulch for weed control in DSR. In DSR fields, a 50% reduction in weed growth (density and biomass) with a concomitant yield increase of 31% was recorded in response to sorghum mulching at 6 t ha^{-1} (Mubeen et al. 2014). Incorporation of wheat residues resulted in a 47% reduction in weed density and biomass in DSR and increased the rice yield by 22.5% (Khaliq et al. 2015). Likewise, rice straw mulching at 15, 17.5, and 20 t ha^{-1} averted the weed density and biomass of grassy and broadleaf weeds in ZT wheat. An increase in mulching rate caused an increase in the suppression of weed biomass (Chaudhary and Iqbal 2013). Mulching of wheat residues reduced weed density and biomass by 43% and 45%, respectively, in DSR (Khaliq et al. 2011). In wheat sown on permanent raised beds and mulched with lawn-mowed *Cynodon dactylon* at 6 t ha^{-1}, density and biomass of noxious *Convolvulus arvensis* were reduced by 84% compared to the bare soil (Fig. 8.3). Application of sorghum straw as surface mulch or its soil incorporation averted *C. rotundus* biomass by 50% and 53% in cornfields (Mahmood and Cheema 2004).

Wheat stubble management options such as residue incorporation, burning, and retention also influenced weed growth in DSR (Haq et al. 2019). Compared to residue incorporation and burning, weedy check plots under wheat residue retention had 43–45% and 49–58% fewer weeds in 2013 and 2014, respectively. Another possibility is to grow *Sesbania* between rows of DSR and then use it as brown manure by killing it at 30 DAS with a broadleaf killer such as MCPA. In the main crop, intercropping of an allelopathic crop can also be opted for weed suppression in conservation tillage systems (Farooq and Nawaz 2014; Farooq et al. 2025).

Stale seedbed or suicidal germination can affect weed pressure in the following crop (Fig. 8.4). Pulverizing the soil 3–6 weeks before planting stimulates weed seed germination. Germinated weed seedlings are then killed by either tillage or using non-selective herbicides such as glyphosate (Matloob et al. 2015a, b). A stale seedbed ensures that the crop emerges and grows in a weed-free environment and has the ability to suppress the late-emerging weeds. However, this method cannot be used in ZT wheat because the land plowing is restricted in ZT. In wheat, inversion tillage with moldboard plow can bury weed seeds into deeper soil layers and can cause a significant reduction in the emergence of photoblastic seeds of annual weeds.

Fig. 8.3 Suppression of noxious *Convolvulus arvensis* due to mulching (4–5 cm thick layer) with lawn-mowed *Cynodon dactylon* at 6 t ha^{-1}, in the field condition. Loose cut straw with large size resulted in better weed suppression. (**a**) Weedy check. (**b**) Mulching with loose cut straw. (**c**) Herbicide treated. (**d**) Mulching with fine cut straw

Narrow row spacing can be another pragmatic approach to reduce weed burden and supplement herbicide efficacy. In wheat, narrow row spacing of 15 cm resulted in a 57% reduction in weed growth with a wheat yield advantage of 29% over wider row spacings of 22.5 and 30 cm (Shah et al. 2005; Abbas et al. 2009). Wider row spacing of 30 cm resulted in 34% more weed biomass in DSR than a 15 cm spaced crop (Chauhan and Johnson 2011). Weedy check plots of DSR sown at a row spacing of 10 cm had 24–27% less weed biomass as compared to 20 cm spacing (Khaliq et al. 2014). In puddled rice, maximum reduction in weed density (22 plants m^{-2}) and biomass (7.6 g m^{-2}) was realized when seedlings were transplanted at 15 cm (Ashraf et al. 2014). Likewise, in corn, reduced densities as well as dry biomass of *T. portulcastrum*, *Cyperus rotundus*, and *Convolvulus arvensis* were recorded for crops sown at 75 cm spaced rows as compared to 85 and 95 cm (Khalil et al. 2010). Reduction in weed growth under narrow row spacing with a concurrent increase in seed yield is well documented in corn (Maqbool et al. 2006; Saeed et al. 2010). Besides averting weed density and biomass, narrow row spacing was also effective in reducing reproductive output of problematic weeds like *E. colona* and *T. portulcastrum* in corn (Fahad et al. 2014). For sugarcane, a narrow row spacing of 60 cm caused a significant reduction in weed density and biomass; nevertheless, yield under this row spacing was 77 t ha^{-1}, which was lower compared to the yields of 92 and 84 t ha^{-1} realized in response to 120 and 90 cm spacing, respectively (Munsif et al. 2015).

Fig. 8.4 Influence of stale seedbed (SSB) practices on dynamics of weed flora for cotton crop in a cotton-wheat rotation. (**a**) SSB under ZT with glyphosate. (**b**) SSB under ZT without glyphosate. (**c**) SSB under shallow tillage without glyphosate

Labrada (2002) postulated the use of high seeding densities within the context of integrated weed management, and the same has been demonstrated for rice (Khaliq et al. 2012a, b; Awan et al. 2014a, b) and wheat (Munsif et al. 2014). Increasing the seed rate of DSR from 20 to 100 kg ha^{-1} reduced weed density by 38–47% (Ahmed et al. 2014). Seeding densities of 120–160 kg ha^{-1} reduced light interception by *S. marianum* to the tune of 87–89% in wheat (Khan and Marwat 2006). In corn, a high planting density of 60,000 plants ha^{-1} in conjunction with pre-emergence application of pendimethalin at 1.03 kg a.i. ha^{-1} resulted in efficient control of weeds along with the highest benefit-cost ratio (Ullah et al. 2008). Diversification of cropping systems and flexibility in rotating crops can reduce the weed menace in cropping systems by altering ecological conditions conducive to weed seed germination and seedling growth. Rotating wheat with sunflower, oilseed rape, berseem clover, and potato decreased the density of *P. minor* significantly (Malik et al. 1998; Brar 2002).

It is unequivocal that crop-weed interference is inevitable; therefore, judicious use of herbicides with diverse modes of action in combination with agronomic interventions can be effective against weeds. Synergistic interaction of cultural practices that provide crops a competitive advantage over weeds by reducing weed burden can significantly contribute toward crop competitive ability and herbicide performance (Khaliq and Matloob 2011). There is a growing thrust in the recent past to manipulate cultural practices in conjunction with herbicides as an integrated weed management tactic (Chauhan and Johnson 2010). Manipulation of stale seedbed (Kumar and Ladha 2011; Matloob et al. 2015a, b, 2020), brown manuring of *Sesbania* (Khalid 2014), allelopathic mulches (Jabran et al. 2015, 2018), retention of crop residues and stubbles (Khaliq et al. 2015; Haq et al. 2019), incorporation of allelopathic crops in rotation (Narwal 2000), intercropping as an allelopathic benefit (Iqbal et al. 2007), row orientation and spacing (Chauhan and Johnson 2011; Khaliq et al. 2014), seeding density (Khaliq et al. 2012a, b; Awan et al. 2014a, b, c), fertilization practices (Di Tomaso 1995; Chauhan and Abugho 2013), and harvest weed seed control (Shaner and Beckie 2013) can help in achieving the goals of sustainable weed management in cropping systems. Additionally, the implementation of strict quarantine legislations will prevent the invasion of alien weed species into the country.

8.5 Conclusions

Development and appraisal of integrated, cost-effective, sustainable, and site-specific weed management systems is a challenging task. Considering the huge agronomic and economic impacts of weeds on various crops and cropping systems in Pakistan (Table 8.1; Fig. 8.5), there is a dire need to devise eco-efficient and economically feasible weed management practices with less emphasis on chemicals or synthetic inputs. In this context, weed ecology and biology can provide practical bases for formulating weed management practices based on various aspects of the weed life cycle, pinpointing potential vulnerabilities and sensitive stages that can be strategically targeted in cropping systems to keep weeds in check. Due emphasis should also be given to the belowground management aspects of weeds, that is, weed seed banks, rather than merely controlling the aboveground emerged weed flushes with little consideration of seed rain and subsequent seed bank buildup. For this, weed management should be based on the eco-biology of the weeds and consider the phyto-sanitary context of the whole agroecosystem. Research on weed eco-biology and related management aspects will assist in the development of ecologically based weed management approaches under diverse agroecological conditions of Pakistan. Moreover, applied research on weed biology and ecology in the backdrop of climate change, especially under long-term system-level experiments, is needed to envisage current and anticipated weed problems.

Fig. 8.5 Weed infestation is a serious production challenge for major crops of Pakistan. (**a**) Wheat. (**b**) Cotton. (**c**) Corn. (**d**) Aerobic rice. (**e**) Sugarcane

References

Aamir M, Khan EA, Baloch MS, Aslam M (2019) Germination ecology of *Lathyrus aphaca*, a problematic weed of wheat crop under semi-arid conditions of Pakistan. Planta Daninha 37:e019185755

Abbas SH, Saleem M, Maqsood M, Mujahid MY, Saleem R (2009) Weed density and grain yield of wheat as affected by spatial arrangements and weeding techniques under rainfed conditions of Pothowar. Pak J Agric Sci 46:242–247

Abbas T, Nadeem MA, Tanveer A, Ali HH, Matloob A (2017) Evaluation and management ACCase inhibitor resistant littleseed canary grass (*Phalaris minor*) in Pakistan. Arch Agron Soil Sci 63:1613–1622

Afghan S (1996) Studies on biological, cultural and chemical control of sugarcane weeds. Pakistan Sugar J 10(2):3–9

Afridi SG, Sultan S, Butt ZA, Ahmad H (2015) Ecological prevalence of the weed species growing in maize and sugarcane fields of district Mardan, Pakistan. J Biodivers Environ Sci 7:1–7

Ahmad Z, Khan SM, Abd-Allah EF, Alqarawi AA, Hashem A (2016) Weed species composition and distribution pattern in the maize crop under the influence of edaphic factors and farming practices: a case study from Mardan, Pakistan. Saudi J Biol Sci 23:741–748

Ahmed S, Salim M, Chauhan BS (2014) Effect of weed management and seed rate on crop growth under direct dry seeded rice systems in Bangladesh. PLoS One 9:e101919

Alam SM (2003) Weeds and their ill-effects on main crops. Available at https://www.dawn.com/news/127357/weeds-and-their-ill-effects-on-main-crops. Accessed on 18 July 2024

Ali HH, Tanveer A, Nadeem MA, Asghar HN (2011) Methods to break seed dormancy of *Rhynchosia capitata*, a summer annual weed. Chil J Agric Res 71:483–487

Ali HH, Tanveer A, Nadeem MA (2012) Evaluation of some seed dormancy breaking methods on germination of *Rhynchosia capitata* (Roth DC). Pak J Weed Sci Res 18:423–432

Ali HH, Peerzada AM, Hanif Z, Hashim S, Chauhan BS (2017) Weed management using crop competition in Pakistan: a review. Crop Protect 95:22–30

Ali S, Hanif MA, Umar M, Manzoor A, Shafique O, Khan BA, Ahmed I (2022) Broad leaved weed's seed bank and flora dynamics in wheat as affected by different herbicides. J Agric Res 35:58–69

Ashiq M, Aslam Z (2014) Weeds and weedicides. Department of Agronomy, Ayub Agricultural Research Institute, Faisalabad & University of Agriculture, Faisalabad, pp 14–142

Ashraf U, Anjum SA, Ehsanullah KI, Tanveer M (2014) Planting geometry-induced alteration in weed infestation, growth and yield of puddled rice. Pak J Weed Sci Res 20:77–89

Asim IZ, Haq F, Iqbal A (2016) Phenology, life form and leaf spectra of the vegetation of Kokarai valley, district Swat. J Biodivers Environ Sci 9:23–31

Aslam F, Khaliq A, Matloob A, Tanveer A, Hussain S, Zahir ZA (2017) Allelopathy in agro-ecosystems: a critical review of wheat allelopathy-concepts and implications. Chemoecology 27:1–24

Awan TH, Chauhan BS, Cruz PCS (2014a) Physiological and morphological responses of *Ischaemum rugosum* Salisb. (Wrinkled Grass) to different nitrogen rates and rice seeding rates. PLoS One 9(6):e98255

Awan TH, Cruz PCS, Chauhan BS (2014b) Ecological significance of rice (*Oryza sativa*) planting density and nitrogen rates in managing the growth and competitive ability of itchgrass (*Rottboellia cochinchinensis*) in direct-seeded rice systems. J Pest Sci 88:427–438

Awan TH, Chauhan BS, Cruz PCS (2014c) Growth plasticity of Junglerice (*Echinochloa colona*) for resource use when grown with different rice (*Oryza sativa*) planting densities and nitrogen rates in dry-seeded conditions. Weed Sci 62:571–587

Bajwa AA, Jabran K, Shahid M, Ali HH, Chauhan BS, Ehsanullah (2015) Eco-biology and management of *Echinochloa crus-galli*. Crop Prot 75:151–162

Bajwa AA, Chauhan BS, Farooq M, Shabbir A, Adkins SW (2016a) What do we really know about alien plant invasion? A review of the invasion mechanism of one of the world's worst weeds. Planta 244:39–57

Bajwa AA, Sadia S, Ali HH, Jabran K, Peerzada AM, Chauhan BS (2016b) Biology and management of two important *Conyza* weeds: a global review. Environ Sci Pollut Res 23:24694–24710

Bajwa AA, Akhter MJ, Iqbal N, Peerzada AM, Hanif Z, Manalil S, Hashim S, Ali HH, Kebaso L, Frimpong D, Namubiro H, Chauhan BS (2017) Biology and management of *Avena fatua* and *Avena ludoviciana*: two noxious weed species of agro-ecosystems. Environ Sci Pollut Res 24:19465–19479

Barberi P, Cozzani A, Macchia M, Bonari E (1998) Size and composition of the weed seedbank under different management systems for continuous maize cropping. Weed Res 38:319–334

Barker B (2022) Weed phenology can impact control. Available at https://www.topcropmanager.com/weed-phenology-can-impact-control/. Accessed on 2 Jan 2024 at 11:05 am

Bhatt R, Kukal SS, Busari MA, Arora S, Yadav M (2016) Sustainability issues on rice–wheat cropping system. Int Soil Water Conserv Res 4:64–74

Brainard DC, Bellinder RR, Hahn RR, Shah DA (2008) Crop rotation, cover crop and weed management effects on weed seedbanks and yields in snap bean, sweet corn and cabbage. Weed Sci 56:434–441

Brar LS (2002) Current status of herbicide resistance in Punjab and its management strategies. In: Proceedings of international workshop on herbicide resistance and zero tillage in rice–wheat cropping system. CCSHAU, Hisar, India, pp 6–10

Chaudhary S, Iqbal J (2013) Weed control and nutrient promotion in zero-tillage wheat through rice straw mulch. Pak J Weed Sci Res 19:465–474

Chauhan BS (2012) Weed ecology and weed management strategies for dry seeded rice in Asia. Weed Technol 26:1–13

Chauhan BS, Abugho SB (2013) Fertilizer placement affects weed growth and grain yield in dry-seeded rice (*Oryza sativa* L.) systems. Am J Plant Sci 4:1260–1264

Chauhan BS, Johnson DE (2009) Influence of tillage systems on weed seedling emergence pattern in rainfed rice. Soil Tillage Res 106:15–21

Chauhan BS, Johnson DE (2010) Implications of narrow crop row spacing and delayed *Echinochloa colona* and *Echinochloa crus-galli* emergence for weed growth and crop yield loss in aerobic rice. Field Crop Res 117:177–182

Chauhan BS, Johnson DE (2011) Row spacing and weed control timing affect yield of aerobic rice. Field Crop Res 121:226–231

Chauhan BS, Opeña J (2012) Effect of tillage systems and herbicides on weed emergence, weed growth, and grain yield in dry-seeded rice systems. Field Crop Res 137:56–69

Chauhan BS, Singh RG, Mahajan G (2012) Ecology and management of weeds under conservation agriculture: a review. Crop Prot 38:57–65

Di Tomaso JM (1995) Approaches for improving crop competitiveness through manipulation of fertilization strategies. Weed Sci 43:491–497

Erenstein O, Laxmi V (2008) Zero tillage impacts in India's rice-wheat systems: a review. Soil Till Res 100:1–14

Fahad S, Hussain S, Saud S, Hassan S, Muhammad H, Shan D, Chen C, Wu C, Xiong D, Khan SB, Jan A, Cui K, Huang J (2014) Consequences of narrow crop row spacing and delayed *Echinochloa colona* and *Trianthema portulacastrum* emergence for weed growth and crop yield loss in maize. Weed Res 54:475–483

Farooq M, Nawaz A (2014) Weed dynamics and productivity of wheat in conventional and conservation rice-based cropping systems. Soil Till Res 141:1–9

Farooq M, Siddique HMK, Rehman H, Aziz T, Lee D, Wahid A (2011) Rice direct seeding: experiences, challenges and opportunities. Soil Till Res 111:87–98

Farooq O, Ul-Allah S, Matloob A, Hussain M, Farooq M (2025) Allelopathic weed control in conservation agriculture systems. In: Basch G, Gonzalez-Sanchez E, Geraghty J, Eslami SV, Duiker SW, Mkomwa S and Bartz M (eds) Weed management in conservation agriculture systems Burleigh Dodds Science Publishing Limited, Cambridge, UK, pp 109–132. https://doi.org/10.19103/AS.2024.0145.05

Fernández Farnocchia RB, Benech-Arnold RL, Mantese A, Batlla D (2021) Optimization of timing of next-generation emergence in *Amaranthus hybridus* is determined via modulation of seed dormancy by the maternal environment. J Exp Bot 72:4283–4297

Finch-Savage WE, Footitt S (2017) Seed dormancy cycling and the regulation of dormancy mechanisms to time germination in variable field environments. J Exp Bot 68:843–856

Forcella F (2003) Debiting the seedbank: priorities and predictions. In: Bekker et al (eds) Seedbanks: determination, dynamics and management. Aspects of applied biology 69. Association of Applied Biologists, Welles Bourne, pp 151–162

Ghersa CM, Holt JS (1995) Using phenology prediction in weed management: a review. Weed Res 35:461–470

Haq M, Khaliq A, Sheng Q, Matloob A, Hussain S, Fatima S, Aslam Z (2019) Weed growth, herbicide efficacy, and rice productivity in dry seeded paddy field under different wheat stubble management methods. J Integr Agric 18:907–926

Haq SM, Lone FA, Kumar M, Calixto ES, Waheed M, Casini R, Mahmoud EA, Elansary HO (2023) Phenology and diversity of weeds in the agriculture and horticulture cropping systems of Indian Western Himalayas: understanding implications for agro-ecosystems. Plants 12:1222

Hassan G, Khan H, Marwat KB, Khan MA (2003) Studies on weed seed dynamics in soil seed bank of rice-based cropping system of Pakistan. Pak J Weed Sci Res 9:259–264

Hassan G, Saeed M, Khan H (2004) Dormancy studies in some major weed seeds of rice based cropping system of Pakistan. Pak J Weed Sci Res 10:185–192

Heap I (2023) Current status of the International Herbicide-Resistant Weed Database. Available at https://www.weedscience.org/home.aspx. Accessed on 28 May 2023 at 2:45 pm

Heap I, Duke SO (2018) Overview of glyphosate-resistant weeds worldwide. Pest Manag Sci 74:1040–1049

Hosseini P, Karimi H, Babaei S, Mashhadi HR, Oveisi M (2014) Weed seed bank as affected by crop rotation and disturbance. Crop Prot 64:1–6

Hussain S, Ramzan M, Akhter M, Aslam M (2008) Weed management in direct seeded rice. J Anim Plant Sci 18:86–88

Hussain F, Shah SM, Fazal-e-Hadi, Asadullah (2009) Diversity and ecological characteristics of weeds of wheat fields of university of Peshawar botanical garden at Azakhel, district Nowshera, Pakistan. Pak J Weed Sci Res 15:283–294

Ikram RM, Tanveer A, Ata Z, Saqib M (2014) Dormancy studies on *Euphorbia dracunculoides* and *Astragalus* spp.: major weeds of arid areas. Planta Daninha 32:747–753

Iqbal J, Cheema ZA, An M (2007) Intercropping of field crops in cotton for the management of purple nutsedge (*Cyperus rotundus* L.). Plant Soil 300:163–171

Iqbal N, Manalil S, Chauhan BS, Adkins SW (2019) Glyphosate-tolerant cotton in Australia: successes and failures. Arch Agron Soil Sci 65:1536–1553

Jabran K (2019) Role of mulching in pest management and agricultural sustainability. Springer, Cham. https://doi.org/10.1007/978-3-030-22301-4

Jabran K (2020) Use of mulches for managing field bindweed and purple nutsedge, and weed control in spinach. Int J Agric Biol 23:1114–1120

Jabran K, Chauhan BS (2015) Weed management in aerobic rice systems. Crop Prot 78:151–163

Jabran K, Ullah E, Hussain M, Farooq M, Zaman U, Yaseen M, Chauhan BS (2015) Mulching improves water productivity, yield and quality of fine rice under water-saving rice production systems. J Agron Crop Sci 201:389–400

Jabran K, Uludag A, Chauhan BS (2018) Sustainable weed control in rice. In: Corres NE, Burgos NR, Duke SO (eds) Weed control: sustainability, hazards, and risks in cropping systems worldwide. CRC Press, Taylor & Francis Group, Boca Raton, pp 276–287

Jarwar AD, Baloch GM, Memon MA, Rajput LS (2005) Efficacy of pre- and post-emergence herbicides in cotton. Pak J Weed Sci Res 11:141–145

Javaid MM, Tanveer A (2014) Germination ecology of *Emex spinosa* and *Emex australis*, invasive weeds of winter crops. Weed Res. https://doi.org/10.1111/wre.12111

Javaid A, Shafique S, Shafique S (2010) Seasonal pattern of seed dormancy in *Parthenium hysterophorus* L. Pak J Bot 42:497–503

Javaid MM, Abbas Z, Waheed H, Mahmood A, Abbas T, Amin MM, Asif M (2020) Studies on germination ecology and seedling characteristics of *Cleome viscosa* as affected by various environmental factors. Pak J Bot 52:1911–1919

Javaid MM, Mahmood A, Alshaya DS, Alkahtani MDF, Waheed H, Wasaya A, Khan SA, Naqve M, Haider I, Shahid MA, Nadeem MA, Azmat S, Khan BA, Balal RM, Attia KA, Fiaz S (2022) Influence of environmental factors on seed germination and seedling characteristics of perennial ryegrass (*Lolium perenne* L.). Sci Rep 12:9522

Khalid W (2014) Effect of various sowing method and weed control practices on growth and yield of direct seeded medium grain rice. M.Sc. (Hons.) thesis. Department of Agronomy, University of Agriculture, Faisalabad, Pakistan

Khalil SK, Mehmood T, Rehman A, Wahab S, Khan AZ, Zubair M, Mohammad F (2010) Utilization of allelopathy and planting geometry for weed management and dry matter production of maize. Pak J Bot 42:791–803

Khaliq A, Matloob A (2011) Weed crop competition period in three fine rice cultivars under direct seeded rice culture. Pak J Weed Sci Res 17:229–243

Khaliq A, Matloob A (2012) Germination and growth response of rice and weeds to herbicides under aerobic conditions. Int J Agric Biol 14:775–780

Khaliq A, Matloob A, Farooq M, Aslam F (2010) Weed occurrence and yield losses in dry direct seeded rice. In: 22nd Conference of Asian Pacific Weed Science Society, Lahore, Pakistan, March 08–12, 2010, p 12

Khaliq A, Riaz MY, Matloob A (2011) Bio-economic assessment of chemical and non-chemical weed management strategies in dry seededfine rice (*Oryza sativa* L.). J Plant Breed Crop Sci 3:302–310

Khaliq A, Mahmood S, Matloob A, Khan MB, Awan IU (2012a) Optimizing seeding density and tank mixture of herbicides help reduce yield losses in dry seeded fine rice. Pak J Weed Sci Res 18:167–181

Khaliq A, Matloob A, Mahmood S, Abbas RN, Khan MB (2012b) Seeding density and herbicide tank mixture furnishes better weed control and improve the growth, yield and quality of direct seeded fine rice. Int J Agric Biol 14:499–508

Khaliq A, Matloob A, Chauhan BS (2014) Weed management in dry seeded fine rice sown under varying row spacing in a rice-wheat system of Pakistan. Plant Prod Sci 17:321–332

Khaliq A, Matloob A, Hussain A, Hussain S, Aslam F, Zamir SI, Chattha MU (2015) Wheat residue management options affect crop productivity, weed growth, and soil properties in direct-seeded fine aromatic rice. Clean Soil Water Air 43:1–7

Khan MA (1991) Studies on germination of *Cressa cretica* L. seeds. Pak J Weed Sci Res 4:89–98

Khan MA, Marwat KB (2006) Impact of crop and weed densities on competition between wheat and *Silybum marianum* Gaertn. Pak J Bot 38:1205–1215

Khan MA, Blackshaw RE, Marwat KB (2009) Biology of milk thistle (*Silybum marianum*) and the management options for growers in north-western Pakistan. Weed Biol Manag 9:99–105

Khan I, Jan AU, Gul F, Ali K, Ali G, Ahmed S, Ahmad N (2011) Weeds as a major constraint in wheat production in district Peshawar. Pak J Weed Sci Res 17:381–386

Khan AM, Hussain Z, Khan I, Khan R, Waqas M, Haroon M, Ali Z, Khan IA (2012) Studies on weed seed bank of new developmental farm, KPK Agricultural University Peshawar, Pakistan. Pak J Weed Sci Res 18:183–189

Khan R, Khan MA, Shah S, Uddin S, Ali S, Ilyas M (2016) Bioherbicidal potential of plant extracts against weeds of wheat crop under agro-climatic conditions of Peshawar-Pakistan. Pak J Weed Sci Res 22:285–294

Khatam A, Khan MZ, Nawab K, Mian IA, Ahmad W (2013) Effect of various herbicides and manual control on yield, yield components and weeds of maize. Pak J Weed Sci Res 19:209–216

Kumar K, Goh KM (2000) Crop residues and management practices: effects on soil quality, soil nitrogen dynamics, crop yield, and nitrogen recovery. Adv Agron 68:197–319

Kumar V, Ladha JK (2011) Direct seeding of rice: recent developments and future needs. Adv Agron 111:297–413

Labrada R (2002) The need for improved weed management in rice. In: Proceedings of 20th session of the International Rice Commission. Sustainable rice production for food security, July 23–26, 2002, Bangkok, Thailand, pp 1–11

Luqman, Hussain Z, Bakht T, Miftah-Ud-Din, Khan H, Khan MR, Ullah A, Shah FA, Khan I, Dawar AMK (2021) Soil weed seed bank status in the agroecological condition of Chitral, Pakistan. Pak J Weed Sci Res 27:253–262

Mahajan G, Chauhan BS, Timsina J (2011) Opportunities for weed control in dry seeded rice in North-Western Indo-Gangetic Plains. In: Alvarez-Farnandez R (ed) Herbicides-environment impact studies and management approaches. InTech Publishers, Rejika, pp 199–208

Mahmood A, Cheema ZA (2004) Influence of sorghum mulch on purple nutsedge (*Cyperus rotundus* L.). Int J Agric Biol 6:86–88

Mahmood S, Hussain A, Malik SA (2012) Persistence of some weed species from wheat (*Triticum aestivum* L.) monoculture via soil seed reserves. Pak J Bot 44:1375–1379

Malik RK, Gill G, Hobbs PR (1998) Herbicide resistance – a major issue for sustaining wheat productivity in rice–wheat cropping systems in the Indo–Gangetic Plains. Rice–Wheat Consortium Paper Series 3. Rice–Wheat Consortium for the Indo–Gangetic Plains, New Delhi, 36p

Manalil S, Coast O, Werth J, Chauhan BS (2017) Weed management in cotton (*Gossypium hirsutum* L.) through weed-crop competition: a review. Crop Prot 95:53–59

Mann RA, Ahmad S, Hassan G, Baloch MS (2007) Weed management in direct seeded rice crop. Pak J Weed Sci Res 13:219–226

Mann RA, Ramzan M, Munir A (2008) Improving the sustainability of wheat production in irrigated areas of Punjab, Pakistan through conservation tillage technology. Int J Agric Biol 10:249–254

Maqbool MM, Tanveer A, Ata Z, Ahmad R (2006) Growth and yield of maize (*Zea mays* L.) as affected by row spacing and weed competition durations. Pak J Bot 38:1227–1236

Maqbool MM, Tanveer A, Ali A, Abbas MN, Imran M, Ahmad M, Abid A (2016) Growth and yield response of maize (*Zea mays*) to inter and intra-row weed competition under different fertilizer application methods. Planta Daninha 34:47–56

Maqsood M, Akbar M, Yousaf N, Mahmood MT, Ahmed S (1999) Studies on weed crop competition in maize. Int J Agric Biol 4:270–272

Maqsood Q, Abbas RN, Khaliq A, Zahir ZA (2018) Weed seed bank dynamics: weed seed bank modulation through tillage and weed management. Planta Daninha v36:e018166706

Marwat KB, Saeed M, Hussain Z, Gul B, Rashid H (2008) Study of various herbicides for weed control in wheat under irrigated conditions. Pak J Weed Sci Res 14:1–8

Matloob A (2014) Weed dynamics and management studies in direct seeded Rice (*Oryza sativa* L.). PhD dissertation. Department of Agronomy, University of Agriculture, Faisalabad, Pakistan

Matloob A, Khaliq A, Chauhan BS (2015a) Weeds of rice in Asia: problems and opportunities. Adv Agron 130:291–336

Matloob A, Khaliq A, Tanveer A, Hussain S, Aslam F, Chauhan BS (2015b) Weed dynamics as influenced by tillage system, sowing time and weed competition duration in dry-seeded rice. Crop Prot 71:25–38

Matloob A, Safdar ME, Abbas T, Aslam F, Khaliq A, Tanveer A, Rehman A, Chadhar AR (2020) Challenges and prospects for weed management in Pakistan: a review. Crop Prot 134:104724

Matloob A, Jabran K, Farooq M, Khaliq A, Aslam F, Abbas T, Ehsanullah, Zaman U, Irshad S, Chauhan BS (2022) Water-wise cultivation of basmati rice in Pakistan. In: Sarwar N, Rehman A, Ahmad S, Hasunzzaman M (eds) Modern techniques of rice crop production. Springer Nature Singapore, pp 197–230

Mubeen K, Nadeem MA, Tanveer A, Jhala AJ (2014) Effects of seeding time and weed control methods in direct seeded rice (*Oryza sativa* L.). J Anim Plant Sci 24:534–542

Munawar S, Naeem M, Ali HH, Jamil M, Iqbal M, Nazir MQ, Balal RM, Safdar ME (2015) Seed dormancy breaking treatments for African purslane (*Zaleya pentandra*). Planta Daninha 33:623–629

Munsif F, Shah S, Samuel AL, Amin M, Shan A, Ahmad A, Ali Z (2014) Integration of weed control methods with seed rates for improving wheat yield. Pak J Weed Sci Res 20:155–165

Munsif F, Ali K, Khalid S, Ali A, Ali M, Ahmad M, Ahmad W, Ahmad I, Basir A (2015) Influence of row spacing on weed density, biomass and yield of chip bud settling of sugarcane. Pak J Weed Sci Res 21:137–144

Mushtaq N, Chema ZA (2007) Weeding out huge crop losses. Available at http://www.pakissan.com/english/news/newsDetail.php?newsid=15752. Accessed on 18 June 2017

Naeem S, Shad RA, Ahmad T (1993) Seed bank of Rabi weeds in the soil of wheat fields of different rainfall zones of rainfed area of Punjab (Pakistan). Pak J Weed Sci Res 6:16–22

Nakabayashi K, Leubner-Metzger G (2021) Seed dormancy and weed emergence: from simulating environmental change to understanding trait plasticity, adaptive evolution, and population fitness. J Exp Bot 72:4181–4185

Narwal SS (2000) Weed management in rice: wheat rotation by allelopathy. Crit Rev Plant Sci 19:246–249

Naseer-ud-Din GM, Shehzad MA, Nasrullah HM (2011) Efficacy of various pre and post-emergence herbicides to control weeds in wheat. Pak J Agric Sci 48:185–190

Nasim G, Shabbir A (2012) Invasive weed species-a threat to sustainable agriculture. In: Ashraf M, Ozturk M, Ahmad MSA, Aksoy A (eds) Crop production for agricultural improvement. Springer, Netherlands, pp 523–556

Nawaz A, Farooq M (2016) Weed management in resource conservation production systems in Pakistan. Crop Prot 85:89–103

Nawaz A, Farooq M, Lal R, Rehman A, Rehman H (2017) Comparison of conventional and conservation rice-wheat systems in Punjab, Pakistan. Soil Till Res 169:35–43

Peerzada AM (2017) Biology, agricultural impact, and management of *Cyperus rotundus* L.: the world's most tenacious weed. Acta Physiol Plant 39:270

Peerzada AM, Bajwa AA, Ali HH, Chauhan BS (2016) Biology, impact, and management of Echinochloa colona (L.) Link. Crop Prot 83:56–66

Peerzada AM, Ali HH, Hanif Z, Bajwa AA, Kebaso L, Frimpong D, Iqbal N, Namubiru H, Hashim S, Rasool G, Manlil S, Meulen A, Chauhan BS (2023) Eco-biology, impact, and management of *Sorghum halepense* (L.) Pers. Biol Invasions 25:955–973

Piskackova RTA, Leon RG (2022) Using weed emergence and phenology models to determine critical control windows for winter-grown carinata (*Brassica carinata*). Weed Sci 70:495–502

Qaderi MM (2023) Environmental regulation of weed seed dormancy and germination. Seeds 2:259–277

Qi Y, Li J, Guan X, Yan B, Fu G, He J, Du L, Zhao C, Zhang D (2020) Effects of herbicides on non-target plant species diversity and the community composition of fallow fields in northern China. Sci Rep 10:9967

Qureshi R (2004) A quantitative account of weeds of sugarcane (*Saccharum officinarum*) crop in District Sukker. Sci Vision 9:1–2

Qureshi H, Arshad M, Bibi Y (2014) Invasive flora of Pakistan: a critical analysis. Int J Biosci 4:407–424

Rahman IU, Afzal A, Iqbal Z, Ijaz F, Ali N, Asif M, Alam J, Majid A, Hart R, Bussmann RW (2018) First insights into the floristic diversity, biological spectra and phenology of Manoor valley, Pakistan. Pak J Weed Sci Res 50:1113–1124

Rajput MT, Tahir SS, Ahmed B, Arain MA (2008) Check list of the weeds found in cotton crops, cultivated in Taluka Ubauro, district Sukkur, Pakistan. Pak J Bot 40:65–70

Riaz M, Jamil M, Mahmood TZ (2007) Yield and yield components of maize as affected by various weed control methods under rain-fed conditions of Pakistan. Int J Agric Biol 9:152–155

Saeed M, Marwat KB, Hassan G, Khan A, Khan IA (2010) Interference of horse purslane (*Trianthema portulacastrum* L.) with maize (*Zea mays* L.) at different densities. Pak J Bot 42:173–179

Safdar ME, Tanveer A, Khaliq A, Riaz MA (2015) Yield losses in maize (*Zea mays*) infested with parthenium weed (*Parthenium hysterophorus* L.). Crop Prot 70:77–82

Safdar ME, Tanveer A, Khaliq A, Maqbool R (2016) Critical competition period of parthenium weed (*Parthenium hysterophorus* L.) in maize. Crop Prot 80:101–107

Safdar ME, Hayyat MS, Alawad HFN, Javaid MM, Mahmood A, Ali B, Nawaz M, Nadeem MA, Ali A, Balal RM, Hashem A, Alfagham A, Parray JA, Abd-Allah EF (2024) Germination ecology of red sprangletop: a problematic weed of direct-seeded rice. Pol J Environ Stud 20:1–10

Salisbury EJ (1961) Weeds and aliens. Collins, London

Shabbir A, Dhileepan K, Adkins SW (2012) Spread of parthenium weed and its biological control agent in the Punjab, Pakistan. Pak J Weed Sci Res 18:581–588

Shabbir A, Zalucki MP, Dhileepan K, Khan N, Adkins SW (2023) The current and potential distribution of parthenium weed and its biological control agent in Pakistan. Plants 12:1381

Shad RA, Siddiqui SU (1996) Problems associated with *Phalaris minor* and other grass weeds in India and Pakistan. Exp Agric 32:151–160

Shad RA, Beg K, Sher M (1989) Germination and growth response of important summer weeds to different light intensities. Pak J Weed Sci Res 2:12–19

Shah NH, Habibullah, Ahmed N, Inamullah (2005) Effect of different methods of weed control on the yield and yield components of wheat. Pak J Weed Sci Res 11:97–101

Shaner DL, Beckie HJ (2013) The future for weed control and technology. Pest Manag Sci 70:1329–1339

Tanveer A (2008) Biology and ecology of weeds. Higher Education Commission, Islamabad, 108p

Tanveer A, Chaudhry NH, Ayub M, Ahmad R (2003) Effect of cultural and chemical weed control methods on weed population and yield of cotton. Pak J Bot 35:161–166

Tanveer A, Arshad MS, Ayub M, Javaid MM, Yaseen M (2012) Effect of temperature, light, salinity, drought stress and seeding depth on germination of *Cucumis melo* var. agrestis. Pak J Weed Sci Res 18:445–459

Tanveer A, Tasneem M, Khaliq A, Javaid MM, Chaudhry MN (2013a) Influence of seed size and ecological factors on the germination and emergence of field bindweed (*Convolvulus arvensis*). Planta Daninha 31:39–51

Tanveer A, Mumtaz K, Mansoor J, Chaudhry MN, Balal RM, Khaliq A (2013b) Effect of ecological factors on germination of horse purslane (*Trianthema portulacastrum*). Planta Daninha 31:587–597

Tanveer A, Sibtain M, Javaid M, Ali HH (2014) Germination ecology of wild onion: a rainfed crop weed. Planta Daninha 32:69–80

Tanveer A, Khaliq A, Ali HH, Mahajan G, Chauhan BS (2015) Interference and management of Parthenium: the world's most important invasive weed. Crop Prot 68:49–59

Teasdale JR, Mangum RW, Radhakrishnan J, Cavigelli AM (2004) Fluctuations of the weed seed bank at the long-term Beltsville Farming Systems Project. USDA-ARS, Beltsville

Ullah W, Khan MA, Arifullah SH, Sadiq M (2008) Evaluation of integrated weed management practices for maize. Pak J Weed Sci Res 14:19–32

Usman K, Ullah I, Khan SM, Khan MU, Ghulam S, Khan MA (2012) Integrated weed management in wheat through tillage and herbicides for wheat production in rice-wheat cropping system in Northwestern Pakistan. J Integr Agric 11:946–953

Wang G, Baskin CC, Baskin JM, Yang X, Liu G, Zhang X, Ye X, Huang Z (2017) Timing of seed germination in two alpine herbs on the southeastern Tibetan plateau: the role of seed dormancy and annual dormancy cycling in soil. Plant Soil 421:465–476

Werth J, Thornby D, Walker S (2012) Assessing weeds at risk of evolving glyphosate resistance in Australian sub-tropical glyphosate-resistant cotton systems. Crop Pasture Sci 62:1002–1009

Werth J, Boucher L, Thornby D, Walker S, Charles G (2013) Changes in weed species since the introduction of glyphosate-resistant cotton. Crop Pasture Sci 64:791–798

Yasin M, Iqbal Z, Safdar ME, Rehman A, Ali A, Asif M, Aziz M, Tanveer A, Pervez MA (2011) *Phalaris minor* control, resistance development and strategies for integrated management of resistance to fenoxaprop-ethyl. Afr J Biotechnol 10:11802–11807

Zafar M, Tanveer A, Cheema ZA, Ashraf M (2010) Weed-crop competition effects on growth and yield of sugarcane planted using two methods. Pak J Bot 42:815–823

Zamljen SA, Rovanˇsek A, Leskovˇsek R (2024) Weed seed bank response during the early conversion period to less intensive tillage systems. Soil Till Res 242:106164

Zhang K, Ji Y, Fu G, Yao L, Liu H, Tao J, Walck JL (2021) Dormancy cycles in *Aquilegia oxysepala* Trautv. et Mey. (Ranunculaceae), a species with non-deep simple morphophysiological dormancy. Plant Soil 464:223–235

Chapter 9
Biology and Management of Weeds in Turkiye

Taseer Ahmad, Khawar Jabran, İlhan Uremis, and Mehmet Nedim Dogan

Abstract The agriculture sector plays a crucial role in Turkiye's economy, providing rural livelihood and economic balance in the country. Overall, Turkiye stands in eighth position in the world and first in Europe by total agricultural output. Cereals, legumes, tubers, and fodder crops are grown in the country for human food, animal feed, and industrial purposes. Among several constraints to crop production, weed problems are prominent ones. The harmful effect of weeds on crop-growing areas of the country is quite high, and these yield losses vary among the regions. The yield losses also vary according to climate conditions and the type of weed flora. It also depends on crop type and the associated weed species as each crop varies in its weed composition and level of competition. The average yield losses due to weeds in Turkiye range from 20% to 35%. Hence, it is important to understand the biology of these weed species and plan an effective weed management program. Different weed management strategies are adopted in the country to tackle the weed problems. These include cultural, mechanical, chemical, biological, and integrated weed control. Herbicide application is the widely used method for controlling weeds, while mechanical tools are also used for managing them. Use of mechanical weeders has also proved effective in managing weeds for crops grown on ridges. Choice of cultivar plays a crucial role in weed management and is a widely used cultural control method for weed control. In conclusion, weed management is essential for sustainable crop production in Turkiye as weeds pose a substantial threat to yield and quality. A sustainable approach by understanding weeds' biology and adoption of chemical and non-chemical methods along with site-specific weed management will be the key to ensuring future food security.

T. Ahmad (✉) · K. Jabran
Department of Plant Production and Technologies, Nigde Omer Halisdemir University, Nigde, Turkiye

İ. Uremis
Department of Plant Protection, Hatay Mustafa Kemal University, Hatay, Turkiye

M. N. Dogan
Department of Plant Protection, Adnan Menderes University, Aydin, Turkiye

B. S. Chauhan, K. Jabran (eds.), *Biology and Management of Weeds in Asia and Africa*, Sustainability Sciences in Asia and Africa,
https://doi.org/10.1007/978-981-95-7314-1_9

Keywords Weeds · Diversity · Cereals · Legumes · Tubers · Management

9.1 Introduction

Agriculture plays a vital role in Turkiye's economy, contributing significantly to the country's GDP, exports, and employment. Cereals and legumes are the most commonly grown crops with 75.5 million tons of production in the country (TUIK 2025). Wheat (*Triticum aestivum* L.) and barley (*Hordeum vulgare* L.) are commonly grown as cool-season crops, while corn (*Zea mays* L.) and rice (*Oryza sativa* L.) are grown as warm-season crops all over the country. Cotton (*Gossypium hirsutum* L.), chickpea (*Cicer arietinum* L.), lentils (*Lens culinaris* Medik.), potato (*Solanum tuberosum* L.), and some forage crops are also grown in the country. In Turkiye, during the year 2024, wheat yield was 20.8 million tonnes, corn was 8.1 million tonnes, rice was 10.19 million tonnes, barley was 8.1 million tonnes, and potato was 6.9 million tonnes (TUIK 2025).

The average yield of these crops is lower compared to agriculturally advanced countries, and several biotic and abiotic factors are involved in this decrease in yield. Among biotic and abiotic constraints to agricultural production, weeds pose severe problems in crop production by competing with crops for water, nutrients, light, and space and producing harmful chemicals (allelopathy) (Ahmad et al. 2023). Moreover, weeds harbor insect pests and diseases, and if left uncontrolled, they can result in total yield loss.

The harmful effects of weeds in cereals, legumes, and tuber growing areas of the country are quite high, and the yield losses vary among the regions. For example, in the Aegean region of Turkiye, yield losses due to weeds were 30% (Tepe 1998), in the Eastern Anatolian region, yield losses were 24% (Guncan 1975), while in the Cukurova region, yield losses were almost 20% (Gungor and Uygur 2005). The average yield losses due to weeds in Turkiye are 20–35% (Sirri et al. 2020).

The yield losses vary according to region and weed types as well as their ecology. There are narrow leaved weeds such as black grass (*Alopecurus myosuroides* Hudson), wild oat (*Avena fatua* L.), canary grass (*Phalaris minor* Retz), winter wild oat (*Avena sterilis* L.), rigid ryegrass (*Lolium rigidum* Gaudin), barnyard grass (*Echinochloa crus-galli* (L.) P. Beauv.), Johnson grass (*Sorghum halepense* (L.) Pers.), and downy brome (*Bromus tectorum* L.), while broadleaf weeds such as lamb's quarter (*Chenopodium album* L.), redroot pigweed (*Amaranthus retroflexus* L.), common purslane (*Portulaca oleracea* L.), field bindweed (*Convolvulus arvensis* L.), wild mustard (*Sinapis arvensis* L.), jimson weed (*Datura stramonium* L.), broomrape (*Orobanche* spp.), and black nightshade (*Solanum nigrum* L.) are the common weeds of cropped areas of Turkiye (Sirri et al. 2020; Ahmad et al. 2021a, b; Ahmad and Jabran 2025). These weeds are commonly found throughout the country, and their management is a challenge. Their impacts vary according to the ecological and biological traits of the weed and their presence in a specific crop. Some of them also have herbicide resistance problems, making it important to understand the biology and ecology of these weeds for proper management. Hence, this chapter aims to review the important weeds in major crops of Turkiye. Yield losses caused by these weeds in different crops in Turkiye have also been explained. Management strategies adopted to control these weeds have also been reviewed.

9.2 Diversity of Weeds and Their Biological Traits in Relation to Specific Crops in Turkiye

9.2.1 *Wheat*

Wheat is an important cereal crop that originated from southeast Turkiye and has been extensively cultivated for 2000 years in the country (Beser et al. 2019). Turkiye is among the countries that are top in bread consumption, enhancing wheat demands in the country. The Central Anatolia region is the highest wheat-producing region of the country, followed by the Southeast Anatolian region (Gummadov et al. 2015). The wheat production in the country is facing decline due to the focus of growers on other cash crops such as cotton, sugar beet (*Beta vulgaris* L.), and stone fruits. Moreover, the average yield of wheat is also lower compared to developing countries (China, India) due to agronomic and management challenges. Among several challenges, the weed problem is also an emerging issue due to the huge flora diversity depending on soil and climatic conditions.

A study on weed diversity in wheat fields of Turkiye observed 40 weed species belonging to 20 different plant families (Pala et al. 2020). Among them, 59 species were dicots with 19 different families, and 12 species were from monocots with one family. The Asteraceae family was identified as the most dominant plant family with 15 weed species, followed by Poaceae with 12 weed species, and Brassicaceae and Fabaceae with 8 weed species (Pala et al. 2020). This shows the dominance of monocot weeds in the wheat field of Turkiye. The list of weeds observed in the wheat field is summarized in Table 9.1.

Among monocots, weeds from *Avena* spp. and *Lolium* spp. are the dominant weeds in wheat-growing areas of Turkiye. Weeds from these two genera are dominant due to their similarity with wheat at early growth stages, sharing the same life span and required climatic conditions. Both these weeds are dominant weeds of arable areas in Mediterranean regions worldwide (Alshallash 2018). Low temperatures (e.g., 20/18 °C) favor the growth of both these species, which helps them to survive even in low temperatures. Both species exhibited some tolerance to salinity stress, which indicates their potential to invade and persist in arable land, particularly in Mediterranean and subtropical regions. Broadleaf weeds from the Brassicaceae and Asteraceae families are also dominant in wheat fields due to their survival in low temperatures. The deep taproot system of weed species from the Brassicaceae family makes them drought-tolerant and possesses highly competitive traits as compared to crops.

Table 9.1 List of important weeds observed in wheat fields of Turkiye

Family	Species	English/common name
Amaranthaceae	*Atriplex patula* L.	Spear saltbush
Apiaceae	*Daucus carota* L.	Wild carrot
	Scandix pecten-veneris L.	Shepherd's-needle
	Turgenia latifolia (L.) Hoffm.	Broadleaf false carrot
Asteraceae	*Anthemis arvensis* L.	Corn chamomile
	Cirsium arvense (L.) Scop.	Thistle
	Conyza canadensis (L.) Cronq.	Horseweed
	Crepis aspera L.	Hawk's beard
	Sonchus asper (L.) Hill.	Prickly sow thistle
	Lactuca serriola L.	Prickly lettuce
Boranginaceae	*Anchusa azurea* Miller.	Italian bugloss
	Buglossoides arvense (L.) Johnst.	Field gromwell
Brassicaceae	*Capsella bursa-Pastoris* (L.) Medik.	Shepherd's purse
	Descurainia sophia (L.) Webb	Tansy mustard
	Sinapis arvensis L.	Wild mustard
	Thlaspi arvense L.	Field pennycress
Caryophyllaceae	*Agrostemma githago* L.	Corncockle
	Stellaria media (L.) Vill.	Chickweed
	Silene conoidea L.	Weed silene
Convolvulaceae	*Convolvulus arvensis* L.	Field bindweed
Fabaceae	*Glycyrrhiza glabra* L.	Liquorice
Geraniaceae	*Erodium hoefftianum* C.A.Mey	Stork's bill
	Geranium dissectum L.	Cutleaf geranium
Lamiaceae	*Lamium amplexicaule* L.	Henbit dead-nettle
Liliaceae	*Allium vineale* L.	Wild garlic
	Muscari neglectum Guss.	Grape hyacinth
Malvaceae	*Alcea pallida* Waldst. & Kit.	Hollyhock
	Malva neglecta Wallr.	Mallow
Papaveraceae	*Papaver rhoeas* L.	Flanders poppy
Plantaginaceae	*Plantago lanceolata* L.	English plantain
Poaceae	*Avena fatua* L.	Common wild oat
	Avena sterilis L.	Winter wild oat
	Alopecurus myosuroides Hudson	Black grass
	Bromus tectorum L.	Cheatgrass
	Hordeum murinum L.	Canary grass
	Phalaris minor Retz	Little seed canary grass
	Poa annua L.	Bluegrass
	Lolium multiflorum Lam.	Italian ryegrass

(continued)

Table 9.1 (continued)

Family	Species	English/common name
Polygonaceae	*Rumex crispus* L.	Curled dock
	Polygonum aviculare L.	Knotgrass
	Polygonum convolvulus L.	Black bindweed
Primulaceae	*Anagallis arvensis* L.	Scarlet pimpernel
Ranunculaceae	*Adonis aestivalis* L.	Summer pheasant's eye
	Ranunculus arvensis L.	Corn buttercup
Rubiaceae	*Galium aparine* L.	Cleavers

Source: Adopted from Pala et al. (2020) and Bozkan and Karaca (2021)

9.2.1.1 Management

Different weed management strategies are adopted in the country to tackle the weed problems in wheat. These include cultural, mechanical, chemical, biological, and integrated weed control. Choice of cultivar plays a crucial role in weed management in wheat and is a widely used cultural control method for weed control. Dwarf cultivars are more prone to weeds as compared to tall ones (Zimdahl 2018). Previously, it has been observed that semi-dwarf wheat genotypes can suppress *Alopecurus myosuroides* and *Avena* spp. more efficiently as compared to dwarf genotypes (Mennan and Isik 2004). Tillage is another important tool that helps in the regulation and reduction of weed seed infestation. For example, tillage practices disturb the weed seed distribution because they affect several soil properties such as penetration resistance, roughness of surface, and bulk density that ultimately change the weed population dynamics (Wang et al. 2022). Tillage helps in burying the weed seeds in the deeper soil layer, thus reducing the weed infestation (Shivran et al. 2020). Zero tillage, along with synthetic carbon-based herbicides, helps in reducing weed emergence and reducing the need for extensive tillage (Loddo et al. 2021). Use of crop rotation is another important tool that helps in weed management. It has been observed that weed numbers, coverage, and frequencies were decreased with crop rotation (Anderson et al. 2007). Wheat rotation with corn, soybean (*Glycine max* (L.) Merr.), sunflower (*Helianthus annuus* L.), and peanut (*Arachis hypogaea* L.) reduced the *A. sterilis* population by 90% in the 4-year trial (Torun and Uygur 2019).

Among all, herbicide application is the fastest, most convenient, and reliable way of weed management in wheat. Herbicides, such as diflufenican, prosulfocarb, pendimethalin, pendimethalin + flufenacet, pyroxsulam + florasulam + cloquintocet-mexyl (safener), pinoxaden + florasulam + cloquintocet-mexyl (safener), clodinafop-propargyl + florasulam + pinoxaden + cloquintocet-mexyl (safener), pinoxaden + pyroxsulam + cloquintocet mexyl (safener), pyroxsulam + cloquintocet-mexyl (safener), sulfosulfuron, chlorsulfuron, dicamba + tritosulfuron, metsulfuron-methyl, diclofop-methyl, clopyralid, 2,4-D + dicamba, and dicamba are the herbicides recommended for weed control in wheat crop in Turkiye (BKU 2025).

Herbicides such as diflufenican, prosulfocarb, pendimethalin, and pendimethalin + flufenacet are recommended and used as pre-emergence for controlling both grass and broadleaf weeds in wheat (BKU 2025). Clopyralid, metsulfuron-methyl, 2,4-D, dicamba, 2,4-D + dicamba, and dicamba + tritosulfuron are applied as post-emergence to manage broadleaf weeds. Diclofop-methyl is recommended as a post-emergence herbicide primarily for controlling grass weeds. Chlorsulfuron, sulfosulfuron, pyroxsulam + florasulam + cloquintocet-mexyl (safener), pinoxaden + florasulam + cloquintocet-mexyl (safener), clodinafop-propargyl + florasulam + pinoxaden + cloquintocet-mexyl (safener), pinoxaden + pyroxsulam + cloquintocet-mexyl (safener), and pyroxsulam + cloquintocet-mexyl (safener), are used as post-emergence herbicides for broad-spectrum control of narrow and broadleaf weeds (BKU 2025).

9.2.2 Rice

In Turkiye, the northwestern (Marmara-Thrace), the Northern (the Black Sea region), and the Southeastern Anatolian regions are producing rice and provinces such as Edirne, Samsun, Balikesir, Sanlıurfa, and Diyarbakır are well known for high cultivation of rice (Özaslan 2015; Yuzbasioglu and Abaci 2023). The existence of weed communities in rice causes a huge decrease in the growth and yield of rice. Weeds also serve as hosts and carriers for disease-causing agents such as rice blast disease. A recent study identified 95 weed species from 26 plant families in rice production areas of Turkiye (Yazlik et al. 2020; Özaslan 2015). Among all weed species, 50 species were terrestrial, 34 were aquatic, and 11 exhibited both aquatic and terrestrial ecosystems. In a survey, it was observed that there were 70 different weed species belonging to 22 families in rice fields of Turkiye (Özaslan 2015). The list of weeds observed in the rice field is summarized in Table 9.2.

Several factors, such as climatic conditions, soil types, and agricultural practices, affect the weed flora of rice. Weeds from the *Echinochloa* genus are the most widely adopted, with huge infestations in the rice-cultivating areas due to shared required growth conditions. Moreover, the species produces a high number of seeds, which ultimately enhances their spread and persistence in the rice-growing areas. The Cyperaceae family was also dominant due to its survival under flooding conditions. For example, *C. difformis* thrives in submerged conditions, similar to rice, hence challenging the control of the species (Özaslan 2015). The eco-biological traits such as fast life cycle, seedbank persistence, and tolerance to flooding make the sedges among the most troublesome weeds in rice ecosystems.

Table 9.2 List of important weeds observed in rice fields of Turkiye

Family	Species	English/common name
Alismataceae	*Alisma canaliculatum* A.Braun & C.D.Bouché	Channeled water plantain
	Alisma plantago-aquatica L.	Common water-plantain
Amaranthaceae	*Amaranthus retroflexus* L.	Redroot pigweed
	Amaranthus albus L.	Tumble pigweed
	Chenopodium album L.	Lamb's quarter
Apiaceae	*Eryngium campestre* L.	Field eryngo
	Eryngium creticum Lam.	Small headed eryngo
Asteraceae	*Anthemis arvensis* L.	Corn chamomile
	Artemisia vulgaris L.	Common mugwort
	Bidens cernua L.	Nodding bur-marigold
	Cirsium arvense (L.) Scop.	Thistle
	Cichorium intybus L.	Chicory
	Conyza canadensis (L.) Cronq.	Horseweed
	Lactuca serriola L.	Prickly lettuce
	Sonchus asper (L.) Hill.	Prickly sow thistle
	Xanthium strumarium L.	Common cocklebur
Boraginaceae	*Heliotropium europaeum* L.	European heliotrope
Convolvulaceae	*Convolvulus arvensis* L.	Field bindweed
	Convolvulus galaticus Rost. ex Choisy	Gray ivy
Cyperaceae	*Cyperus difformis* L.	Variable flat sedge
	Cyperus glomeratus L.	Glomerate flat sedge
	Cyperus longus L.	Sweet cyperus
	Cyperus rotundus L.	Purple nutsedge
	Scirpoides holoschoenus (L.) Sojak.	Roundhead bulrush
Equisetaceae	*Equisetum arvense* L.	Horsetail
Euphorbiaceae	*Chrozophora tinctoria* (L.) Rafin	Dyer's croton
Fabaceae	*Trifolium arvense* L.	Hare's foot clover
	Trifolium resupinatum L.	Reversed clover
Guttiferae	*Hypericum triquetrifolium* Turra.	Curled leaved St. John's wort
Lamiaceae	*Mentha longifolia* (L.) Hudson	Horse mint
	Mentha spicata L.	Spearmint
Lythraceae	*Lythrum hyssopifolia* L.	Hyssop loosestrife
Malvaceae	*Alcea setosa* (Boiss.) Alef.	Bristly hollyhock
	Malva neglecta Wallr.	Mallow
Onagraceae	*Epilobium hirsutum* L.	Hoary willowherb
Plantaginaceae	*Plantago lanceolata* L.	English plantain
	Plantago major L.	Broadleaf plantain

(continued)

Table 9.2 (continued)

Family	Species	English/common name
Poaceae	*Agrostis capillaris* L.	Common bent
	Cynodon dactylon (L.) Pers.	Bermuda grass
	Digitaria sanguinalis (L.) Scop.	Hairy crabgrass
	Echinochloa colonum (L.) Link.	Jungle rice
	Echinochloa crus-galli (L.) P. Beauv.	Barnyard grass
	Poa nemoralis L.	Wood bluegrass
	Setaria viridis (L.) P. Beauv.	Green bristlegrass
	Sorghum halepense (L.) Pers.	Johnson grass
Polygonaceae	*Rumex crispus* L.	Curled dock
	Polygonum aviculare L.	Knotgrass
	Polygonum lapathifolium L.	Pale persicaria
	Polygonum persicaria L.	Lady's thumb
Portulacaceae	*Portulaca oleracea* L.	Common purslane
Solanaceae	*Solanum nigrum* L.	Black nightshade
	Physalis angulata L.	Angular winter cherry
	Physalis philadelphica Lam.	Mexican husk tomato

Source: Adopted from Özaslan (2015) and Yazlik et al. (2020)

9.2.2.1 Management

Weeds are notorious yield reducers in rice crops, so it is crucial to manage them in a timely and efficient manner. Both the direct and indirect control measures are available for managing weeds in rice fields (Ismail and Abdullah 2020). The direct method includes manual weeding, use of mechanical tools and herbicide applications, while in the indirect methods, choice of cultivar, crop rotation, tillage, land preparation, and use of clean weed-free seeds are the major weed management strategies. The use of competitive or resistant rice cultivars also proved its effectiveness in rice; for example, the imidazolinine (IMI)-tolerant rice market is growing in Turkiye (Pala et al. 2023). Weedy rice (*Oryza sativa* f. *spontanea*) is another big problem in rice fields of Turkiye, and keeping the field flooded in wintertime while using pre-emergence herbicides proved effective in controlling this weed (Pala et al. 2023).

The recommended herbicides for managing weeds in rice include cyhalofop-butyl, ipfencarbazone, penoxsulam, quinclorac, halosulfuron methyl, clomazon, tefuryltrone, oxaziclomefone, pentoxazone, imazamox, pretilachlor, benzobicyclon, bispyribac sodium, bensulfuron-methyl, profoxydim, and azimsulfuron (BKU 2025).

Benzobicyclon, pretilachlor, and oxaziclomefone are used as pre-emergence to control a broad spectrum of weeds in rice fields. Clomazone is a pre-emergence recommended herbicide to manage annual grasses and broadleaf weeds, while pentoxazone is applied to target specific broadleaf weeds and sedges.

Cyhalofop-butyl and profoxydim are applied after weed emergence, specifically targeting grass weeds in rice paddies. Imazamox and quinclorac are used as post-emergence herbicides to control various broadleaf and grass weeds. Penoxsulam, tefuryltrione, bispyribac-sodium, and azimsulfuron are employed post-emergence for managing a wide range of weed species. Halosulfuron-methyl and bensulfuron-methyl are used as post-emergence herbicides and are particularly effective against sedges and certain broadleaf weeds.

9.2.3 *Corn*

Corn is a valuable food and feed crop in Turkiye, and it is widely grown in the Mediterranean and Southeast Anatolian region of the country (Cengiz 2014). Weeds are among the limiting factors that affect their growth and yield (Ahmad et al. 2023). Corn fields of Turkiye do share some common weeds, while some weeds are restricted to a specific region due to variation in soil and climate conditions and differences in crop management practices. The presence of these weeds throughout the corn-growing season causes 31–65% yield loss (Doğan and Boz 2005).

A survey on weed species in the corn fields of Turkiye showed 26 weed species from 14 different families. Among them, seven species were from Poaceae, three from Malvaceae, three from Euphorbiaceae, and two from Amaranthaceae, Asteraceae, and Convolvulaceae, while eight species were from other families (Doğru and Kitiş 2023). Among all, *C. rotundus* was the dominant weed species throughout the observed corn fields. The list of weeds observed in the corn field is summarized in Table 9.3.

The dominant weeds in corn fields of Turkiye exhibit several biological traits that contribute to their dominance and competitiveness in the area (Açıkgöz et al. 2023). For example, several corn field weeds, especially weeds from the *Ipomoea* genus, have a fast-growing annual to perennial growth habit, enabling these weeds to colonize the fields rapidly (Hançerli et al. 2018). Moreover, species such as *A. retroflexus* produce a high number of seeds that increase their spread and persistence in a specific area. Also, weed species such as *C. arvensis* have a deep root system that allows this weed to survive under various climatic conditions.

9.2.3.1 Management

Due to the diverse climate of the country, corn fields are affected by several weeds that cause significant yield loss (Sirri et al. 2020). To reduce this yield loss, several chemical and non-chemical weed control strategies have been adopted in the country. Cultural methods such as the use of optimum plant density, crop rotation, and tillage practices proved quite effective in managing weeds in corn. For example, crop rotation of corn with legume or wheat can help in disturbing the weed cycle,

Table 9.3 List of important weeds observed in corn fields of Turkiye

Family	Species	English/common name
Amaranthaceae	*Amaranthus albus* L.	Tumble pigweed
	Amaranthus retroflexus L.	Redroot pigweed
	Amaranthus bilitoides S. Watson	Prostrate pigweed
	Chenopodium album L.	Lamb's quarter
Asclepiadaceae	*Cynachum acutum* L.	Swallow wort
Asteraceae	*Lactuca saligna* L.	Willowleaf lettuce
	Lactuca serriola L.	Prickly lettuce
	Sonchus oleraceus L.	Common sow thistle
	Xanthium strumarium L.	Common cocklebur
Boraginaceae	*Heliotropium europaeum* L.	European heliotrope
Convolvulaceae	*Convolvulus arvensis* L.	Field bindweed
Cyperaceae	*Cyperus rotundus* L.	Purple nutsedge
Euphorbiaceae	*Chrozophora tinctoria* (L.) Rafin	Dyer's croton
Fabaceae	*Glycyrrhiza glabra* L.	Licorice
Guttiferae	*Hypericum triquetrifolium* Turra.	Curled leaved St. John's wort
Lamiaceae	*Mentha longifolia* (L.) Hudson	Horse mint
	Mentha spicata L.	Spearmint
Malvaceae	*Abutilon theophrasti* Medik.	Velvetleaf
	Hibiscus trionum L.	Flower-of-an-hour
Plantaginaceae	*Plantago lanceolate* L.	English plantain
	Veronica officinalis L.	Common gypsyweed
Poaceae	*Cynodon dactylon* (L.) Pers.	Bermuda grass
	Digitaria sanguinalis (L.) Scop.	Hairy crabgrass
	Echinochloa colonum (L.) Link.	Jungle rice
	Echinochloa crus-galli (L.) P. Beauv.	barnyard grass
	Setaria viridis (L.) P. Beauv.	Green bristlegrass
	Sorghum halepense (L.) Pers.	Johnson grass
Polygonaceae	*Rumex crispus* L.	Curled dock
	Polygonum aviculare L.	Knotgrass
	Polygonum persicaria L.	Lady's thumb
Portulacaceae	*Portulaca oleracea* L.	Common purslane
Solanaceae	*Solanum nigrum* L.	Black nightshade
	Physalis angulata L.	Angular winter cherry
	Datura stramonium L.	Jimsonweed
	Physalis philadelphica Lam.	Mexican husk tomato
Zygophyllaceae	*Tribulus terrestris* L.	Puncture vine

Source: Adopted from Alptekin et al. (2023), Açıkgöz et al. (2023) and Doğru and Kitiş (2023)

while deep or zero tillage helps in reducing the weed seed bank inside the soil. The mechanical weeder is also used to control weed density within crop rows.

The use of herbicides is a widely adopted method in the country for managing weeds in the corn field. Herbicides such as nicosulfuron, foramsulfuron, rimsulfuron, halosulfuron methyl, mesotrione, pendimethalin, flumioxazin, pethoxamid, dicamba, dimethenamid-p, s-metolachlor, bromoxynil, 2,4-D, pyridate, and MCPA are recommended for weed control in Turkiye (BKU 2025). Among them, pendimethalin and dimethenamid-P are used as pre-emergence to control broadleaf and grass weeds. Pethoxamid is quite effective against grass weeds as pre-emergence, while s-metolachlor is also used pre-emergence for controlling grass weeds and some broadleaf weeds. Post-emergence herbicides such as foramsulfuron and nicosulfuron proved effective against grass weeds. Herbicides such as 2,4-D, dicamba, pyridate, MCPA, halosulfuron-methyl, mesotrione, and bromoxynil are used as post-emergence herbicides to strictly target broadleaf weeds. Rimsulfuron is recommended as a post-emergence application for controlling both broadleaf and grass weeds.

9.2.4 Barley

Barley possesses significant importance in Turkiye due to its high use in animal feed and the malt industry. It is usually grown in the Southeast and central Anatolian regions of the country (Olgun et al. 2013). A case study of Turkiye showed that the barley field contained 72 weed species from 21 different plant families (Pala 2020). Among them, 59 species were broadleaf and 13 were grass weed species. Asteraceae and Poaceae were the dominant plant families. Another survey conducted in the barley field showed that there were 27 weed species from 13 different plant families (Savci and Gurbuz 2023). This shows that weed density and frequency vary from region to region. The list of weeds observed in the barley field in Turkiye is summarized in Table 9.4.

In barley fields, *S. arvensis* was the dominant weed and was observed in more than 80% of barley fields of Turkiye. *A. sterilis* was the second most dominant weed with a presence in 63% of barley growing areas (Sirri et al. 2020; Pala 2020). Moreover, *A. fatua*, *R. arvensis*, *P. rhoeas*, and *C. arvense* were observed in more than 50% of barley fields of Turkiye. Among all weeds, most of them are annuals, completing their life cycle within a single growing season. This property allows them to rapidly colonize and compete with barley crops. Moreover, many of these weeds have growth requirements similar to barley, i.e., seeds can emerge even at 4–5 °C and optimal growing temperature 12–25 °C that helps them to thrive under the same environmental conditions. Also, the high seed production along with similar seed size and ripening time of *Avena* and *Bromus* spp. make them more problematic and most frequently confused species, and increase their survival rate. Moreover, *A. sterilis* is an annual plant capable of producing up to 200 seeds per plant, with

Table 9.4 List of important weeds observed in barley fields of Turkiye

Family	Species	English/common name
Amaranthaceae	*Salsola kali* L.	Common saltwort
Amaryllidaceae	*Allium nigrum* L.	Black garlic
Apiaceae	*Daucus carota* L.	Wild carrot
	Caucalis platycarpos L.	Bur-parsley
	Bifora radians Bieb	Wild bishop
	Caucalis platycarpos L.	Bur-parsley
Aristolochiaceae	*Aristolochia maurorum* L.	Birthwort
Asteraceae	*Anthemis arvensis* L.	Corn chamomile
	Cichorium intybus L.,	Common Chicory
	Cirsium arvense (L.) Scop.	Thistle
	Chondrilla juncea L.	Rush skeleton weed
	Sonchus asper (L.) Hill.	Prickly sow thistle
	Sonchus oleraceus L.	Common sow thistle
	Lactuca serriola L.	Prickly lettuce
Boraginaceae	*Anchusa azurea* Miller.	Italian bugloss
Brassicaceae	*Capsella bursa-Pastoris* (L.) Medik.	Shepherd's purse
	Cardaria draba (L.) Desv.	Whitetop hoary cress
	Sisymbrium officinale (L.) Scop.	Hedge mustard
	Descurainia sophia (L.) Webb	Tansy mustard
	Sinapis arvensis L.	Wild mustard
	Thlaspi arvense L.	Field pennycress
Caryophyllaceae	*Agrostemma githago* L.	Corncockle
	Silene vulgaris L.	Maidenstears
Convolvulaceae	*Convolvulus arvensis* L.	Field bindweed
	Silene vulgaris L.	Maidenstears
Dipsacaceae	*Cephalaria syriaca* (L.) Schard.	Sryrian cephalaria
Fabaceae	*Lathyrus aphaca* L.	Yellow pea
	Vicia faba L.	Broad bean
Euphorbiaceae	*Euphorbia helioscopia* L.	Sun spurge
Lamiaceae	*Lamium amplexicaule* L.	Henbit dead-nettle
Liliaceae	*Allium vineale* L.	Wild garlic
Malvaceae	*Malva neglecta* Wallr.	Mallow
Papaveraceae	*Papaver rhoeas* L.	Flanders poppy
Plantaginaceae	*Plantago lanceolata* L.	English plantain
	Veronica hederifolia L.	Ivy-leaved speedwell

(continued)

Table 9.4 (continued)

Family	Species	English/common name
Poaceae	*Avena fatua* L.	Common wild oat
	Alopecurus myosuroides Hudson	Black grass
	Avena sterilis L.	Winter wild oat
	Bromus tectorum L.	Cheatgrass
	Hordeum murinum L.	Wall barley
	Phalaris canariensis L.	Canary grass
	Poa annua L.	Bluegrass
	Lolium multiflorum Lam.	Italian ryegrass
Polygonaceae	*Rumex crispus* L.	Curled dock
Primulaceae	*Anagallis arvensis* L.	Scarlet pimpernel
Ranunculaceae	*Ranunculus arvensis* L.	Corn buttercup

Source: Adopted from Pala et al. (2020) and Savci and Gurbuz (2023)

seeds that can remain viable in the soil for several years, making it a persistent competitor in cereal fields (Pala 2020).

9.2.4.1 Management

Weed management in barley includes both chemical and non-chemical weed control methods. Preventive approaches such as the use of false or stale seedbeds can offer effective weed control during crop growth while reducing labor costs (Kanatas et al. 2020). Mechanical tool, such as the use of a spring tine weeder to uproot weed seedlings, is a comparatively efficient non-chemical weed control method (CFS 2025). Crop diversification in a given area has been shown to effectively suppress weed populations, as different crops can limit weed establishment and growth. The inclusion of allelopathic crops in the crop rotation system, i.e., sorghum in a rotation, could provide significant control over weeds compared to a rotation having no inclusion of allelopathic crops.

Herbicides such as 2,4-D, dicamba, prosulforcarb, chlorotoluron, pinoxaden + cloquintocet-mexyl, tritosulfuron + florasulam, prosulfocarb + s-metolachlor, and aminopyralid + florasulam are recommended for controlling weeds in barley (BKU 2025). Among them, prosulfocarb, chlorotoluron and prosulfocarb + s-metolachlor are applied before crop emergence to control a range of grass and broadleaf weeds. Herbicides including 2,4-D, dicamba, tritosulfuron + florasulam, and aminopyralid + florasulam are used as post-emergence to control broadleaf weeds of barley. Pinoxaden + cloquintocet-mexyl is applied as post-emergence to target grass weeds, with cloquintocet-mexyl acting as a safener to protect the barley crop.

9.2.5 Potato

Weeds are among the major limitations in the potato production system in the country (Jabran et al. 2023). Weed infestation in potato crop causes huge losses in tuber yield and quality, and this yield loss varies from 10 to 80% depending upon climate conditions, weed density and diversity, and period of crop-weed competition (Siddiqui et al. 2024). Weeds can harm the potato crop by sheltering disease-causing agents, attracting insect pests, and hindering field operations (Siddiqui et al. 2024). A survey conducted on weed problems in potato showed that weeds from Poaceae (9 species), Polygonaceae (5 species), Brassicaceae (3 species), and Asteraceae (3 species) families were dominant in the potato fields Turkiye (Ozdemir 2018). In spring-grown potatoes, *C. album* was dominant before cultivation, while *P. oleracea* was widely found after potato cultivation, i.e., on lateral growth stages of potato. In summer cultivation, *E. crus-galli* were detected as the most dominant weed, while in autumn, *P. oleracea* were dominant before sowing, while *C. album* was densely found in almost all the regions of Turkiye. Another study showed a total of 60 weed species belonging to 21 families in potato fields of Nigde province, Turkiye (Kilic 2016). The list of weeds observed in potato fields is summarized in Table 9.5.

Weed species such as *C. album*, *A. retroflexus*, and *S. kali* are also widely observed in the potato field of Turkiye (Ahmad and Jabran 2025). These species belong to the same family and share several traits, such as the same growth habit and growing conditions that make them competitive against the potato. Also, these species produce a high number of seeds, and an efficient way of seed dispersal facilitates their existence in a specific area. The yield loss varies due to weeds and is usually weed-specific as well as their density. For example, *A. retroflexus* causes 9–49%, *C. album* 9–48%, *Salsola kali* 32–63% and *Solanum nigrum* 22–43% in potato depending on low to high competition, respectively (Ahmad and Jabran 2025).

9.2.5.1 Management

Weed management in potato includes both non-chemical and chemical control practices. Non-chemical control measure includes cultural and mechanical weed control. Cultural methods such as the choice of competitive cultivar and adoption of crop rotation play a significant role in declining weed infestation in the potato growing areas. For example, seeding competitive cultivars such as Russet Burbank, which has a wide canopy structure, can suppress the weed growth very easily (Hutchinson et al. 2004). Another study showed that cultivar "Lady Olympia" caused the highest weed suppression, reducing weed dry biomass by 60% or more, making it suitable for organic farming systems, while "Agria" cultivar exhibited superior weed tolerance with minimal yield loss, making it ideal for conventional high-yield farming systems (Siddiqui and Jabran 2025). Crop rotation of legumes, pulses, and oilseed crops with potato helps in breaking the weed growth cycle and proves effective in weed management (Khakbazan et al. 2019). Hand hoeing is

Table 9.5 List of important weeds observed in potato fields of Turkiye

Family	Species	English/common name
Amaranthaceae	*Amaranthus retroflexus* L.	Redroot pigweed
	Amaranthus viridis L.	Slender amaranth
	Salsola kali L.	Common saltwort
	Chenopodium album L.	Lamb's quarter
Apiaceae	*Daucus carota* L.	Wild carrot
Asteraceae	*Matricaria chamomilla* L.	Chamomile
	Cirsium arvense (L.) Scop.	Thistle
	Lactuca serriola L.	Prickly lettuce
Boraginaceae	*Heliotropium europaeum* L.	European heliotrope
Brassicaceae	*Sinapis arvensis* L.	Wild mustard
	Diplotaxis tenuifolia L.	Wall rocket
	Raphanus raphanistrum L.	Wild radish
Convolvulaceae	*Convolvulus arvensis* L.	Field bindweed
Cyperaceae	*Cyperus rotundus* L.	Purple nutsedge
Equisetaceae	*Equisetum arvense* L.	Horsetail
Malvaceae	*Malva sylvestris* L.	Common mallow
Papaveraceae	*Papaver rhoeas* L.	Flanders poppy
Primulaceae	*Anagallis arvensis*	Scarlet pimpernel
Poaceae	*Avena fatua* L.	Common wild oat
	Cynodon dactylon (L.) Pers.	Bermuda grass
	Cyperus rotundus L.	Purple nutsedge
	Digitaria sanguinalis (L.) Scop.	Hairy crabgrass
	Lolium rigidum Gaudin	Rigid ryegrass
	Echinochloa crus-galli (L.) P. Beauv.	Cockspur grass
	Poa annua L.	Bluegrass
	Setaria viridis (L.) P. Beauv.	Green bristlegrass
Polygonaceae	*Rumex crispus* L.	Curled dock
	Polygonum aviculare L.	Knotgrass
	Polygonum amphibium L.	Water knotweed
	Polygonum persicaria L.	Lady's thumb
Portulacaceae	*Portulaca oleracea* L.	Common purslane
Solanaceae	*Solanum nigrum* L.	Black nightshade
	Datura stramonium L.	Jimsonweed

Source: Adopted from Kilic (2016), Ozdemir (2018) and Jabran et al. (2023)

common practice for weed management in potatoes as potatoes usually grow on ridges and are generally adopted by small farmers.

Chemical weed control is the most widely used method for weed control in potato, and it can increase yield by more than 50% as compared to untreated plots (Mayerova et al. 2018). Both pre- and post-emergence herbicides are available for controlling weeds in potato. The recommended herbicides include metribuzin, fluazifop-butyl, prosulfocarb, clethodim, cycloxydim, flurochloridone, cycloxydim, pendimethalin, pethoxamid, clomazone, metobromuron, diquat dibromide, and flufenacet + metribuzin (BKU 2025). Preemergence herbicides such as metolachlor and S-metolachlor, trifluralin, pendimethalin, linuron, fomesafen, flumioxazin, and EPTC proved highly effective in managing weeds (Borzouei et al. 2013). Postemergence herbicides include sethoxydim, clethodim, fluazifop-P-butyl, and propaquizafop (Uremis and Uludag 2020), and these are usually used for grass weeds, while some of their mixtures are used for both narrow and broadleaf weeds.

9.3 Conclusions and Future Prospects

Agriculture is a vital sector in Turkiye, and weeds are the main challenge faced by this sector that ultimately impacts sustainable crop production. It is important to determine the problematic weed species in each crop, and their successful control in that specific area in the country. Studying the biology of these weed species provides significant insight into the growth and management of specific weed species and their survival in a specific area. Studies focusing on aspects such as seed dormancy, germination ecology, emergence patterns, competitive ability, and seedbank dynamics are particularly valuable. For instance, knowing the seasonality of weed emergence or the longevity of seeds in the soil can inform the timing of cultural practices like tillage, sowing, and crop rotation. Similarly, understanding the adaptability to abiotic stress and reproductive traits of emerging invasive weeds can help predict their spread and dominance under changing climatic conditions. Such ecological studies can also guide the design of crop diversification strategies, planting density adjustments, and the use of cover crops to naturally suppress weeds. Currently, herbicides are the widely used method for controlling these weeds, but excessive use of these herbicides has led to resistance in several weed species. Nonchemical weed control or herbicides with a different mode of action should be considered while managing weeds. In future, weed management in Turkiye should be based on the selection of improved cultivars along with AI-powered weed detection and the adoption of site-specific weed management. These precision tools can allow real-time identification of weed species and densities, enabling variable-rate herbicide application or selective mechanical weeding, thereby reducing costs and environmental impact. Ultimately, a combination of ecological research, integrated management practices, and precision technologies will be key to developing sustainable, efficient, and adaptable weed management strategies in Türkiye.

References

Açıkgöz Ç, Gürbüz R, Çoruh İ (2023) Determination of the weed flora and the efficacy of some herbicides on weeds and yield in maize fields of Iğdır Province, Türkiye. Turk J Weed Sci 26(1):26–37

Ahmad T, Jabran K (2025) Inter- and intra-specific competition of potato with four common weed species and their impact on tuber yield. Potato Res. https://doi.org/10.1007/s11540-025-09879-5

Ahmad T, Olowu E, Jabran K, Tepe I (2021a) Weed dynamics and control in legumes under changing climate. In: Erman M (ed) Grain legumes under abiotic stress: yield, quality, enhancement and acclimatization. Iksad Publications, Ankara, pp 163–184

Ahmad T, Jabran K, Moss SR (2021b) *Alopecurus myosuroides*. Biology and management of problematic crop weed species, pp 1–19

Ahmad T, Jabran K, Cheema ZA, Bajwa AA, Farooq MA (2023) Global perspective of education in weed science. Weed Sci 71:536–548

Alptekin H, Ozkan A, Gurbuz R, Kulak M (2023) Management of weeds in maize by sequential or individual applications of pre- and post-emergence herbicides. Agriculture 13(2):421

Alshallash KS (2018) Seed germination of rigid ryegrass (*Lolium rigidum*) and sterile oat (*Avena sterilis*) under water salinity conditions at constant or alternating temperatures. Egypt J Bot 58(3):539–545

Anderson RL, Stymiest CE, Swan BA, Rickertsen JR (2007) Weed community response to crop rotations in western South Dakota. Weed Technol 21:131–135

Beser N, Kaya Y, Civi G, Gumus T (2019) The evaluation of performance of some Imi herbicide resistant cultivars in wheat in Trakya Region. International Biological, Agricultural and Life Science Congress, pp 478–485

BKU (2025) Department of plant protection products. Available at: https://bku.tarimorman.gov.tr/Bitki/Details/271. Accessed 10 Feb 2025

Bozkan N, Karaca M (2021) The species and intensities of weed seeds obtained from wheat flour mill plants in Turkey. Selcuk J Agric Food Sci 35(2):150–157

Borzouei M, Izadi E, Rashed Mohassel MH, Hassanzadeh M, Rastgoo M (2013) Effect of Trifluralin herbicide residues in soil on the growth of some plants. Proceedings of the 4th Iranian Weed Conference, Ahvaz University, 6–8 (In Persian with English abstract)

Cengiz R (2014) Country report maize in Turkey. In: Maize for food, feed, nutrition and environmental security, p 352

CFS (Crop Fertility Services) (2025) Tine weeders. Available at: https://www.cropfertilityservices.com/organic-farming-equipment/hatzenbichler-tine-weeder/. Accessed 7 Sept 2025

Doğan MN, Boz Ö (2005) Comparison of weed problems in main and second crop maize (*Zea mays* L.) growing areas in Turkey. Asian J Plant Sci 4(3):220–224

Doğru AS, Kitiş YE (2023) The weed species in corn fields in Antalya Province of Türkiye. Ege Üniversitesi Ziraat Fakültesi Dergisi 60(3):429–436

Gummadov N, Keser M, Akin B, Cakmak M, Mert Z, Taner S, Ozturk I, Topal A, Yazar S, Morgounov A (2015) Genetic gains in wheat in Turkey: winter wheat for irrigated conditions. Crop J 3(6):507–516

Guncan A (1975) Erzurum çevresinde bulunan yabancı otlar ve önemlilerinden bazılarının yazlık arpa ve buğdayda mücadele imkanları üzerinde araştırmalar. Atatürk Üniversitesi Ziraat Fakültesi Araştırma Serisi No: 135, Erzurum, 79 p

Gungor M, Uygur FN (2005) Adana ili mısır ekim alanlarında yabancı otlara karşı uygulanan kimyasal mücadelenin önemi ve ortaya çıkan sorunların araştırılması. M.Sc. Thesis, Çukurova Üniversitesi Fen Bilimleri Enstitüsü, Adana, 171 p

Hançerli L, Ayata MU, Çakan H, Uygur FN, Uygur S (2018) A new weed species record for the flora of Turkey: Ipomoea hederifolia L. (Convolvulaceae). Turk. J Weed Sci 21(2):36–38

Hutchinson PJ, Ransom CV, Tonks DJ, Eberlein CV (2004) Russet Burbank potato tolerance to dimethenamid-p. Weed Technol 18(3):850–852

Ismail SNF, Abdullah AS (2020) Recent developments of weed management in rice fields. Rev Agric Sci 8:343–353

Jabran K, Ahmad T, Siddiqui AO, Üremiş İ, Doğan MN (2023) Weed management in potato. In: Çalişkan ME et al (eds) Potato production worldwide. Academic, pp 121–131. ISBN 9780128229255.

Kanatas PJ, Travlos IS, Gazoulis J, Antonopoulos N, Tsekoura A, Tataridas A, Zannopoulos S (2020) The combined effects of false seedbed technique, post-emergence chemical control and cultivar on weed management and yield of barley in Greece. Phytoparasitica 48(1):131–143

Khakbazan M, Mohr RM, Huang J, Xie R, Volkmar KM, Tomasiewicz DJ et al (2019) Effects of crop rotation on energy use efficiency of irrigated potato with cereals, canola, and alfalfa over a 14-year period in Manitoba, Canada. Soil Tillage Res 195:104357

Kilic OK (2016) Determination of weed species, distribution and density in potato fields (*Solanum tuberosum* L.) in Nigde Province. Bitki Koruma Bulteni 56(4):417–428

Loddo D, McElroy JS, Giannini V (2021) Problems and perspectives in weed management. Italian J Agron 16:1854. https://doi.org/10.4081/ija.2021.1854

Mayerova M, Madaras M, Soukup J (2018) Effect of chemical weed control on crop yields in different crop rotations in a long-term field trial. Crop Prot 114:215–222

Mennan H, Isik D (2004) The competitive ability of *Avena* spp. and *Alopecurus myosuroides* Huds. influenced by different wheat (*Triticum aestivum*) cultivars. TURJAF 28(4):245–251

Olgun M, Erdoğan S, Kutlu İ, Ayter NG, Başçıftçı ZB (2013) Evaluation acreage, production and yield on barley by krigging method in Turkey. Anadolu Tarım Bilimleri Dergisi 28(3):141–149

Özaslan C (2015) Determination of weeds in rice fields of South Eastern Anatolia region of Turkey. Sci Pap Ser A Agron 58:260–265

Ozdemir N (2018) Determination of weed problems in three different growing seasons in potato fields (*Solanum tuberosum* L.) in Odemis District of Izmir. Intl J Scient Technol Res 4(9):7–17

Pala F (2020) Observation of weed species, frequency and density in common barley (*Hordeum vulgare* L.) fields of Diyarbakir, Turkey: a case study. J Agric Sci 26(2):164–172. https://doi.org/10.15832/ankutbd.500963

Pala F, Erman M, Cig F, Dilmen H (2020) A study on weed flora and importance value index of weeds in wheat crop. Int J Sci Technol Res 6(1):49–59

Pala F, Mennan H, Jabran K (2023) Competitive ability of imidazolinone-tolerant rice (cv. Luna) with different weedy rice (*Oryza sativa* f. spontanea) biotypes. Phytoparasitica 51(5):1161–1172

Savci H, Gurbuz R (2023) Determination of density and frequency of weeds in commercial crops of Hamur District of Ağrı, Türkiye. Turk Tarimsal Arastirmalar Derg 10(1):38–50

Shivran H, Yadav RS, Singh SP, Godara AS, Bijarniya AL, Samota SR (2020) Tillage and weed management effect on productivity of wheat in North-West Rajasthan. Indian J Weed Sci 52(2):127–131

Siddiqui AO, Jabran K (2025) Weed tolerance and suppressive ability of potato cultivars to natural weed infestations. J Crop Health 77(1):47

Siddiqui AO, Işık D, Jabran K (2024) Impact of weed competition on morphological and biochemical traits of potato: a review. Potato Res 67(2):451–462

Sirri M, Farooq S, Ozaslan C (2020) Weed species of winter cereals and their management in Turkey. In: Cereal grain: productions and improvement. Iksad Publishing House, Ankara, pp 275–318

Tepe I (1998) Türkiye'de tarım ve tarım dışı alanlarda sorun olan yabancı otlar ve mücadeleleri. Yüzüncü Yıl Üniversitesi Yayınları No: 32, Ziraat Fakültesi Yayınları No: 18, Van, 237 p

Torun H, Uygur FN (2019) Effect of crop rotations on winter wild oat (*Avena sterilis* L.) populations in Osmaniye province wheat sown areas. J Res Weed Sci:345–357

TUIK (2025) Crop production statistics 2024. Available at: https://data.tuik.gov.tr/Bulten/Index?p=Crop-Production-Statistics-2024-53447. Accessed 7 Jan 2025

Uremis I, Uludag A (2020) Patateste yabancı otlar ve mücadelesi. In: Çalışkan ME (ed) Patates. Bey Medya San. Tic., Kanyılmaz Matbacılık, Izmir, p 174

Wang R, Ma L, Lv W, Li J (2022) Rotational tillage: a sustainable management technique for wheat production in the semiarid Loess Plateau. Agriculture 12(10):1582
Yazlik A, Bör AR, Eroğlu E (2020) Evaluation of weed status in rice production in Turkey. Black Sea J Agric 3(4):290–300
Yuzbasioglu R, Abaci NI (2023) Paddy rice production costs and factors affecting paddy rice productivity: case study in Sinop Province, Turkey. Custos Agronegocio Online 19(1):147–162
Zimdahl RL (2018) Fundamentals of weed science. Academic Press, USA

Chapter 10
Weed Biology and Management in Vietnam

H. L. Thi, N. H. Son, T. D. Khanh, and N. T. Cuong

Abstract Weeds remain one of the most persistent constraints to agricultural productivity and sustainability in Vietnam. Their high genetic diversity, adaptive reproductive strategies, and capacity for rapid evolution under intensive management regimes enable them to outcompete crops for light, nutrients, water, and space. In rice-based systems, particularly wet direct-seeded rice (WDSR), yield losses from dominant species such as *Echinochloa crus-galli*, *Leptochloa chinensis*, and weedy rice (*Oryza sativa f. spontanea*) can reach 30–80% if unmanaged. Beyond direct yield reduction, weeds aggravate pest and disease incidence, contaminate grain, and drive escalating production costs. The increasing prevalence of herbicide resistance linked to overreliance on a narrow set of chemical modes of action further challenges conventional control strategies. This chapter synthesizes current knowledge of weed biology and ecology in Vietnam, emphasizing seed dormancy, soil seedbank dynamics, vegetative propagation, and adaptive mechanisms underpinning resistance. We highlight region-specific weed diversity across rice, upland, orchard, and industrial cropping systems, and evaluate the ecological and economic impacts of weed infestations. Building on this biological foundation, we critically examine integrated weed management (IWM) strategies combining cultural, mechanical, chemical, and biological tools. Particular attention is given to allelopathic crops (e.g., rice, cucumber, cosmos) and plant-derived herbicides as emerging sustainable alternatives. Advances in digital agriculture, including artificial intelligence, remote sensing, and precision herbicide application, are also discussed as future pathways.

H. L. Thi (✉)
Can Tho University, Can Tho, Vietnam
e-mail: hlthi@ctu.edu.vn

N. H. Son
Genetic Agricultural Institute, Hanoi, Vietnam

T. D. Khanh
Vietnam Academy of Agricultural Sciences, Hanoi, Vietnam

N. T. Cuong
Cuu Long Delta Rice Research Institute, Can Tho, Vietnam

B. S. Chauhan, K. Jabran (eds.), *Biology and Management of Weeds in Asia and Africa*, Sustainability Sciences in Asia and Africa,
https://doi.org/10.1007/978-981-95-7314-1_10

By integrating ecological insights with innovative technologies, Vietnam has the potential to transition from chemically dependent weed control toward resilient, climate-adaptive, and environmentally sound management systems.

Keywords Herbicide resistance · Integrated weed management · Allelopathy · Rice-based systems · Sustainability, Vietnam agriculture

10.1 Introduction

Vietnam is one of the leading rice-producing countries in Asia, with agricultural systems that span a wide range of agroecological zones from mountainous uplands in the North to intensively cultivated floodplains in the Mekong River Delta. Among these, rice is the most important staple crop, grown on approximately 7.1 million hectares annually and forming the backbone of food security and rural economies across the nation (Gianessi 2013). In recent decades, Vietnam has rapidly transitioned from traditional transplanted rice cultivation toward wet direct-seeded rice (WDSR), particularly in the lowland deltas, as farmers seek to reduce labor costs and adopt mechanized practices (Chauhan and Johnson 2010; Chauhan 2013). However, this shift in crop establishment methods has unintentionally created ideal conditions for weed proliferation. Unlike transplanted rice, which involves puddling and a head start for crop seedlings, WDSR systems eliminate soil disturbance and allow weeds to germinate simultaneously with the crop. These conditions favor rapid weed emergence, persistence, and competition, particularly during the early stages of crop growth when rice is most vulnerable (Chin 2001; Chauhan 2013). Weeds in rice fields compete aggressively for essential resources such as sunlight, water, nutrients, and rooting space, often leading to significant reductions in crop yield, resource-use efficiency, and overall profitability (Chauhan and Johnson 2010; Busi et al. 2016).

The threat posed by weeds in Vietnam is exacerbated by the widespread presence of aggressive and difficult-to-control species. Dominant weed species such as *Echinochloa crus-galli* (barnyardgrass), *Leptochloa chinensis* (Chinese sprangletop), *Fimbristylis miliacea*, *Cyperus difformis*, *Cyperus iria*, *Monochoria vaginalis*, and *Ludwigia adscendens* are well adapted to flooded rice environments, while invasive species such as *Mikania micrantha* and *Imperata cylindrica* persist year-round in perennial orchard systems (Viet et al. 2022). Among these, *Oryza sativa* f. *spontanea* (weedy rice) has emerged as a major challenge due to its close genetic relationship to cultivated rice, which allows it to mimic crop phenotypes and evade both mechanical and chemical control (Chauhan and Johnson 2010; Busi et al. 2016). These weed species possess a suite of biological and ecological adaptations, including prolific seed production, strong dormancy, vegetative propagation, rapid phenological development, and allelopathic interactions, that enhance their competitiveness and resilience under intensified cropping systems (Chauhan 2013; The 2021). For example, a single *Echinochloa crus-galli* plant can produce tens of thousands of seeds, while *Cyperus rotundus* spreads through underground tubers and rhizomes that are resistant to conventional tillage (Chauhan 2013). These traits not only enable weed persistence across seasons but also contribute to the buildup of soil seedbanks, increasing future weed pressure (Chauhan and Johnson 2010; Thi et al. 2025d).

A growing concern in Vietnam is the development of herbicide resistance in weed populations, particularly under systems of continuous rice cultivation and overreliance on a limited number of herbicide modes of action. Herbicides such as bispyribac-sodium, cyhalofop-butyl, and florpyrauxifen-benzyl are frequently used in rice production without rotation or integration with other methods, thereby imposing strong selection pressure on weed communities (Duy et al. 2018a, b; Thuy et al. 2022). Cases of resistance have been reported in *Echinochloa crus-galli*, *Leptochloa chinensis*, and weedy rice populations across multiple regions (The 2021; Duy et al. 2018a, b; Thuy et al. 2022). Mechanisms of resistance include both target-site mutations and enhanced metabolic detoxification pathways, such as those involving cytochrome P450 monooxygenases (The 2021). When combined with labor shortages and limited access to training, the escalating resistance problem threatens the long-term sustainability of Vietnam's rice production systems (Gianessi 2013; Chauhan 2013).

In this chapter, we synthesize current knowledge on the biology, ecology, and management of weeds in Vietnam's diverse cropping systems, with a particular emphasis on rice-based agriculture. Section 10.1 provides an overview of weed diversity and biological traits that confer competitive advantages under Vietnamese agroecosystems. Section 10.2 discusses the ecological and agronomic impacts of weed infestations on crop productivity, resource-use efficiency, and ecosystem function. Section 10.3 reviews conventional and emerging weed management strategies, including cultural, chemical, and biological approaches. Section 10.4 highlights recent scientific advances in the use of plant-derived allelochemicals, bioherbicides, and artificial intelligence tools for weed identification and precision management. The final section outlines key research priorities and future directions for transitioning to climate-smart, integrated weed management (IWM) systems that align with national sustainability goals and international best practices (Thi et al. 2022; Vang et al. 2024; Hieu and Hien 2020).

10.2 Weed Diversity and Key Biological Traits

Vietnam harbors an exceptionally rich and ecologically complex weed flora, a consequence of its tropical monsoon climate, steep ecological gradients, and diverse agricultural practices. The country's agroecosystems range from irrigated lowland rice fields and upland rainfed systems to perennial orchards, plantations, and aquatic margins. These varying ecosystems create niches for a wide range of weed species

Table 10.1 Major weed groups and species richness in Vietnam

Weed group	Dominant families	Number of species	Representative species
Grasses (Poaceae)	Poaceae	53 species (pp. 1–54 in the flora)	*Echinochloa crus-galli, Leptochloa chinensis, Ischaemum rugosum, Eleusine indica, Cynodon dactylon*
Sedges (Cyperaceae)	Cyperaceae	53 species (pp. 55–108 in the flora)	*Cyperus difformis, Cyperus iria, Cyperus rotundus, Fimbristylis miliacea, Fuirena umbellata*
Broadleaf weeds	Multiple families (Asteraceae, Amaranthaceae, etc.)	312 species (pp. 110–420 in the flora; the largest and most diverse group)	*Ageratum conyzoides, Eclipta prostrata, Ludwigia adscendens, Monochoria vaginalis, Bidens pilosa, Chromolaena odorata, mimosa pudica*

Adapted from Viet et al. (2022)

with distinct life histories, ecological strategies, and impacts on crop productivity and agroecosystem sustainability (Son 2011).

To date, more than 400 weed species have been identified across Vietnam (Viet et al. 2022). These species are typically categorized into three major groups based on their morphology and taxonomy: grasses (Poaceae), sedges (Cyperaceae), and broadleaf weeds (representing diverse dicot families) (Table 10.1). This grouping is essential not only for identification but also for understanding functional dynamics related to their phenology, reproduction, competitiveness, and response to management interventions (Chauhan 2013).

10.2.1 Broad Overview of Weed Groups

Among these groups, broadleaf weeds are the most species-rich, with over 300 species across families such as Asteraceae, Fabaceae, Amaranthaceae, Convolvulaceae, and Onagraceae. These weeds range from creeping annuals to robust perennials with significant plasticity, enabling them to adapt to both disturbed and undisturbed habitats.

In contrast, grasses and sedges, though each accounting for 53 species, are more prevalent in disturbed environments such as rice paddies, canal edges, and levees. These groups often possess traits such as C4 photosynthesis, rapid seedling emergence, and prolific seed production that make them particularly successful in intensively managed systems. Their physiological resilience allows them to survive under variable moisture regimes, light conditions, and anthropogenic disturbance (Son 2000a, b; Viet et al. 2022).

10.2.2 Weed Distribution Across Agroecosystems

Weed communities in Vietnam are strongly shaped by the country's agroecological zoning and cropping patterns. In lowland irrigated rice ecosystems—such as those found in the Mekong River Delta (MRD), Red River Delta (RRD), and coastal plains—weed flora is dominated by grasses and sedges. Key species include *Echinochloa crus-galli*, *Leptochloa chinensis*, *I. rugosum*, *F. miliacea*, *Cyperus difformis*, and *Ludwigia adscendens* (Fig. 10.1). These species possess traits such as dormancy, variable germination temperatures, and tolerance to flooding that allow them to persist despite seasonal variations and agronomic practices (Son 2011; Chin 2001; Chauhan 2013).

Perennial orchard systems, including durian, jackfruit, and citrus plantations, tend to harbor different weed assemblages. Here, species such as *Imperata cylindrica*, *Ageratum conyzoides*, and *Mikania micrantha* dominate. These weeds, particularly problematic during the dry season, exhibit aggressive ground cover, inhibit seedling establishment, and compete for nutrients and light. Their dominance in orchards is partly due to reduced tillage and limited use of selective herbicides (Thi et al. 2025e, f).

Aquatic and semiaquatic habitats, such as paddy bunds and water margins, support unique weed species with specialized morphological and physiological traits

Fig. 10.1 Morphological diversity of weedy rice and associated weed species in rice ecosystems of Vietnam. (**a**) Thirty-eight accessions of weedy rice (*Oryza sativa* f. *spontanea*) collected from Thap Muoi, Vietnam. (**b**) Chinese sprangletop (*Leptochloa chinensis* (L.) Nees). (**c**) Barnyardgrass (*Echinochloa crus-galli* (L.) P. Beauv.). (**d**) Sedge (*Fimbristylis miliacea* (L.) Vahl). (**e**) Rice field heavily infested by mixed weed populations. (Photos by Kieu Cong Vinh and Ho Le Thi, Weed science ATL 5.19 Lab, Can Tho University, Vietnam)

for water tolerance. Examples include *Monochoria vaginalis*, *Sagittaria sagittifolia*, and *Marsilea minuta*. These species illustrate the convergence of aquatic adaptations with weed traits, such as floating leaves, air-filled tissues (aerenchyma), and rapid clonal propagation (Viet et al. 2022).

10.2.3 Biological Traits of Dominant Weed Species

Understanding the biological traits of dominant weeds is crucial for effective and ecologically grounded weed management. Traits such as seed dormancy, vegetative propagation, allelopathy, C4 photosynthesis, and phenotypic plasticity significantly influence weed persistence and crop-weed competition (Table 10.2).

These species exhibit a range of life cycles (annual vs. perennial), reproductive modes (sexual vs. vegetative), habitat preferences (upland, lowland, aquatic), and yield loss potentials (often >50% under high densities). For instance, *E. crus-galli* may cause complete yield loss in poorly managed fields, while species like *Marsilea minuta* and *Bolboschoenus maritimus* thrive in flooded, oxygen-deficient environments.

10.2.4 Weed Diversity by Soil Type and Crop

Weed species richness and composition also vary by soil type and cultivated crop. Surveys across key agricultural zones have identified 71 species from 22 families, with Poaceae and Cyperaceae being the most dominant (Son 2011).

By crop type, cornfields had the highest weed richness (51 species), followed by upland rice (49 species), and soybean (41 species). Broadleaf weeds tend to dominate in upland systems, especially on red soils with reduced competition from rice monoculture (Table 10.3).

10.2.5 The Weedy Rice Challenge

Among Vietnam's weed flora, weedy rice (*Oryza sativa* f. spontanea) represents and emerges as a unique threat due to its close genetic relationship with cultivated rice. In direct-seeded rice systems, particularly under water-saving practices, weedy rice emerges and establishes alongside the crop, exploiting its similar morphology to escape detection (Fig. 10.1). Studies (Cuong et al. 2022, 2023) have characterized local weedy rice lines as having high early vigor, diversity, prolonged seed dormancy, severe seed shattering, and synchronous germination with the crop. These traits contribute to significant yield losses (up to 50%) and grain quality degradation due to contamination with red pericarp seeds.

Table 10.2 Key weed species in Vietnam and their biological traits

No.	Scientific name	Life cycle	Max height (cm)	Reproduction	Habitat preference	Yield loss potential	References
1	*Echinochloa crus-galli* (L.) P. Beauv.	Annual	150	40,000 seeds/plant, dormant all year	Moist soil, poorly grows in deep water	Up to 60% (high density)	Chin (2001), Khanh et al. (2018), Thuy et al. (2022), and Thi et al. (2021)
2	*Echinochloa glabrescens* Munro ex Hook. f.	Annual	50–100	Thousands of seeds/plant	Moist soil	7–47% (infected area 5–50%)	Chin (2001) and Suk et al. (2021)
3	*Echinochloa colona* (L.) Link	Annual	90	Year-round seed set	Moist soil, poor in continuous flooding	>45% yield loss	Son (2011) and Thi et al. (2022)
4	*Ischaemum rugosum* Salisb.	Annual	120	Seeds with dormancy	Moist soil and wetlands	15–42%	Chin (2001) and Chauhan (2013)
5	*Leersia hexandra* Swartz	Perennial	100	Seeds, rhizomes, stolons	Continuously flooded wetlands	Up to 40% yield loss	Ho (1993) and Chinh et al. (1978)
6	*Leptochloa chinensis* (L.) Nees	Annual/ Perennial	120	Seeds or basal shoots	Weak swamp soils when flooded	>40% yield loss	Chin (2001), Busi et al. (2016), and Thi et al. (2021)
7	*Paspalum distichum* L.	Perennial	30–60	Stolons or seeds	Wet soils	Up to 55% yield loss	Son et al. (2005), Son (2011), and Suk et al. (2021)
8	*Cyperus difformis* L.	Annual	75	50,000 seeds/plant	Wet, shallowly flooded soils	12–50% yield loss	Son (2000a), The (2021), and Thuy et al. (2022)

(continued)

Table 10.2 (continued)

No.	Scientific name	Life cycle	Max height (cm)	Reproduction	Habitat preference	Yield loss potential	References
9	*Fimbristylis quinquangularis* (Vahl) Kunth	Annual	60	10,000 seeds/plant, year-round	Moist soil with fluctuating water	Up to 50% yield loss	Thi et al. (2021) and Suk et al. (2021)
10	*Cyperus iria* L.	Annual	~60	5000 seeds/plant, dormancy possible	Wet soils	~40% yield loss	Son (2000a, b) and The (2021)
11	*Bolboschoenus maritimus* (L.) Palla	Perennial	90–100	Rhizomes, tubers	Muddy, waterlogged soils	40–60% yield loss	Ho (1993) and Chinh et al. (1978)
12	*Marsilea minuta* var. incurva (A. Braun) Launert	Perennial	–	Rhizomes, underground stems	Aquatic, lowland soils	>40% yield loss	Ho (1993) and Son (2011)
13	*Monochoria vaginalis* (Burm.f.) C. Presl	Annual/ Perennial	50	Mainly seeds	Aquatic and semiaquatic	>45% yield loss	Chin (2001) and Viet et al. (2022)
14	*Ludwigia octovalvis* (Jacq.) P.H. Raven	Annual	100	Seeds, some dormancy	Moist, flooded soils	30–60% yield loss	Son (2011) and Suk et al. (2021)
15	*Sphenoclea zeylanica* Gaertn.	Annual	150	Seeds, year-round	Wet soils	>45% yield loss	Son (2000b) and Chinh et al. (1978)

Table 10.3 Weed diversity on different soil types (Son 2011)

Soil type	Number of weed species	Dominant and problematic species
Red ferralitic	57	*Ageratum conyzoides, Rypholium trilobatum, Borreria latifolia*
White-gray	46	*Imperata cylindrica, Leptochloa chinensis*
Fallow land	68	*Saccharum spontaneum, Paspalum conjugatum, Panicum bisulcatum, Imperata cylindrica, Mimosa pudica, Borreria latifolia*

10.2.6 Herbicide Resistance and Adaptive Potential

Intensive herbicide use in rice systems, particularly the repeated application and overreliance on post-emergence herbicides such as bispyribac-sodium and cyhalofop-butyl, has selected for herbicide-resistant biotypes of *Echinochloa spp. and Leptochloa spp.* Resistance can arise through enhanced metabolic detoxification mediated by cytochrome P450 enzymes as well as target-site mutations. In addition, phenotypic plasticity and crop mimicry further complicate management (Chauhan and Johnson 2010; The 2021).

The biological success of these weeds is underpinned by traits such as prolific seed production (e.g., >40,000 seeds in *E. crus-galli*), dormancy that enables multi-season persistence, and vegetative propagation in species like *Cyperus rotundus*, which produces tubers and rhizomes that survive tillage and regenerate (Son 2000b; Son et al. 2005; The 2021).

Vietnam's weed flora presents both a biodiversity reservoir and a significant agronomic constraint. Effective weed management requires an integrated approach tailored to specific agroecosystems and weed biology. Strategies should combine preventive (certified seed use, crop rotation), cultural (water management, tillage), and technological (precision herbicide application, herbicide-tolerant varieties) measures (Khanh 2003; Son et al. 2005; Son 2011).

The spatial heterogeneity in weed distribution necessitates localized strategies rather than a one-size-fits-all model. Understanding weed diversity and functional traits underpins the design of sustainable, resilient weed management programs. Future research should focus on predictive weed modeling, resistance evolution, and the integration of agroecological principles into weed control frameworks across Vietnam's diverse agricultural landscapes.

10.3 Biological and Ecophysiological Traits of Weeds in Vietnam

Weeds possess a remarkable set of biological and ecophysiological adaptations that allow them to persist in disturbed habitats, compete effectively with crops, and thrive under varied environmental conditions. These adaptations are central to their survival, reproductive success, and interference with cropping systems.

10.3.1 Reproductive Strategies: Sexual and Asexual Modes

Weed species can reproduce through sexual reproduction (seed-based) and/or asexual (vegetative) means. Annual weeds predominantly propagate via seed, while many perennial species rely on vegetative structures such as rhizomes, stolons, bulbs, corms, or tubers in addition to seed reproduction (Chinh et al. 1978; Ho 1993).

Seed production is typically prolific. For example, *E. crus-galli* may yield up to 40,000 seeds per plant, contributing to rapid seedbank buildup and widespread dispersal (Son 2000b).

Vegetative reproduction, while slower in terms of dispersal range, offers survival advantages under adverse conditions. Structures like rhizomes or bulbs store nutrients that support regrowth even under abiotic stress (e.g., drought, flooding). However, vegetative propagules are generally more vulnerable to flooding, tillage, and high temperatures compared to seeds. This tradeoff explains why annuals dominate in seasonally disturbed fields, while perennials dominate in continuously wet or less disturbed areas.

10.3.2 Seed Dormancy and Germination Physiology

Dormancy enables weeds to delay germination until conditions are favorable. It is a key trait contributing to temporal avoidance of weed control efforts.

In species like *E. crus-galli*, dormancy is genetically regulated, whereas in others, such as *F. miliacea* or *Imperata cylindrica*, it is environmentally modulated by temperature and moisture (Chinh et al. 1978).

Thermal cues: The optimal germination range for *E. crus-galli* is 25–35 °C, with dormancy induced below 10 °C or above 45 °C. Similarly, dormancy in *F. miliacea* shortens at high temperatures and prolonged moisture, allowing synchronization with the onset of rains (Chinh et al. 1978).

Moisture regulation: Optimal soil moisture for germination lies at 80–90% field capacity; below 70%, germination sharply declines (Ho 1993). This sensitivity underpins the importance of soil water management in weed suppression strategies.

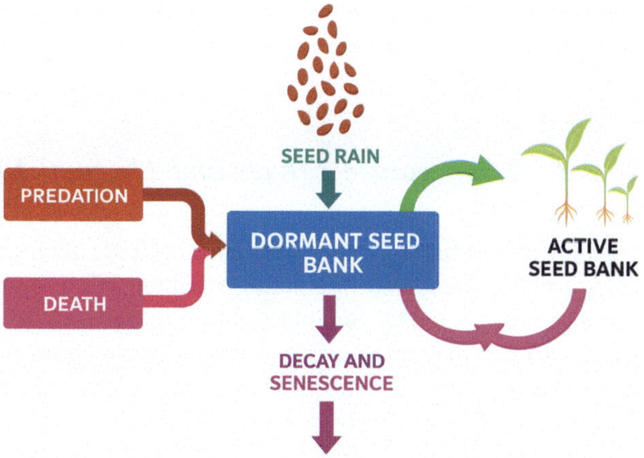

Fig. 10.2 Conceptual model of the weed seed bank dynamics in rice agroecosystems. (Modified from Anju et al. (2022))

These dormancy traits contribute to seed bank longevity, a major challenge in integrated weed management. Some seeds may remain viable for 5–10 years, with staggered germination across seasons. Therefore, in Vietnamese rice production systems, particularly in wet- and dry-direct seeded rice of the Mekong River Delta, the soil seed bank is the critical reservoir sustaining weed populations across seasons. Seed rain from previous crops contributes to the dormant seed bank, where seeds of key species such as weedy rice, Chinese sprangletop, and grasslike fimbristylis may persist for several years due to strong dormancy and protective seed coats. Under favorable conditions of light, temperature, soil moisture, and oxygen, a fraction of these seeds transitions into the active seed bank and germinates, competing directly with rice. Conversely, substantial seed losses occur through predation, mortality, and decay (Fig. 10.2). Understanding these dynamics is essential for designing integrated weed management (IWM) strategies, including stale seedbed practices, reduced tillage, competitive rice cultivars, crop rotation, and the use of allelopathic extracts from rice or Asteraceae species. Effective management of the soil seed bank is therefore a cornerstone for reducing long-term weed pressure and mitigating herbicide resistance in Vietnamese rice agroecosystems.

10.3.3 Role of Water Regime and Flooding

In flooded rice systems, typical of the Mekong River Delta, water depth becomes a critical determinant of weed composition and emergence timing: Research shows that maintaining flood water at 2.5–5 cm after sowing can significantly inhibit emergence of *E. crus-galli*, *L. chinensis*, *F. miliacea*, and *C. iria* (Son 2000b). At 7 cm water depth, germination of *L. chinensis* and *Fimbristylis* species is fully suppressed. *Cyperus* spp. are also strongly inhibited at 5 cm (Son 2000b). This hydrological barrier is particularly effective against seed-propagated annuals. However, it may unintentionally favor perennials that reproduce vegetatively (e.g., *Marsilea minuta*, *Bolboschoenus maritimus*).

10.3.4 Temperature Effects and Thermal Tolerance

Weeds often exhibit greater thermal amplitude than crops, enhancing their resilience to heat stress and fluctuating weather. High temperatures accelerate growth and shorten life cycles, enabling multiple generations per season. Tropical weed species like *E. crus-galli* and *A. conyzoides* can germinate and complete life cycles rapidly at 30–40 °C (Son et al. 2005). Cold-tolerant species such as *Senecio vulgaris* (introduced) exhibit early spring germination, allowing early resource capture. Temperature-mediated dormancy and plasticity allow weed populations to persist even when climatic conditions shift due to seasonal or interannual variability.

10.3.5 Photosynthetic Pathways and Water-Use Efficiency

Many grasses and sedges use the C4 photosynthetic pathway, which confers advantages under high light intensity, temperature, and low moisture conditions (Son 2011). Species such as *E. crus-galli*, *L. chinensis*, and *I. rugosum* are C4 weeds with high water-use efficiency and competitive growth in direct-seeded rice fields. In contrast, broadleaf weeds like *Monochoria vaginalis* or *Ludwigia* spp. rely on C3 photosynthesis, favoring moist or semiaquatic niches (Table 10.4). C4 weeds tend to dominate upland and aerobic conditions, while C3 species are more common in lowland and flooded fields (Table 10.4).

10.3.6 Herbicide Selectivity and Ecophysiology

Weed ecophysiology also governs herbicide performance (Table 10.4). Herbicides such as metsulfuron-methyl (ALS-inhibitor) are highly selective against broadleaf species, but ineffective against grasses and some sedges (Son 2000a). The reduced efficacy in flooded fields is often linked to: (i) poor herbicide translocation in submerged leaves; (ii) dominance of perennials with non-susceptible vegetative structures; and (iii) timing mismatches between herbicide application and weed growth stages. Therefore, ecophysiological traits must be integrated into herbicide choice and application timing.

The biological and ecophysiological complexity of Vietnam's weed flora underlies the persistence and adaptive potential of these species across cropping systems. Traits such as seed dormancy, vegetative propagation, thermal tolerance, and C4 physiology enable weeds to exploit management gaps and environmental variability.

Table 10.4 Ecophysiological profiles of selected weed species in Vietnam (Viet et al. 2022)

Species	Life cycle	Reproduction mode	Germination range (°C)	Moisture tolerance	Flood tolerance	Photosynthetic pathway	Management note
Echinochloa crus-galli	Annual	Seeds (40,000/plant)	10–45 °C (opt: 25–35)	High	Poor in deep flooding	C4	Dormant seeds; resistant biotypes present
Leptochloa chinensis	Annual	Seeds	20–38 °C	Medium	Tolerant to shallow flooding	C4	Difficult to control post-emergence
Fimbristylis miliacea	Annual	Seeds	20–35 °C	High	Inhibited at >5 cm flooding	C3	Shallow flooding suppresses emergence
Cyperus iria	Annual	Seeds	Broad	Medium	Suppressed >5 cm water	C4	Emerges after field drainage
Imperata cylindrica	Perennial	Rhizomes	–	Low to medium	Low	C4	Aggressive in orchards, drylands
Monochoria vaginalis	Annual	Seeds	~25–35 °C	Very high	High	C3	Aquatic adaptation; thrives in puddled fields

Effective weed management requires a nuanced understanding of these traits to guide integrated weed management (IWM) approaches. Cultural practices such as stale seedbed, flooding control, and crop rotation can target specific life histories. Chemical control should be based on ecophysiological profiles, and herbicide timing should align with weed emergence and growth patterns. Combining ecological knowledge with precision agronomy will be essential to sustainably control weeds in Vietnam's evolving agricultural landscapes.

10.4 Impacts of Weeds on Cropping Systems and Agroecosystems

Weeds are a major constraint to agricultural productivity, ecological balance, and economic sustainability in Vietnam. Their impacts extend beyond direct yield losses to influence crop competition, nutrient depletion, pest and disease dynamics, and long-term soil degradation. Given the increasing intensification of rice-based systems in the Mekong River Delta (MRD), Red River Delta (RRD), and upland cropping zones, the pressure exerted by weeds has intensified, particularly in direct-seeded rice systems where the absence of water during germination favors early weed establishment.

10.4.1 Yield Losses Caused by Competitive Weed Species

Weeds reduce yields by competing with crops for essential growth resources such as light, water, nutrients, and physical space, especially during early developmental stages (Table 10.2). In rice-based systems, species like *E. crus-galli*, *Leptochloa chinensis*, and *I. rugosum* often germinate synchronously with rice seedlings and exhibit rapid early vigor, shading crops and reducing tiller number, panicle formation, and grain filling (Chauhan and Johnson 2010; Thi et al. 2025d).

Direct-seeded systems are particularly vulnerable. Trials in the Mekong Delta have shown that *E. crus-galli* and *L. chinensis* can reduce rice yield by 30–80% under unmanaged infestations (Busi et al. 2016). Historic data from the 1970s demonstrated that rice yields could drop to just 19.5% of their potential under unchecked weed growth (Chinh et al. 1978).

10.4.2 Nutrient and Water Competition

Weeds often extract soil nutrients more aggressively than crops. *L. chinensis* and *C. iria* are highly competitive for nitrogen, reducing crop uptake during critical stages such as tillering and grain filling (Krupnik et al. 2012). In experiments in southern Vietnam, nitrogen application in the absence of weed control led to increased yield losses from 16.6% to 38.9% as nitrogen input rose from 0 to 150 kg ha^{-1} due to enhanced weed growth (Son 2000a, b).

Water competition is also severe, especially in upland crops or orchards during the dry season. *Imperata cylindrica* and *Mikania micrantha* develop extensive root systems, depleting soil moisture and reducing water-use efficiency in fruit trees (Thi et al. 2025d).

10.4.3 Physical and Structural Impacts

Dense weed growth alters field microclimates and obstructs agronomic operations. Species such as *Mikania micrantha*, *Ageratum conyzoides*, and *Imperata cylindrica* form thick mats that impede manual weeding, fertilizer application, and irrigation. In orchards and plantations, such weeds harbor rodents and serve as refuges for harmful insects, further increasing management burdens.

In rice fields, tall-growing weeds such as *Sphenoclea zeylanica* and *Ludwigia octovalvis* can physically overshadow the crop, reducing photosynthetic light interception. Additionally, weeds like *Cyperus rotundus* spread via underground tubers and rhizomes, regenerating even after mechanical disturbance.

10.4.4 Weed–Pest–Pathogen Interactions

Several common weed species serve as alternate hosts for pathogens and insect vectors. *Ageratum conyzoides* hosts begomoviruses, which threaten tomato, chili, and tobacco crops, while *Ludwigia adscendens* and *Monochoria vaginalis* can harbor planthoppers and leafhoppers that vector rice tungro virus and other diseases (The 2021). The persistence of these weeds in canals, field margins, and unmaintained plots supports year-round pest survival and amplifies damage during cropping seasons.

10.4.5 Persistent Weed Seed Banks

Weed seeds with dormancy, particularly *Echinochloa*, *Fimbristylis*, *Cyperus*, and weedy rice, can survive in the soil for multiple seasons, regenerating populations even after good control in the current season (Chauhan 2013; Cuong et al. 2022). Seed bank persistence is a major issue in WDSR systems, where reduced soil disturbance and absence of a water layer allow rapid emergence (Table 10.5). In an analysis of the soil seed bank across MRD rice fields, weed densities of up to 14,000 seeds m^{-2} were recorded, with weedy rice accounting for up to 18% of the total (Son 2011).

Table 10.5 Common weed species with persistent soil seed banks (Son 2011)

Species	Dormancy time
Weedy rice (*O. sativa f. spontanea*)	12–24 months dormancy
Echinochloa crus-galli	8–36 months
Fimbristylis miliacea	6–12 months
Cyperus difformis	Up to 3 years
Ammannia spp.	Long dormancy under dry conditions

10.4.6 Herbicide Costs and Economic Burden

Weed control is one of the most expensive inputs for smallholder rice farmers in Vietnam. On average, herbicide expenditure ranges from 2.3% to 3.2% of gross income but may rise to 10–15% in fields with resistance issues (The 2021). Manual weeding is even costlier, accounting for 20–30% of total labor costs, or approximately 1.2 million VND ha^{-1}.

Herbicide resistance is compounding the issue. Biotypes of *Echinochloa*, *Leptochloa*, and weedy rice now show resistance to common post-emergence herbicides (e.g., cyhalofop-butyl, florpyrauxifen-benzyl), driven by repeated exposure and lack of rotation (Viet et al. 2022; Thi et al. 2025d). Farmers are forced to use higher doses or more expensive combinations, increasing financial risk.

10.4.7 Weedy Rice: A Unique and Persistent Threat

Weedy rice (*Oryza sativa f. spontanea*) is among the most damaging weeds in Vietnam's WDSR systems (Fig. 10.1a). It mimics cultivated rice morphologically during early stages, making manual detection difficult. Once established, it competes for nutrients, sheds seeds prematurely, and contaminates harvested grain with red pericarp kernels, reducing milling quality and market value (Chauhan and Johnson 2010; Cuong et al. 2024).

Studies by Cuong et al. (2022, 2023) identified key traits: high early vigor, strong shattering ability, and long dormancy. These traits enable the weed to dominate fields rapidly and persist across seasons via soil seed banks.

10.4.8 Ecological and Environmental Impacts

Weeds affect agroecosystem health beyond crops. Dominance of invasive species like *Mikania micrantha* can displace native flora, reducing biodiversity and disrupting ecological processes like pollination or nutrient cycling (Khanh et al. 2006c).

Aquatic weeds such as *Monochoria vaginalis* reduce oxygen in irrigation canals, affecting fish and beneficial aquatic insects.

Overreliance on chemical control also raises environmental risks. In a study of runoff patterns in Can Tho and An Giang provinces, residues of metsulfuron-methyl and pretilachlor were detected in water bodies adjacent to paddy fields, suggesting a pathway for contamination and biodiversity loss (The 2021).

10.5 Weed Management Practices in Vietnam

Weed management in Vietnam has evolved significantly over the past decades, transitioning from labor-intensive hand weeding toward herbicide-based programs, and more recently toward integrated weed management (IWM). Given the complexity of Vietnam's agroecosystems, the diversity of weed species, and the increasing challenge of herbicide resistance, no single method is universally effective (Chauhan 2013; Viet et al. 2022; Thi et al. 2025d). Instead, sustainable solutions must combine conventional, cultural, biological, and innovative practices tailored to local conditions.

10.5.1 Conventional Approaches: Chemical and Manual Control

Chemical Control

Chemical herbicides remain the most widely adopted strategy in rice-based systems due to their rapid effectiveness and labor-saving advantages. In wet direct-seeded rice (WDSR), both pre-emergence and early post-emergence herbicides are extensively used. Common active ingredients include pretilachlor, butachlor, cyhalofop-butyl, bispyribac-sodium, and, more recently, florpyrauxifen-benzyl (Duy et al. 2018a, b). While these products provide satisfactory control of dominant grasses and sedges, repeated use of the same modes of action has driven resistance in *Echinochloa crus-galli*, *Leptochloa chinensis*, and weedy rice populations. Resistance mechanisms include both target-site mutations in ALS and ACCase enzymes, and metabolic resistance mediated by cytochrome P450 activity (The 2021; Duy et al. 2018b). These findings highlight the urgent need for herbicide rotation, mixtures, and stewardship programs.

Manual Weeding

Manual hand weeding continues to play an important role, especially in smallholder fields or where herbicide use is constrained due to cost or crop sensitivity. It is most effective during the critical weed-free period (15–35 days after sowing) in rice

(Chauhan and Johnson 2010). However, rural labor shortages and the labor-intensive nature of manual weeding reduce its long-term viability. In perennial orchard systems, mowing or slash-and-burn practices are also common, but these typically provide only temporary suppression and often stimulate regrowth of persistent perennials such as *Imperata cylindrica* (Vang et al. 2024).

Evidence from Corn Production

Experiments in vegetable corn demonstrated that manual weeding combined with increased planting density significantly improved weed suppression and yield. Thiem et al. (2020) reported that densities of 111,111–138,889 plants ha^{-1}, coupled with one or two hand-weedings, substantially reduced weed density and dry weight, resulting in higher cob yield (Table 10.6). These results highlight how integrating planting density with hand weeding can optimize crop competitiveness, minimize labor inputs, and reduce dependence on chemical herbicides.

10.5.2 Cultural and Biological Innovations

10.5.2.1 Cultural Practices

Cultural management practices form a cornerstone of sustainable weed control. Techniques such as land leveling, optimizing sowing dates, and deploying competitive rice cultivars suppress early weed establishment. Dense planting reduces light penetration, thereby inhibiting germination, while early flooding suppresses upland weed species. The stale seedbed technique, where irrigation stimulates weed germination before planting followed by destruction of emerged seedlings, is also effective in depleting the soil seed bank (Chauhan 2013; Khanh et al. 2006a).

Table 10.6 Effect of planting density and hand weeding on weed suppression and vegetable corn yield (Thiem et al. 2020)

Treatment	Weed density (plants m^2)	Weed dry weight (g m^2)	Weed control efficiency (WCE, %)	Cob yield (tons ha^{-1})	Green biomass (tons ha^{-1})
D1 + NW	192.9	118.8	–	1.35	18.78
D1 + HW1	133.5	70.5	40.7	1.47	19.78
D1 + HW2	92.9	27.7	76.7	1.61	20.46
D3 + NW	170.8	99.2	–	1.61	25.67
D3 + HW1	112.8	54.1	45.5	2.01	27.66
D3 + HW2	72.1	12.7	87.2	2.23	27.63
D4 + HW1	96.2	49.9	47.4	2.12	33.36
D4 + HW2	60.9	12.5	86.8	2.15	33.22

NW No weeding, *HW1, HW2* Hand weeding once and twice
Note: D1, D3, D4: Planting densities (79,365, 111,111, 138,889 plants ha^{-1})

10.5.2.2 Biological Control

Biological control efforts in Vietnam have achieved notable successes. Classical introductions include the stem-borer moth *Carmenta mimosa* for *Mimosa pigra* and the weevil *Neochetina bruchi* for *Eichhornia crassipes* (Cam 1998). These efforts are part of a broader strategy to address the heavy reliance on synthetic herbicides in key sectors such as rice cultivation, which has caused environmental and health concerns (Khanh et al. 2006b). Vietnamese research is therefore promoting integrated biological management, including exploiting plant allelopathy to develop weed-suppressing rice cultivars. In rice ecosystems, beneficial fungi such as *Exserohilum monoceras* have been evaluated; when applied at 10^6 spores mL^{-1}, this pathogen caused up to 100% mortality of *E. crus-galli* seedlings at the 2–4 leaf stage without injuring rice (Tuat 1997).

10.5.2.3 Allelopathy and Bioherbicides

Allelopathy has emerged as a key area of innovation in Vietnam's weed management research. Rice (*Oryza sativa*) has been identified as a potent allelopathic crop. Studies by Thi et al. (2014, 2021, 2022) and Vang et al. (2024) demonstrated that phenolic acids and *N-trans-cinnamoyltyramine* (NTCT) isolated from Vietnamese cultivars significantly inhibit the germination and growth of *E. crus-galli*, *Leptochloa chinensis*, and weedy rice. Rice by-products, such as straw and husk extracts, also reduce seedling survival of sedges like *Fimbristylis miliacea*.

Other crops and wild plants also exhibit allelopathic activity. Sweet potato (*Ipomoea batatas*) suppresses *Imperata cylindrica*, *Ageratum conyzoides*, and *Bidens pilosa*, while *Cosmos sulphureus* extracts (Asteraceae) inhibit germination of *E. crus-galli* and *F. miliacea* (Xuan et al. 2016; Trang et al. 2024). Further studies have identified allelopathic compounds from *Cucumis sativus* (cucumber), *Passiflora edulis* (passion fruit), and *Castanea crenata*, including cinnamic acids, phenolic acids, flavonoids, and coumarins, which disrupt weed seed germination and seedling growth (Thi et al. 2008; Khanh et al. 2006a; Tuyen et al. 2018).

These findings are summarized in Tables 10.7 and 10.8, which provide an overview of allelopathic plants, target weed species, observed effects, and identified compounds. Together, they demonstrate that allelopathy offers a promising pathway toward novel herbicide development and sustainable crop rotations.

10.5.3 Toward Integrated Weed Management (IWM)

Integrated weed management (IWM) is increasingly promoted as the most viable solution to sustain productivity while minimizing ecological harm. IWM is built on several guiding principles:

Table 10.7 Some studies on allelopathic plants in Vietnam

Allelopathic plant	Target weed species	Observed effects	Methodology	Reference
Higher plants from Southeast Asia	Various paddy weeds	Weed suppression	Screening of allelopathic potential	Xuan et al. (2005)
Rice (*Oryza sativa*)	Various paddy weeds	Reduction in weed biomass	Field trials assessing rice allelopathy	Khanh et al. (2007)
Sweet potato (*Ipomoea batatas*)	Cogongrass (*Imperata cylindrica*), Bidens (*Bidens pilosa*), Goatweed (*Ageratum conyzoides*)	Suppression of weed emergence and growth	Field evaluation of sweet potato varieties	Xuan et al. (2016)
Barnyardgrass (*Echinochloa crus-galli*)	Rice (*Oryza sativa*)	Allelopathic interaction affecting rice growth	Assessment of allelopathic effects	Khanh et al. (2018)
Rice (*Oryza sativa*)	Barnyardgrass (*Echinochloa crus-galli*)	Inhibition of germination and growth	Isolation and identification of allelopathic compounds	Thi et al. (2014)
Cucumber (*Cucumis sativus*)	Barnyardgrass (*Echinochloa crus-galli*)	Suppression of seedling growth	Extraction and bioassay of allelopathic substances	Thi et al. (2008)
Rice (*Oryza sativa*)	Barnyardgrass (*Echinochloa crus-galli*), weedy rice	Growth inhibition, suppression of invasive weeds	Isolation and testing of N-trans-cinnamoyltyramine	Vang et al. (2024)
Rice by-products	*Echinochloa crus-galli, Leptochloa chinensis, Fimbristylis miliacea*	Reduced seed germination and seedling survival	Use of rice by-products and field testing	Thi et al. (2021)
Rice aqueous methanolic extracts	Junglerice (*Echinochloa colona*)	Phytotoxicity and inhibition of germination	Aqueous methanol extraction and bioassay	Thi et al. (2022)
Asteraceae plant extracts	*Echinochloa crus-galli, Echinochloa colona*	Phytotoxicity and biological control potential	Extraction and bioassay analysis	Trang et al. (2024)

- Herbicide rotation and mixtures to delay resistance evolution.
- Healthy crop establishment through competitive cultivars, optimal density, and balanced fertilization.
- Economic threshold-based interventions, ensuring control is justified only when yield loss exceeds intervention costs.
- Natural regulation, leveraging beneficial microbes, pathogenic fungi, and allelopathic crops.

Table 10.8 List of critical allelochemicals identified from plant extracts explored in Vietnam

Botanical name	Target weed species	Identified compound	Authors
Billygoat-weed (*Ageratum conyzoides*)	*Echinochloa crus-galli* *Monochoria vaginalis* *Aeschynomene indica* *Raphanus sativus*	Catechin *(Stem)*[a] Coumaric acid Gallic acid Protocatechuic acid *(Whole plant)*[a] Benzoic acid p-Coumaric acid Sinapic acid *(Leaf only)*[a]	Xuan et al. (2004)
Rice variety OM 5930	*Echinochloa crus-galli* *Leptochloa chinensis* *Amaranthus palmeri* *Phleum pratense* *Oryza sativa* f. *spontanea*	N-trans-Cinnamoyltyramine (NTCT)	Vang et al. (2024)
Castanea crenata	*Echinochloa crus-galli* *Lactuca sativa* *Raphanus sativus*	*Gallic acid* *Protocatechuic acid* *p-Hydroxybenzoic acid* *Caffeic acid* *Ferulic acid* *Ellagic acid* *Cinnamic acid* *(GC-MS and HPLC analyses)*[b] *$2\alpha,3\beta,7\beta$,23-Tetrahydroxyurs-12-ene-28-oic acid* *(^{1}H- and ^{13}C-NMR, IR, and LC-MS)*[b]	Tuyen et al. (2018)

(continued)

Table 10.8 (continued)

Botanical name	Target weed species	Identified compound	Authors
Cucumis sativus	*Echinochloa crus-galli*	Coumaric acid Protocatechuic acid p-Hydroxybenzoic acid Caffeic acid Syringic acid Sinapic acid Trans-Coumaric acid Trans-Cinnamic acid *(HPLC analyses)*[b]	Thi et al. (2008)
Passiflora edulis	*Echinochloa crus-galli* *Monochoria vaginalis*	Courmarin Myristic acid Palmitic acid Stearic acid Kavain Yagonin Dihydromethysticin	Khanh et al. (2006a)
Cucumis sativus L. cv.	*Lepidium sativum* *Echinochloa crus-galli*	(S)-2-Benzoyloxy-3-phenyl-1-propanol 9-Hydroxy-4,7-megastigmadien-9-one (6S,7E,9S)-6,9,10-Trihydroxy-4,7-megastigmadien-3-one Sisymbrifolin	Kato-Noguchi et al. (2012)

Oryza sativa L.cv. OM 2395, 5451, 6976, 380, 5930, 4498, 3536, N406, and 7347	*Echinochloa crus-galli*	2,4-Dihydroxybenzaldehyde 2,6-Dimethoxybenzoic acid 3,4-Dihydroxybenzoic acid 3,4-Dihydroxyphenylacetic acid 3-Hydroxybenzoic acid 4-Hydroxybenzoic acid 5-Methoxysalicylic acid 7-Oxostigmasterol Benzoic acid 2,4-Dimethoxybenzoic acid 2,5-Dihydroxybenzoic acid 3,4-Dimethoxybenzoic acid 3,5-Dihydroxybenzoic acid 3,5-Dimethoxybenzoic acid Cinnamic acid Coumarin Ergosterol peroxide p-Hydroxycinnamic acid Salicylic acid Vanillic acid *(UHPLC-MS analysis)*[b]	Thi et al. (2020)
Cucumis sativus	*Lepidium sativum* *Lactuca sativa* *Medicago sativa* *Lolium multiflorum* *Phleum pratense* *Digitaria sanguinalis* *Echinochloa crus-galli* *Echinochloa colonum*	(S)-2-Benzoyloxy-3-phenyl-1-propanol	Thi et al. (2008)
	Lepidium sativum *Echinochloa crus-galli*	9-Hydroxy-4,7-megastigmadien 9-one (HMO) (6S,7E,9S)-6,9,10-Trihydroxy-4,7-megastigmadien-3-one (THMO) *(HPLC analyses)*[b]	Kato-Noguchi et al. (2011)

Note: [a]Plant part
[b]Analytical chromatography technique

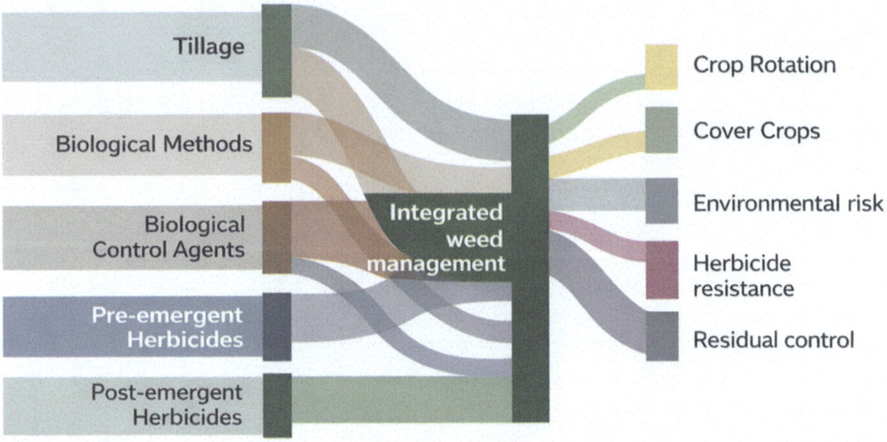

Fig. 10.3 Integrated weed management (IWM) framework showing the interactions of different control tactics (tillage, biological methods, biological control agents, pre- and post-emergence herbicides) and their outcomes (crop rotation, cover crops, residual control, herbicide resistance, and environmental risk)

In practice, successful IWM models in the Mekong Delta have combined pre-emergence herbicides, stale seedbed preparation, competitive rice varieties, timely manual weeding, and allelopathic plant extracts, achieving superior suppression while reducing chemical inputs (Vang et al. 2024).

Despite documented benefits, adoption of IWM remains limited due to farmer awareness gaps, limited access to technologies, and perceptions of complexity (Thi et al. 2025d, e, f). Scaling up IWM will require participatory farmer training, demonstration projects, policy incentives, and sustained investment in research on bio-herbicides, precision application, and herbicide resistance monitoring. With proper institutional support, IWM can become the foundation of resilient, climate-smart weed management systems in Vietnam's rice-based agroecosystems (Fig. 10.3).

10.6 Recent Scientific Advances and Future Research Directions

10.6.1 Advances in Weed Biology and Management Research in Vietnam

In recent years, research on weed biology and management in Vietnam has made notable progress, particularly in the exploration of plant-derived allelochemicals and other sustainable control measures. One major development has been the identification and characterization of bioactive compounds from indigenous plant species with allelopathic potential. For example, the rice cultivar OM5930 has been shown to produce *N-trans*-cinnamoyltyramine (NTCT), a phenolic amide with strong inhibitory effects on weedy rice, barnyardgrass, Chinese sprangletop, and other invasive species when applied at appropriate concentrations (Thi et al. 2008, 2014; Vang et al. 2024). This compound disrupts seed germination and seedling growth without causing damage to the rice crop, demonstrating significant potential for bioherbicide development.

Other studies have focused on allelopathic plants from the Asteraceae family, such as *Cosmos sulphureus*, whose extracts have been shown to reduce germination rates and biomass accumulation of *E. crus-galli*, *F. miliacea*, and other dominant rice-field weeds (Trang et al. 2024; Thi et al. 2025c). In greenhouse and laboratory bioassays, these extracts displayed consistent phytotoxicity across multiple weed species, suggesting that they could be integrated into weed management programs as natural growth inhibitors. The use of allelopathic cover crops, such as *Lantana camara*, Zingiberaceae species, and sweet potato (*Ipomoea batatas*), has also been investigated, showing the ability to suppress weed emergence while improving soil health (Thi et al. 2025a, b; Xuan et al. 2016).

Advances in molecular and physiological studies have shed light on the mechanisms underlying herbicide resistance in Vietnamese weed populations. Research on barnyardgrass and sprangletop biotypes has identified both target-site mutations, such as those affecting acetolactate synthase (ALS) and acetyl-CoA carboxylase (ACCase) and enhanced metabolic resistance mediated by cytochrome P450 monooxygenases (Duy et al. 2018a, b; The 2021). These findings highlight the urgent need for routine resistance monitoring and for integrating herbicide stewardship principles into farmer training programs.

Technological innovations are beginning to influence weed research and management. The application of deep learning models for weed recognition has achieved promising accuracy rates. For example, The (2021) trained the MobileNetV2 architecture on a dataset of over 28,000 plant images representing 109 Vietnamese species, achieving a classification accuracy of 83.9%. Such advances open possibilities for integrating image-based weed detection into unmanned aerial vehicles (UAVs) or mobile applications, enabling real-time, site-specific weed control. Precision spraying systems, guided by such detection tools, could significantly reduce herbicide use and environmental impact.

10.6.2 Future Research Priorities and Policy Implications

Despite these promising developments, many advances remain at the experimental stage. Future research should prioritize scaling up promising bioherbicide candidates, such as NTCT and *Cosmos sulphureus* extracts, through field trials across different agroecological zones. There is also a need to explore synergistic effects between allelopathic compounds and reduced herbicide rates, potentially lowering chemical inputs while maintaining efficacy. Additionally, incorporating molecular breeding approaches to develop rice cultivars with enhanced allelopathic traits could offer a long-term, built-in strategy for weed suppression (Khanh et al. 2006c).

Research should also focus on the integration of biological and technological innovations within comprehensive IWM frameworks. For example, AI-assisted detection systems could be combined with cultural and allelopathic control measures to create adaptive, data-driven weed management programs. Moreover, the rapid technological advancements in AI, unmanned aerial vehicles (UAVs), imaging, and sensor systems over the past decade necessitate the integration of smart weed management strategies designed to enhance control efficiency while reducing methane (CH_4) emissions (Xuan et al. 2025). In parallel, establishing herbicide resistance monitoring networks and standardized weed distribution databases will be essential to inform timely and targeted interventions.

Finally, policy and institutional support will be critical to sustaining innovation in weed management. Government and private sector investment in bioherbicide research, precision agriculture technologies, and farmer capacity building can help accelerate the transition from chemically dependent control to more diversified, environmentally sustainable approaches. With Vietnam's growing scientific capacity and increasing recognition of environmental concerns, the next decade presents a window of opportunity to reshape weed management toward models that balance productivity, sustainability, and climate resilience (Vang et al. 2024; Thi et al. 2025a, b, c, d, e, f).

10.7 Conclusions

Weed management in Vietnam is both a persistent challenge and a major lever for sustainable intensification. In rice-based systems where national food security and rural livelihoods depend on stable yields, the dominance of highly competitive species such as weedy rice (*Oryza sativa* f. *spontanea*), barnyardgrass (*Echinochloa crus-galli*), and Chinese sprangletop (*Leptochloa chinensis*) continues to erode productivity and grain quality. Heavy reliance on herbicides has delivered short-term labor savings but has also accelerated resistance evolution and raised environmental concerns. The body of evidence synthesized in this chapter points to integrated weed management (IWM) as the most viable pathway to balance productivity, profitability, and environmental stewardship over the long term.

Encouraging advances strengthen this trajectory. Allelopathic discoveries (e.g., rice-derived NTCT and extracts from *Cosmos sulphureus*) and cover-crop systems offer crop-compatible suppression of dominant weeds and a pipeline toward novel herbicides. At the same time, precision tools, including UAV-based mapping and AI-enabled recognition enable site-specific interventions that reduce inputs and minimize off-target impacts. When these biological and digital innovations are embedded within adaptive IWM program, coupling competitive cultivars, optimal planting density, water management (including stale seedbed and timely flooding), and judicious herbicide rotation/mixtures, the result is more reliable control with lower ecological cost and improved resilience to climate variability.

To translate promise into practice, Vietnam's strategy should prioritize coordinated, farmer-centered adoption supported by policy and institutions. Community-based actions (e.g., synchronized weedy-rice control at the village/canal system scale) amplify field-level gains, while national resistance surveillance ensures timely, evidence-based recommendations. Investment in R&D and extension is essential to mature bioherbicide candidates, breed allelopathic/competitive crop ideotypes, and mainstream precision technologies in smallholder settings.

Actionable Priorities

1. Scale up field validation of bioherbicides and allelopathic crops across agroecological zones; establish product-development pipelines with safety and efficacy standards.
2. Build a national herbicide-resistance monitoring and reporting network (standardized diagnostics, open dashboards) to guide rotations/mixtures and stewardship.
3. Integrate IWM into crop improvement (breeding for competitiveness and allelopathic potential) and agronomy packages (density, sowing windows, water regimes).
4. Deploy precision weed management (UAV mapping, AI recognition, variable-rate application) with service models accessible to smallholders.
5. Institutionalize community-based campaigns for landscape-level problems (e.g., weedy rice), synchronized by season calendars and water management.
6. Align incentives and training—extension curricula, performance-based subsidies, and private-sector partnerships—to lower adoption barriers and ensure continuity.

Vietnam is well placed to move from a predominantly chemical paradigm to a biologically informed, technologically enabled, and ecologically sustainable weed-management model. Done well, this transition will deliver immediate on-farm benefits (higher, more stable yields; lower costs; safer production) and broader public goods improved environmental quality, better water and methane management in rice, and a more climate-resilient agricultural sector. IWM is not a single tactic but a continuous improvement process; with sustained innovation and strong stakeholder engagement, Vietnam can become a regional exemplar of ecologically balanced weed management.

References

Anju MV, Warrier RR, Kunhikannan C (2022) Significance of soil seed bank in forest vegetation—a review. Seeds 1(3):181–197. https://doi.org/10.3390/seeds1030016

Busi R, Chauhan BS, Powles SB (2016) Weedy rice in the Philippines and Vietnam – final report. Australian Centre for International Agricultural Research (ACIAR)

Cam TT (1998) Biological control of invasive weeds in Vietnam: experiences with *Mimosa pigra* and water hyacinth. In: Proceedings of the international workshop on biological control and management of Mimosa pigra. CSIRO, Darwin, Australia, pp 115–120

Chauhan BS (2013) Strategies for effective and sustainable weed management in dry-seeded rice. Crop Prot 42:18–23. https://doi.org/10.1016/j.cropro.2012.07.002

Chauhan BS, Johnson DE (2010) The role of seed ecology in improving weed management strategies in the tropics. Adv Agron 105:221–262. https://doi.org/10.1016/S0065-2113(10)05006-6

Chin DV (2001) Biology and management of barnyardgrass, red sprangletop, and weedy rice in rice. Weed Biol Manag 1(1):37–41. https://doi.org/10.1046/j.1445-6664.2001.00004.x

Chinh PD, Tuyen DH, Truong L (1978) Weeds and control measures. NXB. Nong Nghiep, Hanoi. (Printed in Vietnamese)

Cuong NT, Lem DT, Thoan NT, Ninh LH, Anh NVT, Dat VP, Thi HL (2022) Research on the morphological and biological characteristics of some weedy rice lines in the Mekong Delta. J Agric Rural Dev 18:10–17. (In Vietnamese)

Cuong NT, Lem DT, Anh NVT, Dat VP, Thi HL (2023) Description of morphological and biological characteristics of some common weedy rice lines in the Mekong Delta. Version B Vietnam J Sci Technol 65(2):33–40. https://doi.org/10.31276/VJST.65(2).33-40. (In Vietnamese)

Cuong NT, Vinh KC, Chac TV, Kha NK, Thi HL (2024) Integrated approaches for weedy rice (*Oryza sativa* L. spontanea) management in Vietnam and beyond: challenges and proposed strategies. English vesion. Vietnam J Agric Rural Dev 6(1):39–50. (In Vietnamese)

Duy L, Chon MN, Bobba VNK, Richard KM (2018a) Weed management practices to control herbicide-resistant *Echinochloa crus-galli* in rice in Mekong Delta, Vietnam. Res Crops 19(1):20–27. https://doi.org/10.5958/2348-7542.2018.00003.7

Duy L, Chon NM, Mann RK et al (2018b) Efficacy of Rinskor™ (florpyrauxifen-benzyl ester) on herbicide resistant barnyardgrass (*Echinochloa crus-galli*) in rice fields of Mekong Delta, Vietnam. J Crop Sci Biotechnol 21:75–81. https://doi.org/10.1007/s12892-017-0142-0

Gianessi LP (2013) The value of herbicides in Vietnamese rice production. CropLife Foundation Report

Hieu NV, Hien NLH (2020) Automatic plant image identification of Vietnamese species using deep learning models. Int J Eng Trends Technol 68(4):25–28. https://doi.org/10.14445/22315381/IJETT-V68I4P205

Ho PH (1993) Weeds of Vietnam, volume III, book 2. NXB, Montréal, pp 810–812. (Printed in Vietnamese)

Kato-Noguchi H, Thi LH, Teruya T, Suenaga K (2011) Two potent allelopathic substances in cucumber plants. Sci Hortic 129(4):894–897. https://doi.org/10.1016/j.scienta.2011.04.031

Kato-Noguchi H, Le TH, Sasaki H, Suenaga K (2012) A potent allelopathic substance in cucumber plants and allelopathy of cucumber. Acta Physiol Plant 34:2045–2049. https://doi.org/10.1007/s11738-012-0997-8

Khanh NQ (2003) Study on weed composition and herbicide efficacy on some crops on sloping land in northern mountainous Vietnam. Master's thesis. University of Agriculture I, Hanoi

Khanh TD, Chung IM, Tawata S, Chung IM (2006a) Weed suppression by Passiflora edulis and its potential allelochemicals. Weed Res 46(4):296–303. https://doi.org/10.1111/j.1365-3180.2006.00512.x

Khanh TD, Xuan TD, Chin DV, Chung IM (2006b) Current status of biological control of paddy weeds in Vietnam. Weed Biol Manag 6(1):1–9. https://doi.org/10.1111/j.1445-6664.2006.00189.x

Khanh TD, Chung IM, Xuan TD, Tawata S (2006c) The exploitation of crop allelopathy in sustainable agricultural production. J Agron Crop Sci 192(3):172–184. https://doi.org/10.1111/j.1439-037X.2006.00299.x

Khanh TD, Xuan TD, Chung IM (2007) Rice allelopathy and possibility for weed management. Ann Appl Biol 151(3):325–339. https://doi.org/10.1111/j.1744-7348.2007.00183.x

Khanh TD, Trung KH, Anh LH, Xuan TD (2018) Allelopathy of barnyardgrass (*Echinochloa crus-galli*) weed: an allelopathic interaction with rice (*Oryza sativa*). Vietnam J Agric Sci 1(1):97–116. https://doi.org/10.31817/vjas.2018.1.1.10

Krupnik TJ, Six J, Ladha JK, Paine MJ, Kessel CV, Linquist BA (2012) Water use efficiency in rice production: agro-ecological and agronomic considerations for direct-seeded rice. Adv Agron 115:75–138. https://doi.org/10.1016/B978-0-12-394276-0.00003-8

Son NH (2000a) Evaluation of the status and economic efficiency of manual weeding and herbicide use on transplanted rice in the Red River Delta at present. J Agric Food Ind (3/2000) (Printed in Vietnamese)

Son NH (2000b) Some research results on weeds in transplanted rice fields and control measures in the Red River Delta. Doctoral dissertation. University of Agriculture I, Hanoi

Son NH (2011) Weeds in lowland rice fields in northern Vietnam and their control measures [Monograph]. Agriculture Publishing House, Hanoi

Son NH, Tuat NV, Tan NT (2005) Weed component and control practice in some annual upland crops in the North of etnam. In: Proceedings of the 20th Asian-Pacific Weed Science Society conference. Agriculture Publishing House, Hanoi, pp 595–600

Suk T, Le VD, Nguyen TT, Thi HT (2021) Current status of weed composition in major rice-producing regions of the Mekong Delta, Vietnam. Vietnam J Agric Sci 4(3):522–534

The ND (2021) Herbicide-resistant weeds: The case of resistance to glyphosate. Vietnam J Sci Technol Eng 63(2):74–80. https://doi.org/10.31276/VJSTE.63(2).74-80

Thi HL, Teruya T, Suenaga K, Chin DV, Kato-Noguchi H (2008) Allelopathy and the allelopathic activity of a phenylpropanol from cucumber plants. Plant Growth Regul 56(1):1–5. https://doi.org/10.1007/s10725-008-9283-6

Thi HL, Lin C-H, Smeda RJ, Leigh ND, Wycoff WC, Fritschi FB (2014) Isolation and identification of an allelopathic phenylethylamine in rice. Phytochemistry 108:109–121. https://doi.org/10.1016/j.phytochem.2014.08.019

Thi LH, Nguyen TTC, Vu DC, Nguyen NY, Nguyen TTT, Phong TNH, Nguyen CT, Lin C-H, Lei Z, Sumner LW, Le VV (2020) Allelopathic potential of Rice and identification of published allelochemicals by cloud-based metabolomics platform. Metabolites 10(6):244. https://doi.org/10.3390/metabo10060244

Thi HL, Nguyen CT, Vu DC, Nguyen TTC, Nguyen VQ, Smeda RJ (2021) Rice by-products reduce seed and seedlings survival of *Echinochloa crus-galli, Leptochloa chinensis* and *Fymbristylis miliacea*. Agronomy 11(4):776. https://doi.org/10.3390/agronomy11040776

Thi LH, Nguyen VL, Phan LK, Nguyen CT, Nguyen THD, Van VL, Reid SJ (2022) Phytotoxicity in aqueous methanolic extracts of rice against junglerice and total activities of identified phytotoxic compounds. Ann Appl Biol 180(2):196–210. https://doi.org/10.1111/aab.12746

Thi HL, Bao PDL, Hoang TV, Dao XTT, Ngan NT (2025a) Allelopathic effects of essential oils from Zingiberaceae species on weed growth. Trop J Nat Prod Res (TJNPR) 9(4):1457–1463. https://doi.org/10.26538/tjnpr/v9i4.11

Thi HL, Trang NTT, Huy NG, Ngan NT, Cuong NT (2025b) Laboratory assessment of *Lantana camara* L. extracts for selective inhibition of rice weeds: phytotoxicity, crop response, and phenolic composition. J Environ Sci Health B 60(5):208–218. https://doi.org/10.1080/03601234.2025.2494190

Thi HL, Trang NTT, Men TT, Khang DT (2025c) Allelopathic effects of yellow cosmos (*Cosmos sulphureus* var. *hirsuticaulis*): phytochemicals composition and its weed suppression potential. Allelopath J 65(1):35–50. https://doi.org/10.26651/allelo.j/2025-65-1

Thi HL, Toan NM, Truong LN, Huy NG, Chac TV (2025d) Investigation of current cultivation practices and weed composition in rice fields in Hau Giang province, Vietnam. Biodiversity 26(2):132–145. https://doi.org/10.1080/14888386.2025.2453458

Thi HL, Vu NT, Huy NG (2025e) Current status of cultivation, weed management, and solutions to improve jackfruit (*Artocarpus heterophyllus* Lam.) production efficiency in Vinh Long province. J Agric Sci Technol – Hue Univ Agric For 9(2):4807–4819. https://doi.org/10.46826/huaf-jasat.v9n2y2025.1235

Thi HL, Huy NG, Khoa TA (2025f) Current cultivation status and weed composition in durian orchards in Cai Be district, Tien Giang province. Vietnam J Sci Technol-Part B 67(6):1–11. https://doi.org/10.31276/VJST.2024.3046

Thiem TT, Thu TTP, Loan N (2020) Effect of plant density and hand weeding on weed control and yield of the vegetable corn. Vietnam J Agric Sci 3(4):784–797. https://doi.org/10.31817/vjas.2020.3.4.02

Thuy NTT, Tuyet TTA, Kristiansen P, Truong NV (2022) Investigating the resistance of barnyard-grass populations to pretilachlor in direct-seeded rice in the Central Region of Vietnam. Afr J Agric Res 18(7):542–554. https://doi.org/10.5897/AJAR2022.16016

Trang NTT, Cuong NT, Van Vang L, Le Thi H (2024) Evaluation of phytotoxic potential in Asteraceae plant extracts for biological control of *Echinochloa crus-galli* and *Echinochloa colona*. Plant-Environment Interactions 5(5):e70009. https://doi.org/10.1002/pei3.70009. PMCID: PMC11384620

Tuat NV (1997) Pathogenic fungi associated with rice weeds and their potential as bioherbicides in Vietnam. Vietnam Agric Sci Technol J 10(3):45–52

Tuyen PT, Xuan TD, Tu Anh TT, Mai Van T, Ahmad A, Elzaawely AA, Khanh TD (2018) Weed suppressing potential and isolation of potent plant growth inhibitors from *Castanea crenata* Sieb. et Zucc. Molecules 23(2):345. https://doi.org/10.3390/molecules23020345

Vang LV, Nguyen AV, Luu DT, Fritschi FB, Nguyen CT, Thi HL (2024) Inhibitory effects of *N*-trans-cinnamoyltyramine on growth of invasive weeds and weedy rice. Plant Environ Interact 5(6):e70017. https://doi.org/10.1002/pei3.70017. PMCID: PMC11384620

Viet H, Thi HL, An DTK, Phen TV, Hoa NX, Binh HTT, Hoang VD, Truong NV, Liem NV (2022) In: Koo LSJ, Kwoon YW, Duong VC, Hoang AC (eds) Common weeds in Vietnam. Agriculture Publishing House, Hanoi, 445 pp, ISBN: 978-604-60-3624-1. (Printed in bilanguages: Vietnamese and English)

Xuan TD, Shinkichi T, Hong NH, Khanh TD (2004) Assessment of phytotoxic action of *Ageratum conyzoides* L.(billy goat weed) on weeds. Crop Prot 23(10):915–922. https://doi.org/10.1016/j.cropro.2004.02.005

Xuan TD, Shinkichi T, Khanh TD, Chung IM (2005) Biological control of weeds and plant pathogens in paddy rice by exploiting plant allelopathy: an overview. Crop Prot 24(3):197–206. https://doi.org/10.1016/j.cropro.2004.08.004

Xuan TD, Tsuzuki E, Terao H, Matsuo M, Khanh TD, Chung IM (2016) Sweet potato (*Ipomoea batatas*) allelopathy for weed management in sustainable agriculture. Allelopath J 37(2):231–246

Xuan TD, Khanh TD, Minh TTN (2025) Implementation of conventional and smart weed management strategies in sustainable agricultural production. Weed Biol Manag 25:e70000. https://doi.org/10.1111/wbm.70000

The manufacturer's authorised representative in the EU is Springer Nature Customer Service Centre GmbH, Europaplatz 3, 69115 Heidelberg, Germany. If you have any concerns regarding our products, please contact ProductSafety@springernature.com

Printed and bound by CPI Group (UK) Ltd, Croydon, CR0 4YY
07/10/2026
02229349-0011